The Scientist as Philosopher

Friedel Weinert

The Scientist as Philosopher

Philosophical Consequences of Great Scientific Discoveries

With 52 Figures

Dr. Friedel Weinert
University of Bradford
Department of Social Sciences and Humanities
Bradford BD7 1DP
Great Britain
e-mail: F.Weinert@Bradford.ac.uk

Cover figures reproduced from Faraday, M.: A Course of 6 Lectures on the Various Forces of Matter and Their Relation to Each Other. Ed. by W. Crookes (Griffin, London, Glasgow 1860)

ISBN 3-540-21374-0 Springer Berlin Heidelberg New York

Library of Congress Control Number: 2004112076

This work is subject to copyright. All rights are reserved, whether the whole or part of the material is concerned, specifically the rights of translation, reprinting, reuse of illustrations, recitation, broadcasting, reproduction on microfilm or in any other way, and storage in data banks. Duplication of this publication or parts thereof is permitted only under the provisions of the German Copyright Law of September 9, 1965, in its current version, and permission for use must always be obtained from Springer. Violations are liable for prosecution under the German Copyright Law.

Springer is a part of Springer Science+Business Media
springeronline.com

© Springer-Verlag Berlin Heidelberg 2005
Printed in Germany

The use of general descriptive names, registered names, trademarks, etc. in this publication does not imply, even in the absence of a specific statement, that such names are exempt from the relevant protective laws and regulations and therefore free for general use.

Typesetting: Data conversion by LE-TEX Jelonek, Schmidt & Vöckler GbR, Leipzig, Germany using a Springer TEX macro package
Cover design: Erich Kirchner, Heidelberg

Printed on acid-free paper SPIN 10988442 57/3141/tr 5 4 3 2 1 0

The Scientist as Philosopher

Science has always had (...) a metaphoric function – that is, it generates an important part of a culture's symbolic vocabulary and provides some of the metaphysical bases and philosophical orientations of our ideology. As a consequence the methods of argument of science, its conceptions and its models, have permeated first the intellectual life of the time, then the tenets and usages of everday life. All philosophies share with science the need to work with concepts such as space, time, quantity, matter, order, law, causality, verification, reality.

G. Holton, *Einstein, History and Other Passions* (2000), 43

Preface

This is not a philosophy of physics book. It is a book about how philosophy inspires physics and how physics influences philosophy. One section touches on biology. The reader will find in the following pages a study of the interaction between science and philosophy, with particular emphasis on physics. Philosophy offers science very general notions or presuppositions. Through scientific discoveries, these notions often become questionable and may undergo revision. Scientific discoveries therefore have consequences for philosophical thought. The great scientists are often aware of these lines of influence. This explains my title and subtitle. The Introduction elaborates that the interaction concerns the notions of Nature, Physical Understanding, Time, Causation and Determinism.

This volume was written in the reverse order from the way it is presented now. This was an accident of discovery. I first became aware of the scientists' involvement with philosophy in discussions regarding the understanding of quantum physics.

During the time of the composition of this book I have benefited from the unwavering support of my colleagues and friends Anthony O'Hear and Roger Fellows. Peter Galison invited me to the Department of the History of Science at Harvard University as a research fellow. Lawrence Sklar received me for a research visit at the University of Michigan. Both offered encouragement and general advice on the manuscript, for which I am very grateful. Gerald Holton at Harvard read part of the manuscript and made some useful suggestions. I was the recipient of a generous grant for my research stay in the United States. I would like to thank the Leverhulme Trust for supporting me with a Leverhulme Study Abroad Fellowship. Thank you to Peter Galison, Lawrence Sklar and Nancy Cartwright for acting as referees.

Steven French of Leeds University and Brigitte Falkenburg of the University of Dortmund read the whole manuscript. They offered valuable advice, which I acknowledge here. I would like to thank them for their constructive criticism and generosity. As usual any remaining errors will have to be my responsibility. Finally, my editor at Springer-Verlag, Angela Lahee, has been very helpful and supportive.
I hope the reader will enjoy reading this study as much as I enjoyed writing it.

Bradford, January 2004 *Friedel Weinert*

Contents

1

Introduction

The aim of this book is to investigate the philosophical consequences of great scientific discoveries. Philosophers have often insisted that there is an interaction between science and philosophy. For instance they have claimed that science has *metaphysical* foundations. This approach has led to a number of classic studies, which have highlighted the philosophical presuppositions built into the scientific enterprise. The disadvantage of this approach is that the actual scientific discoveries are ignored or dealt with in most general terms, in favour of the metaphysical insights. To redress the balance, we may want to turn to the *history of science*. In books on the history of science the scientific findings are spelt out in great detail but now at the expense of the philosophical dimensions of science. This is also true of many popular science books. In recent years, however, there have been a number of excellent studies on the *interrelation* between science and philosophy. In these studies close attention is paid to both scientific detail and philosophical sophistication. Often the narrative moves from philosophy to some representative piece of science and back. In other cases the story moves from scientific discoveries to their philosophical consequences.

The present book falls within this latter category. But there is a difference. This study focuses on the numerous philosophical discussions scientists themselves have offered in the wake of significant discoveries. It aims to provide a comprehensive study of the thinking of research active scientists who consider the *philosophical* consequences, which their discoveries entail. This book will look at how *fundamental notions*[1], with the aid of which humans attempt to construct a coherent scheme of nature, have undergone radical changes as a direct result of scientific discoveries. The book will develop the topics from the way scientists perceive the tight connection between science and philosophy. There seems to be a widespread assumption

[1] These notions are: Nature and physical understanding; time; causation and determinism. Others could be added, e.g. substance. Although this notion is discussed *en passant*, my concentration on the stated notions is due to the scientists' primary interest in them. In *Image and Logic* (1997) Peter Galison has shown that the notion of 'experiment' is not a constant in the history of physics either.

that the scientist turns to philosophy when his or her active research career is over. Yet the pages of the great scientific journals of the 20th century – *Die Naturwissenschaften, Nature, Physikalische Zeitschrift, The Philosophical Magazine, The Physical Review, Zeitschrift für Physik*, – abound with discussions of philosophical consequences, which arose directly from scientific developments, such as quantum mechanics and relativity theory. Besides, physicists like Bohr, de Broglie, Eddington, Einstein, Frank, Heisenberg, Jeans, Langevin wrote book-length studies during the active years of their scientific research to draw attention to the philosophical consequences of the latest discoveries.

Recent years have seen a continuation of this tradition. Scientists like Bernard d'Espagnat, Costa de Beauregard, François Jacob, Ilya Prigogine, Julian Barbour, David Bohm, Paul Davies, James Cushing, Peter Medawar, Roger Penrose, Stephen Hawking, Peter Mittelstaedt, Carl Friedrich von Weizsäcker, Hans-Dieter Zeh, to mention but some, have drawn attention, in various ways, to the ways, in which our view of the material world has changed and will continue to change as a consequence of new discoveries. Today we seem to stand on the threshold of a new scientific revolution.

To characterize the *aim* of the present book in this way is to emphasise two points.

First, many of the scientists involved in the great scientific revolutions realize that their discoveries have philosophical consequences. They become aware that scientific discoveries may change the fundamental notions with which we conceptualise the world around us. They invite an active exchange between science and philosophy. But they do not wait till they are past the prime of their scientific abilities. Philosophical reflections are part and parcel of their scientific work. This either takes the form of philosophical *asides* in otherwise more mathematical work. Or it expresses itself in more sustained reflections in essays or whole books. In either case: *Philosophy appears as a consequence of science.*

In both forms scientists hold that new findings associated with great scientific revolutions must lead to a *revision* of old ideas and concepts.

> *A direct questioning of nature by experiment has shown the philosophical background hitherto assumed by physics to have been faulty.*[2]

This philosophical background is constituted by what we regard as *Nature* and *Reality* and by the changes in these notions due to reorientations in the fundamental

[2] Jeans, *Physics and Philosophy* (1943), 2, 190, 216; contrast this view, prevalent amongst scientists, with the view that the metaphysical principles of science 'are *assumed* to be true independently of any scientific experience', Dilworth, *The Metaphysics of Science* (1996), 71. It should be stressed that many physicists of a later generation have also seen this connection between philosophy and physics, as formulated by Jeans. See Bohm, *Special Theory of Relativity* (1966), 122; Mittelstaedt, *Philosophische Probleme* (1972), Introduction; d'Espagnat, *A la Recherche* (1981), *Une Incertaine Réalité* (1985), *Penser la Science* (1990); Redhead, *From Physics to Metaphysics* (1995) and from a Marxist point of view Harig/Schleifstein *eds*, *Naturwissenschaft* (1960), Introduction; see also Elkana, *Discovery* (1974) who proposes a 'concepts-in-flux' thesis.

notions of *causation* and *determinism*, *space* and *time*. Revisions of these notions, in turn, have implications for what constitutes *physical understanding*, that is, new ways of comprehending Nature. Such notions are not constant across the history of science. But it requires the conceptual upheavals associated with the various scientific revolutions to make scientists and philosophers aware, as Max Born put it that 'old notions are dissolved by new experiences.'[3] This is quite a common feeling among physicists. The revision of the old concepts has to happen under the constraint of new experiences.[4] It is really a general problem, which emerges at many points during this study. It is not contested that scientific discoveries have a bearing on philosophical notions. What is debatable is the *extent* of this influence. The concepts used by humans are not eternal. They must be adapted to new empirical discoveries. This is where philosophy comes to the fore. For the new empirical findings suggest adaptations of the conceptual network. Many great scientists were aware of this impact of new discoveries on the fundamental notions. This awareness made them *philosopher-scientist*. The present study will provide numerous illustrations of this awareness up to the present day. But the philosopher must ask a question of *evaluation*: To which *extent* are these conceptual shifts really philosophical consequences of the empirical discoveries?

In this study we find that there is a *true interaction* between science and philosophy, which is best described as a *dialectic*. Philosophy does not simply exist in the margins of science. Science does not straightforwardly lead to philosophical consequences. The philosopher will agree with Max Born that 'physics free from metaphysical hypotheses is impossible' but will disagree with his further assessment that 'these assumptions have to be distilled out of physics itself and continuously adapted to the actual empirical situation.'[5] The philosophical consequences of sciences are negotiable: the scientific discoveries do not pinpoint *one* consequence with iron logic. Rather, the philosopher must evaluate to which extent the logic of the scientific argument compels the logic of the philosophical consequences.

We can see this *dialectic* between science and philosophy at every turn of this study. We will find that scientists have often borrowed philosophical ideas from the philosophical tradition. The philosophical tradition is part and parcel of the general culture, by which scientists are influenced. Philosophical ideas enter the scientists' attempt to understand their findings and to construct a coherent interpretation of Nature. The general culture is also influenced by the scientific tradition,

---

[3] Born, *Natural Philosophy* (1949), 75; Cassirer, 'Determinismus und Indeterminismus' (1936), 273; Heisenberg, *Physical Principles* (1930), Chap. IV, §3; 'Prinzipielle Fragen der modernen Physik' (1936); d'Espagnat, *Une Incertaine Réalité* (1985), 19–26

[4] Heisenberg, 'Prinzipielle Fragen der modernen Physik' (1936), 110, cf. 'Über die Grundprinzipien' (1927), 21, 'Die Plancksche Entdeckung' (1928), 205, *Physical Principles* (1930), 62ff, 'Philosophische Grundlagen' (1958), 'Philosophische Probleme' (1967), 422; Bohr, 'Wirkungsquantum' (1929); Eddington, *Philosophy of Physical Science* (1939), 33; Cassirer, 'Determinismus und Indeterminismus' (1936), 273; Einstein/Infeld, *Evolution of Physics* (1938), 18f; Langevin, *La Physique* (1923), 265; Rosenberg, 'Idea of Causality' (1942/1979), 446–7

[5] Born, 'Physics' (1951), 630; see also de Regt, 'Philosophy' (1996)

which supplies it with striking images. Two such famous images are the Clockwork Universe and Nature's Quantum Jumps. The scientific tradition is marked by magnificent revolutions. They change the images of the general culture and influence the philosophers in their evaluations of the philosophical consequences of science.

The *first* point to emphasize is that scientists are aware of the impact of their discoveries on philosophy. The *second* point is that the great scientific revolutions, which have taken place over the last 400 years, have led to transformations in many of the fundamental concepts with which humans conceptualise the material world around them. In Chap. 2 we shall see that the very concept of *Nature* has been reshaped and redefined ever since the Scientific Revolution of the 17th century.

The revolutionary theories of the 20th century, the relativity theory and the quantum theory, also impacted on the notion of *physical understanding*. Roughly speaking, physical understanding is the scientist's interpretation of the empirical data. It is an attempt to construct a coherent view of Nature. New ideas about physical understanding are connected with new views of Nature. The main vehicle of physical understanding, as both Maxwell and Hertz emphasized, is the scientific model (Chap. 3).

In Chap. 4, we turn our attention to another of the fundamental notions, which straddles science, philosophy and everyday thinking: the notion of *time*. The Special theory of relativity redefined our notion of time. One of the fundamental innovations of special relativity is that events, as seen from different reference frames, are not judged to have the same temporal dimensions. Classical physics granted time an independent existence in the nature of things. All observers agreed on time and space. All observers agreed on the duration of events and the lengths of objects. The Special theory of relativity showed that this was mistaken. In fact, as Wolfgang Pauli put it, there 'are as many times and spaces as there are Galilean reference systems.'[6] Time is relegated, so it seems, to the perspective of the observer. Stationary and moving observers disagree on 'what time it is'. The passage of time, so it appeared to Einstein and others, was just a human illusion. The physical universe just is, it is a *block universe*. And a scientific theory – the Special theory of relativity – was there to prove it! Many physicists believed that the Special theory of relativity had confirmed the Kantian view that time was nothing but a form of human intuition and absent from the physical world. Yet today time is increasingly seen as an emergent property.

More was to come, as we shall see in Chap 5. The quantum theory – the physics of the atom – challenged traditional assumptions about *causation* and *determinism*. The French physicist Pierre Laplace invented a demon whose intellectual powers were so vast that neither the past nor the future was opaque to him. The Laplacean demon perceived the world as a long causal chain of events: the present state of the universe was the *effect* of its previous state and it was the *cause* of a later state of the universe. From the point of view of the Laplacean demon, past and future have as much reality as the present. The present contains the future. The whole universe is like a *map*. All the routes are already traced. The Laplacean

[6] Pauli, 'Relativitätstheorie' (1921/1981), 15

demon sees the world stretched out like a *filmstrip*. It is a block universe. For the demon there is no novelty. One frame causes the next frame, and is caused by the previous frame. For the Laplacean demon causation and determinism are the same. Many classical and even modern physicists were influenced by the Laplacean ideal of causation and determinism. It formed the philosophical background, against which the strange, seemingly acausal behaviour of atoms was considered. Quantum theory poured cold water on the Laplacean assumption of causal determinism. The present state of an atomic system does not allow us to predict its future state with deterministic certainty. Only its *probable* future state can be known. According to quantum theory the world is probabilistic. It is therefore indeterministic. It does not prescribe a precise trajectory for its future evolution in space and time. There is room for novelty. We shall see how new empirical discoveries about the behaviour of atoms still challenge our assumptions about causation and determinism.

In evaluating the philosophical consequences of great scientific discoveries, philosophers may disagree with physicists. But all will take the findings of the scientific revolutions on board. In this study we consider the philosophical presuppositions, which scientists make and how these are affected by scientific discoveries. Then we move from the scientific discoveries to the philosophical consequences, as seen by the scientists, and from there to an evaluation of the extent to which scientific discoveries have philosophical consequences. In this sense this book seeks a *new* angle in the assessment of the interrelation between science and philosophy.

Philosophy-of-science books often argue in favour of philosophical positions and seek support in scientific case studies. The present study argues from fundamental discoveries to fundamental notions. Throughout we pursue a dual perspective. The *historical* exercise looks at the physicists' 'philosophies' concerning the above-mentioned notions. The *systematic* exercise focuses on the bearing of science on philosophy.

Why should an audience interested in science and philosophy read this tome? It is important to draw attention to these connections between science and philosophy because science, since the Scientific Revolution, has occupied a dominant role in the conception of rationality and our worldviews. Today the excessive specialization and technical sophistication of science tend to hide the fact that science has philosophical underpinnings and consequences.

It is important for both scientists and non-scientists to remember that science is much more than a set of equations: it has both philosophical and cultural consequences for the understanding of the wider world around us. Scientists and philosophers know that science has philosophical dimensions. They are embedded in the actual discoveries. This context is the focus of this study. In particular: how the great scientists saw the impact, which purely scientific discoveries had on our most cherished philosophical notions. We must emphasize that the interaction between science and philosophy is an ongoing concern. Revolutionary discoveries in quantum gravity and quantum cosmology are about to reshape the physicist's view of Nature. Yet again.

Part I

The Philosopher Scientist

2

The Concept of Nature

A new look at nature will throw new light on many fundamental concepts, which dominate science and guide the progress of research.

A.W. Whitehead, *The Concept of Nature* (1920), 25

2.1 Introduction

What humans understand by *Nature* has undergone important changes for two thousand years. We will be concerned with the changes since the Scientific Revolution of the 17th century. The Scientific Revolution itself changed the way Nature was perceived. The founding fathers of the scientific revolution transformed the Greek *organismic* universe into a modern *mechanistic* universe. But the notion of Nature has not remained fixed. Our understanding of the notion of Nature has undergone important changes since the Scientific Revolution. These changes in our notion of Nature occurred as a result of philosophical speculation and empirical findings. The philosophical speculations at first mirror a growing awareness that Nature forms a *system* with many *interconnections*. Immanuel Kant and Pierre Laplace speculated that the whole cosmos had a history. Its present shape was the result of mechanical forces, which had acted from its very origin. Important discoveries in the physical sciences in the 19th century – physical fields and thermodynamic laws – then begin to confirm the philosophical representation of Nature as an interrelated system. Darwin discovered that the biological world had a history. A long evolutionary process had shaped the existence and the form of species. Our understanding of Nature is made up of fundamental notions, which themselves undergo revisions as a result of scientific discoveries. Each revision in our understanding of *time*, *causation* and *determinism* has in turn affected the concept of Nature. And with a changing notion of Nature, the idea of a *physical understanding* of the material world evolves too.

The notion of Nature has an ongoing history. In this chapter we will trace its major evolution since the Scientific Revolution. Some of these conceptual traces are well known; others have not been worked out in any systematic fashion. We will sketch the better-known conceptual steps in the evolution of the concept of Nature and devote more time to the lesser-known developments. We start with a reminder of the tremendous transition from the *organismic* conception of the universe to the *mechanistic* conception of the universe, which occurred during the Scientific Revolution.[1]

2.2 From the Organismic to the Mechanistic Universe

> Nature was no longer a sphinx asking man riddles; it was man that did the asking, and Nature, now, that he put to the torture until she gave him the answer to his questions.
>
> R. Collingwood, *Autobiography* (1939), 78

The traditional Greek conception of the universe attributed to 'brute' matter biological functions. Physical objects possess inherent, natural tendencies. Objects strive towards certain ends. Stones fall to the ground because they strive to regain their natural position on earth, which occupies the centre of the universe. Smoke rises into the air because it strives to regain its natural position in the atmosphere above the earth. These movements are natural because they belong to the *nature* or essences of things. The whole cosmos was seen on the analogy of a biological organism. The analogy at least applies to the tendency of biological organisms to work towards certain ends and goals. The Greek cosmos is an *organismic universe*. Aristotle distinguished four kinds causes, amongst which was the *teological* or final cause. It attributed to natural things goal-seeking behaviour. Leaves grow to provide shade and roots grow to provide food. Equally the behaviour of inanimate objects is explained by relating their observable behaviour to their inherent natures. This 'explains' the downward movement of stones and the upward movement of smoke. The *organismic universe* is fundamentally split between a sphere of perfection and a sphere of imperfection. The sphere of imperfection comprises the central earth, reaching up to the moon. This *sublunar* sphere is a sphere of change, decay and corruption. The sphere of perfection comprises the orbits of the planets lying beyond the moon, including the sun, and reaching all the way to the fixed stars. This

[1] This transition has often been described – see for instance Burtt, *The Metaphysical Foundations of Modern Science* (1924); Cassirer, *Das Erkenntnisproblem in der Philosophie und Wissenschaft der neueren Zeit* (^{3}1922); Collingwood, *The Idea of Nature* (1945); Butterfield, *The Origins of Modern Science* (1949); Dijksterhuis, *The Mechanization of the World Picture* (1961); Koyré, *From the Closed World to the Infinite Universe* (1957); Koyré, *Newtonian Studies* (1965); Hesse, *Forces and Fields* (1961); Čapek, *The Philosophical Impact of Modern Science* (1962); Blumenberg, *Die Genesis der kopernikanischen Welt* (1975); Prigogine/Stengers, *La Nouvelle Alliance* (1979); Pais, *Inward Bound* (1986); d'Espagnat, *Penser la Science* (1990), Chap. 7; Crombie, *Styles of Scientific Thinking in the European Tradition* (1994); Spielberg/Anderson, *Seven Ideas* (1995); Cushing, *Philosophical Concepts in Physics* (1998)

supralunar sphere is a sphere of perfection. In it the planets move in perfect circles, since the circle is the perfect geometric figure. The firmament constitutes the outer boundary of the *organismic cosmos*. It is the panoply of fixed stars. The Greek cosmos is a closed world. Beyond the firmament resides the Unmoved Mover – a Deity who provides the energy to keep the planets moving in their circular orbits. The *organismic universe* is energy-deficient. To keep it running a Deity must constantly pour energy into it. The *organismic universe* requires two different kinds of physics. *Terrestrial* physics for the sublunar sphere of change and corruption and *celestial* physics for the supralunar sphere of perfection.

The description and explanation of the organismic universe relied on *qualitative* concepts and *metaphysical* principles. The very idea that the observable behaviour of physical objects is natural, following the commands of their natures or essences, involves a use of language whose terms escape quantitative precision (Box 2.1). 'Objects strive towards ends': neither the term 'strive' nor the term 'end' lends itself to quantitative analysis and empirical testing. The *principles*, on which the explanation of physical behaviour was based, were themselves of a metaphysical, qualitative kind. Such metaphysical principles were conceptually meant to fulfil *explanatory* roles. And to the Greek mind they may have satisfied this role.

Natura abhorret vacuum – Nature abhors the vacuum
Natura nihil facit frustra – Nature does nothing in vain
Natura non facit saltus – Nature does not make jumps.

Thus, the principle that *nature abhors the vacuum* may appear to explain why there is no void in the physical universe. On closer inspection, however, and with hindsight, these principles fall foul of quantification. Terms like 'nature' and 'abhorrence' cannot be quantified and therefore escape the possibility of empirical testing.

It was in the transition to the *mechanistic* worldview, that the vacuity of these principles was exposed. We shall see in this connection that the philosophical work of Robert Boyle was of particular importance. The transition to the *mechanistic universe* was the collective work of the generations of natural philosophers who developed the heliocentric hypothesis into the clockwork universe. This process matured from the time of the publication of Copernicus's *On the Revolutions of Heavenly Spheres* (1543) to the publication of Newton's *Mathematical Principles of Natural Philosophy* (1687). During this transition, central concepts, which belonged to the *organismic universe*, were slowly replaced by new concepts, which the *mechanistic universe* required. It was an ordered, reasoned transition from *qualitative* to *quantitative* concepts. The philosophical arguments of Boyle and some of his contemporaries have left the conceptual traces, which allow us to reconstruct this transition.

Let us first state the basic elements, out of which the *mechanistic universe* is constructed. It is a universe, constituted out of *matter* and *motion*. Matter is made up of ultimate corpuscles or indivisible atoms. Mechanical laws govern all motion. The *mechanistic universe* is written in the language of mathematics. Its analogy is a clockwork, not an organism. It is a deterministic universe. There is only one kind of physics and the distinction between sublunar and supralunar sphere disappears. Final causes are banished from the *mechanistic universe.* Physical objects posses no natural tendencies. They possess no essential natures. Many of the inherent properties of physical objects are replaced by relational ones. For instance, gravity is not an inherent property of an object, but a phenomenon due to an interaction between two bodies. In the 20th century, as we shall see, physics, as a rational account of Nature, will become relational. In the mechanistic view, physical objects are subject to mechanical causes. In physical objects we have to separate the *primary* qualities from the *secondary* qualities. The *primary* qualities – extension, matter, and motion – are inherent in the physical universe. They exist irrespective of our knowledge of them. The *secondary* qualities – feelings of warmth and cold, perception of colours and sounds – depend on human sensations. They are the response of our perceptual apparatus to the experience of primary qualities, which cause these sensations in us. Finally, the *mechanistic* universe is not a closed energy-deficient world but an infinite energy-sufficient cosmos. The energy-sufficiency of the clockwork universe is maintained in different ways. For Boyle, Leibniz, Kant and Laplace, the universe once set in motion by God, needs no further divine attention. The cosmic clock will tick with perfect regularity. Newton was less certain. God set the universe in motion but the cosmic clockwork needed occasional divine adjustments to keep its regularity.

These basic constituents of the *mechanistic universe* lend themselves to quantitative analysis. Newton emphasises the importance of defining the fundamental notions used in his analysis. He defines the notions of space and time to be used in the new physics and introduces the fundamental laws of motions as the axioms of his system. From this basis further laws can be mathematically deduced.

The birth of the *mechanical worldview* is the most striking philosophical consequence of the first formulation of scientific laws by Kepler and the empirical discoveries, which supported the heliocentric worldview. In the conceptual revolution of the 17th century, *mechanization* and *mathematization* have often been discussed. But part of this conceptual re-orientation was an exclusion of fundamental concepts and principles, which were not accessible to quantification, i.e. concepts like 'essence' and 'form' and 'Natures' and principles like *Natura abhorret vacuum*. These scholastic concepts and principles were not simply dropped. They became the target of philosophical arguments, which attempted to demonstrate their vacuity. The shift from qualitative to quantitative concepts was a reasoned transition. It moved the emphasis from *Natures* to *Nature*.[2] The natural philosophers of the 17th century played an essential part in the argumentative shift. It is one of the many examples in this book of the dialectic connection between physical and philosophical considerations.

---

[2] N. Cartwright has recently reintroduced the term *natures* into the discussion, but not in the sense of scholastic essences; rather they are testable capacities or dispositions; see Cartwright, *Nature's Capacities* (1989) and *Dappled World* (1999)

2.3 From Natures to Nature

> People believe they can compensate for a lack of knowledge by words.
>
> P.T. d'Holbach, *Système de la Nature* (1770), Chap. 6 [translated by the author]

It is clear from this brief characterization that Newtonian physics is much more than a set of equations. It is accompanied by philosophical interpretations, which give rise to a model of the universe compatible with the mathematical equations. In the transition to a new worldview conceptual groundwork needs to be done over and above the formulation of mathematical equations. In this conceptual groundwork fundamental concepts of the old worldview come into review and, if found inadequate, are replaced by new concepts, which are backed by new empirical discoveries. In subsequent chapters we will see this at work in the notions of time and causation. In this chapter we find that one of the most important conceptual transformations, which accompanied the transition to the mechanical worldview, is a redefinition of the concept of Nature.

Many 17th century scientists contributed to this conceptual effort but it is in the work of Robert Boyle that the conceptual trace from *Scholastic Natures* to *Mechanistic Nature* is most clearly delineated. Boyle's philosophical criticism of the notion of *natures* consists of two parts. First, he uses a *nominalist* attack on the scholastic notion of natures in order to show that they play *no* explanatory role (Box 2.1). He then reformulates the notion of Nature, which he considers adequate for a mechanistic worldview. In this constructive part of his argument, Boyle states the *corpuscular philosophy* in all its clarity. But Boyle's formulation contains new elements: in particular he anticipates conceptual developments, which will become of increasing importance in the evolution of the concept of Nature. This is that Nature is a *system* with many *interconnections*. As we shall see, 19th century science – both in physics and biology – finally embraces the idea of *Nature* as an *interrelated system*.

2.3.1 A Nominalist Critique of Scholastic Natures – Robert Boyle

Robert Boyle (1627–1691)

Boyle reconsiders the explanatory usefulness of scholastic principles like *Nature abhors the Vacuum*. When such a principle is employed to explain such a mundane process as water rising in a water pump, does this principle really do explanatory work? Should we be satisfied that water rises in a water pump *because* 'Nature abhors the vacuum'? Boyle's answer is unambiguous: this principle serves no useful explanatory purpose and it skips over the real mechanical processes, which lead to the water rising in a water pump. Boyle is not so much disturbed by our inability to quantify terms like 'Nature' and 'abhor'. His criticism sets in at a more fundamental level. Is it not possible that 'Nature' is just a convenient term – a name – and not a causal agent – a thing –, as it is treated in the scholastic principle?

> I mean, whether it be a real existent thing, or a notional entity somewhat akin to those fictitious terms that men have devised that they might compendiously express several things together by one name (...)

Boyle then goes on to argue, as much as the later Wittgenstein was to do, that the form of language deceives us into thinking that a term is a real physical agent, as when we say that 'law punishes murder' or 'Nature abhors the vacuum'.

> (...) it came to my mind that the naturalists might demand me how, without admitting their notion, I could give any tolerable account of those most useful forms of speech, which men employ when they say that 'nature does this or that' or that 'such a thing is done by nature' or 'according to nature', or else happens 'against nature'? (...) such phrases as that 'nature' or 'faculty' or 'suction' 'does this or that' are not the only ones wherein I observe that men ascribe to a notional thing that which indeed is performed by real agents.

To treat Nature as a physical agent is not to provide a physical explanation of the natural process under scrutiny.

> (...) when a man tells me that 'nature does such a thing', he does not really help me to understand or to explicate how it is done. For it seems manifest enough that whatever is done in the world, at least wherein the rational soul intervenes not, is really effected by corporeal causes and agents, acting in a world so framed as ours is according to the laws of motion settled by the omniscient author of things.

Anticipating his constructive account of the corpuscularian philosophy, Boyle tells his readers that only mechanical explanations will satisfy the natural philosopher.

> [If a man can explain a phenomenon mechanically] he has no more need to think or to say that nature brought it to pass than he that observes the motions of a clock has to say that it is not the engine, but it is art, that shows the hour.[3]

The inefficiency of the scholastic principle can be gleaned from its failure to explain the rising of water in a water pump. The scholastics claim that Nature lifts 'the heavy body of water' against the gravitational pull to prevent a vacuum. Yet when a glass tube is 'but a foot longer than 34 or 35 feet' it will leave 'about a foot of deserted space, which they call the vacuum, at the top of the glass.' This Boyle regards as an anomaly in the scholastic explanation. To avoid it, a mechanical explanation will suffice: the water rises in the water pump 'by the pressure of the atmosphere acting upon the water according to statistical rules or the laws of the equilibrium of liquors settled by God among fluids.'[4]

---

[3] Boyle, *A Free Enquiry* (1686), 32–5

[4] Boyle, *A Free Enquiry* (1686), 65, 106

Box 2.1: Nominalism and Essences

Nominalism is a philosophical position, which originated in the Middle Ages. It had an able defender in the person of Abelard (1079–1142). It was developed to great subtlety by the English monk William of Ockham (*circa* 1285–1349). From its very inception, Nominalism was an opposing force. It attacked the entrenched position of the Middle Ages: Realism.

Medieval Realism is the metaphysical belief that over and above the empirical world of facts, change and chance lay an intelligible world of essences, immutability and necessity. From Platonism derived the idea that *real* knowledge (as opposed to sense impressions) could only be gained from the intelligible world of essences. From Aristotle the Middle Ages had inherited a formal procedure: syllogism. This logical method was used by the Scholastics to guide the intellect's conquest of the world of essences.

For example, both Plato and Aristotle were humans, not animals, because there is an *essence* (also called *form* or *idea*) 'humanity' of which all humans are instances. The essence comes first; the individual thing is its copy. The essence was a worthy object of inquiry, since empirical facts were considered to be too changeable, too imperfect to lead to *real* knowledge. Given the absence of scientific method in the Middle Ages and thus the ability to penetrate beneath the surface of phenomena, this disregard for the empirical world still expressed a serious desire to arrive at structures and give explanations.

But the whole spirit of the age, endorsed by what Ockham called "the authorities" was not factual and objective, like the modern scientific spirit. Knowledge of essences was to bring the medieval mind closer to knowledge of the absolute, i.e. God. Philosophy and theology were interlocked in their search for truth. The tradition, which Ockham attacked, had conceived the essences as entities or realities of a higher order, located in the divine.

This existential claim created the dispute. The question was whether one could hold that besides concrete things, like Fido and Fred, there existed abstract entities like 'dog' or 'animal' and 'man'. These terms denoting genera and species which, in more neutral vocabulary are described as *universals*, became the bone of contention. Such notions had been discussed in Aristotle's *Categories*. The problem was set when the Neo-Platonic philosopher Porphyry in the 3rd century AD wrote an introduction to Aristotle's text in which he refused to commit himself as to whether genera and species were concepts, ideas or essences. In addition to this type of universal, Scholastic philosophy had to wrestle with the further problem, inherited from Platonism, of whether or not universals like *beauty* or *goodness* could be said to exist. The dominant view was that of *Realism*: each individual object and thing in the empirical world partook of an essence, which had a higher and more perfect reality.

Abelard in the 12th century, and Ockham in the 14th century were two of the outstanding *nominalist* thinkers who pleaded for a principle of economy in all explanations. In philosophical jargon this has been termed *Ockham's razor*: 'one should not multiply explanations and causes unless it is strictly necessary;

Box 2.1 (continued)

everything is explained using a smaller number of causes. In the novel *The Name of the Rose* (1980), William of Baskerville offers this just quoted direct translation of the famous formula *entia non multiplicanda preater necessitatem*. It is the best-known formulation (by a 17th century scholar) of Ockham's nominalist principle.

Nominalists denied that universals were essences. Only individual things existed and universality was a property of signs. The nominalists interpreted words like 'man', 'dog' and 'animal' as terms predictable of a number of things. Their universality resided in their meaning. A *universal*, they said, was the result of a mental process of abstraction, a sign which denoted that which several things had in common (being beautiful or human) without referring to extra mental realities (essences). If you can manage with a simple explanation ("universals are signs"), do not look for a difficult one ("universals are realities, essences").

If this sounds very scholastic, Ockham's attack on medieval realism had far-reaching consequences:

- The intelligible world of essences, which had been the preoccupation of the traditionalists, could be dealt with in a logic of language. Universals were signs, not essences. The realm of forms became the object of linguistic research and was no longer that of a 'scientific' inquiry.
- The authorities had explained the existence of the empirical world and the objects, which populate it by means of a principle of individuation, a sort of derivation from the general essences, which were prior to the individual things. Ockham reversed the situation: the individual things are first, the universals being the result of mental activity. The empirical world becomes the foundation of *all* knowledge. All knowledge begins with sensual experience of individual things. There is only one world, and a multiplicity of signs to refer to it.

When Boyle and his contemporaries, who did so much to establish the new mechanistic worldview, attacked the scholastic *natures* they used nominalist arguments introduced by Abelard and Ockham. In the view of the 17th century nominalist scientists the term *Nature* is not a universal in the sense of medieval realism. It is a mere sign, which is used to refer to a multiplicity of empirical things. These empirical objects were the real focus of scientific investigation.

Boyle's *nominalist* critique of the scholastic principles then consists in pointing out that the term *Nature* is erroneously used as a causal agent when in fact it is just a convenient term, 'a compendious form of speech'. There is a tendency in humans to regard what is merely a name for a myriad of individual processes as a thing in itself. *Men ascribe to a notional thing that which indeed is performed by real agents.* Thus the term Nature has come to be identified with an entity, which is supposed to exist over and above the physical objects in the universe. The notion of Nature becomes 'personified', as Paul Thiry d'Holbach was to complain in his *Système de la Nature* a hundred years later.

However, it is more than just a matter of linguistic analysis. If Nature is a superfluous explanatory term than natural philosophy, as practiced by the scholastics, is hampered by their tendency 'to assign imaginary things or arbitrary names as true causes of phenomena.'[5] The problem with the scholastic doctrine of *substantial forms* is that it gives scholastics the impression that they have explained natural processes, when in fact their true mechanical causes are still to be investigated. The doctrine is a hindrance to natural philosophy. There is no place for substantial forms in a corpuscular philosophy. Boyle rejects substantial forms, because the properties of matter are sufficient to explain the phenomena of nature. The Aristotelian doctrine of substantial forms is part of the problem rather than part of the solution.[6]

Before we turn to Boyle's *corpuscular philosophy* as the new paradigm of Nature, we should consider that nominalist thinking was an important conceptual ingredient of the Scientific Revolution. Many of the most important proponents of the new natural philosophy adopted a nominalist viewpoint. Nominalism became an important conceptual tool in the hands of the new natural philosophers to dismantle and re-interpret qualitative scholastic concepts, like 'Natures', 'quality', 'substance', 'substantial forms'. Nominalism denies that such abstract general terms or *universals* have a counterpart in the *particulars* of the physical universe. The external world consists of particulars; universals are only admitted as names. Nominalism therefore encourages a certain orientation towards Nature as an object of scientific study. Nominalism acts as a philosophical presupposition that guides the search for a new view of Nature. We shall see in later chapters that philosophical presuppositions also played their part in considerations of time and causation. Boyle uses nominalist arguments to reorient the study of Nature from the scholastic preoccupation with metaphysical, untestable notions and principles towards an empirical study of the particular physical agents at work in concrete physical situations.

Boyle was not the only nominalist at the time of the Scientific Revolution. A brief review of the practice of nominalist thinking amongst the new natural philosophers will show that Nominalism was an important, if neglected aspect of the conceptual revolution, which spelt out the Scientific Revolution. Although the scholarly emphasis has been on the effects of mechanization and mathematization, the philosophical doctrine of Nominalism played an equally important part in transforming the conceptual system of Scholasticism into a new conceptual network. This is not to say that the natural philosophers were nominalists in the sense of the Middle Ages. Rather, they used nominalist strategies for the particular purpose of remoulding a scholastic system of concepts, which was rapidly failing to provide explanatory patterns of the empirical phenomena.

Francis Bacon retains some of the Aristotelian concepts, like 'form' and 'quality', but invests them with a new meaning.

[5] Boyle, *Free Inquiry* (1686), VIII
[6] Boyle, *The Origin of Forms and Qualities* (1666), VIII

> When we speak of forms, we mean simply those laws and limitations of pure act which organise and constitute a simple nature, like heat, light or weight, in every kind of susceptible material and subject. The form of heat therefore or the form of light is the same thing as the law of heat or the law of light, and we never abstract or withdraw from things themselves and the operative side.[7]

This shift towards the notion of laws of nature has to be seen in connection with Bacon's plea for an experimental natural philosophy. The nominalist trait in Bacon's thinking is clearly manifest.

> The human understanding is carried away to abstractions by its own nature, and pretends that things which are in flux are unchanging. But it is better to dissect nature than to abstract; as the school of Democritus did, which penetrated more deeply into nature than the others. We should study matter, and its structure (*schematismus*), and structural change (*metaschematismus*), and pure act, and the law of act or motion; for *forms* are figments of the human mind, unless one chooses to give the name of *forms* to these laws of act.[8]

With his influential philosophical ideas, Francis Bacon paved the way for a new view of Nature and for the establishment of the new experimental philosophy. On the Continent, René Descartes became an advocate of the mechanistic worldview. Descartes, too, resorts to Nominalism in his defence of a new view of Nature. 'Numbers and universals depend on our mind', he declares.[9] And just like Boyle he refuses to regard Nature as 'quelque Diesse' [some Goddess]. Rather by Nature Descartes understands matter and its properties. The changes, which takes place in matter and Nature, Descartes calls 'les lois de la Nature', because these changes happen according to rules.[10] Other 17th and 18th century philosophers – Gassendi, Malebrache, Locke and d'Holbach – were also nominalists or used nominalist arguments in the establishment of the mechanistic worldview. Robert Boyle was not alone amongst the founding fathers of the new science to choose nominalist positions on the question of what Nature is and how it is to be investigated. William Harvey is famous for his discovery of the circulation of blood (1628). But Harvey also produced philosophical ideas on how 'Nature is herself to be addressed'. Harvey emphasises, like Robert Hooke was to do a little later, the importance in scientific research of combining theory and observation. All science begins with sensual experience but proceeds, by way of reasoning, to general principles and universals.

> Although there is but one road to science, to wit, in which we proceed from things more known to things less known, from matters more manifest to matters more obscure; and universals are principally known to us, science springing by reasonings from universals to particulars; still the comprehension of universals by the understanding is based upon the perception of individual things by the senses.

---

[7] Bacon, *The New Organon* (1620), Book II, §17

[8] Bacon, *The New Organon* (1620), Book I, §51; italics in original

[9] Descartes, *Les Principes de la Philosophie* (1644), I, §§58–9: 'les nombres et universaux dépendent de notre pensée'; similarly d'Holbach, *Système de la Nature* (1770) ends Chap. I by stating that 'Nature is an abstract thing', which must not be personified: we invent words to put them in place of things.

[10] Descartes, *Le Monde et Le Traité de L'Homme* (1664), in *Œuvres* **XI**, 370

In empiricist mode, Harvey argues that abstract ideas follow the sensible ideas given to the mind in sensations:

> (…) sensible things are of themselves and antecedent; things of intellect, however, are consequent, and arise from the former, and indeed, we can in no way attain to them without the help of the others. And hence it is, that without the due admonition of the senses, without frequent observations and reiterated experiments, our mind goes astray after phantoms and appearances.[11]

What Harvey proposes is principally an inductive method. Most proponents of the Scientific Revolution accepted some form of inductivism. Robert Hooke, a contemporary of Newton, constructs his *Philosophical Algebra* on the succession of sense, memory and reason. Unlike many of his contemporaries, Hooke insists on the importance of 'instruments, engines and contrivances to aid the senses'. The advantage of using instruments like the microscope and the telescope to assist the senses is that 'everything is reduced to Regularity, Certainty, Number, Weight and Measure.'[12] Hooke attempts to construct 'a true method of building a solid philosophy' because scholasticism has failed as a scientific method.

> For 'tis not to be expected from the Accomplishments the Creator has endowed Man withal, that he should be able to leap, from a few particular Informations of the Senses, and those very superficial at best, and for the most part fallacious, to be the general Knowledge of Universals or abstracted Natures, and thence be able, as out of an inexhaustible Fountain, to draw out a perfect Knowledge of all particulars to deduce the Causes of all Effects and Actions form this or that Axiome or Sentence, and as it were intuitively, to know what Nature does or is capable of effecting: And after what manner and Method she works; and yet this Method supposes little less.[13]

Skilful experiments and instrument-assisted observations, rather than a 'Belief in implanted Notions', provided the best method of 'discovering the hidden ways of Nature'. Scholastic logic is to be replaced by the method of induction. Universals are generalisations from particulars. Universals are abstract notions to which no scholastic forms correspond. Newton too emphasised this combination of the inductive method with the rejection of scholastic entities. In his *Opticks* Newton deals with magnetic, electric and gravitational forces. He does not regard them as occult properties, in the scholastic sense, but as lawful regularities, the effects of which can be observed in the phenomena. Only the causes of these effects are, according to Newton, still obscure or 'occult'.

> To tell us that every Species of Things is endow'd with an occult specific Quality by which it acts and produces manifest Effects, is to tell us nothing: But to derive two or three Principles of Motion from Phaenomena, and afterwards to tell us how the Properties and Actions of all corporeal Things follow from those manifest principles, would be a very great step in Philosophy, though the Causes of those Principles were not yet discover'd: And therefore I scruple not to propose the

[11] Harvey, 'Anatomical Exercises' (1651), in *Works* (1847), 154, 157

[12] Hooke, 'A General Scheme' (1705), §IV; for more quotes, in a similar vein, see Shapin/Schaffer, *Leviathan* (1985), 36–8

[13] Hooke, 'A General Scheme' (1705), 6

> Principles of Motion above mention'd, they being of very Extent, and leave their Causes to be found out.[14]

The cause of gravitation could, according to Newton, remain a metaphysical speculation, for which there is no room in science. What cannot be inductively inferred from phenomena remains for Newton a mere *hypothesis*, which has no place in experimental philosophy. It is true that natural phenomena, like the attraction between the sun and the earth, can be explained through gravitational forces.

> But hitherto I have not been able to discover the cause of those properties of gravity from phenomena, and I frame no hypotheses; for whatever is not deduced from the phaenomena is to be called an hypothesis; and hypothesis, whether metaphysical or physical, whether occult qualities or mechanical have no place in experimental philosophy. In this philosophy particular propositions are inferred from the phaenomena, and afterwards rendered general by induction. Thus it was that the impenetrability, the mobility, and the impulsive force of bodies, and the laws of motion and of gravitation, were discovered. And to us it is enough that gravity does really exist, and act according to the laws which we have explained, and abundantly serves to account for all the motions of the celestial bodies, and of our sea.[15]

In the 17th century a tight connection exists between a) the empirical investigation of natural phenomena as well as the inductive generalization of empirical data to lawful regularities and b) the rejection of scholastic entities due to nominalist arguments. Modern science arises out of an engagement with the scholastic tradition. In the ensuing conceptual transition – Bacon's reinterpretation of Aristotelian forms to laws or Boyle's redefinition of the notion of Nature – nominalist arguments played a decisive role. They help transform the outdated scholastic paradigm of Nature into a new view, which is compatible with experimental philosophy.

2.3.2 Corpuscular Philosophy

After the nominalist dismantling and remoulding of old concepts comes a constructive rebuilding of a new paradigm of Nature. Robert Boyle calls this new paradigm the *corpuscular philosophy*. Many 17th century philosophers had helped to build the mechanistic worldview. What distinguishes Boyle is that he was a chemist, an experimental philosopher, to whom we owe Boyle's law. Boyle proposes a new view of Nature, nourished by a philosophical commitment to Nominalism and an empirical commitment to Investigation. Nature is not an agent, but a system of rules [laws] for material bodies. If we speak of Nature generally, we mean by it an aggregate of material bodies, with their laws of motion. In this sense Nature is a 'cosmical mechanism'. If we speak of the particular nature of things, we mean by it the individual mechanism of a particular body.[16] This also answers the question as to what is *real* under the Corpuscular Philosophy: material bodies and their motions, and the primary qualities inherent in these bodies comprise *reality* for

[14] Newton, *Opticks* (^{3}1721), 377

[15] Newton, *Mathematical Principles* (1687), Vol. II, Bk. III, General Scholium, 314

[16] Boyle, *A Free Inquiry* (1686), IV, VII

this view of Nature. By contrast, Aristotelian forms are not real. The corpuscular hypothesis encompasses the following principles:

I. There is a universal matter common to all bodies; a body is a 'substance extended, divisible and impenetrable'
II. To differentiate this universal matter 'into a variety of natural bodies, it must have motion in some or all its designable parts'
III. Matter must be divided into sensible corpuscles or particles, which have size and shape; corpuscles are either at rest or in motion.[17]

According to this corpuscular philosophy, the world is *a self-moving engine*, comparable to a *rare clock*. This *clockwork image of the universe*, which became a standard analogy for the mechanistic universe in the 17th century, replacing the organismic analogy of the Greeks, appears first in the writings of the 14th century natural philosopher Nicolas Oresme. [18] Boyle is much impressed with the clock analogy. It should replace the scholastic notion of the universe as a puppet on a string, guided by God, the immovable mover, residing outside the closed universe. The universe, he writes

> is like a rare clock, such as may be that at Strasbourg, where all things are so skilfully contrived that the engine being once set a-moving, all things proceed according to the artificer's first design, and the motions of the little statues that at such hours perform these or those things do not require (like those of the puppets) the peculiar interposing of the artificer or any intelligent agent employed by him, but perform their functions upon particular occasions by virtue of the general and primitive contrivance of the whole engine.[19]

The world as a self-moving engine or as gigantic clockwork: this is a well-established analogy of the mechanistic universe. Boyle adds an afterthought, however, which anticipates an understanding of the notion of Nature whose importance became dominant only in the 19th century. It is the idea that the world is a *great system of things corporeal.*

> For we must consider each body, not barely as it is in itself, an entire and distinct portion of matter, but as it is a part of the universe, and consequently placed among a great number and variety of other bodies, upon which it may act, and by which it may be acted on in many ways.[20]

As we shall see, the *idea of Nature* as an *interconnected system* at first appears as a philosophical thesis in the 17th and 18th centuries. At that time its significance is still eclipsed by the main ingredients of a corpuscular philosophy, matter and

[17] Boyle, *The Origin of Forms and Qualities* (1666), 460–2; see also Burtt, *Metaphysical Foundations* (1924), Chap. VI; Hesse, *Forces and Fields* (1961), Chap. V; Holton, *Thematic Origins* (1973), Chap. I; Shapin/Schaffner, *Leviathan* (1985)

[18] Wendorff, *Zeit und Kultur* (^{3}1985), 144

[19] Boyle, *A Free Enquiry* (1686), 13; *The Origin of Forms and Qualities* (1666), 474

[20] Boyle, *The Origin of Forms and Qualities* (1666), 464. With the title of his *Treatise of the System of the World* (1728), Newton also anticipates the idea of interrelatedness, although it is restricted to planetary motion.

motion. But with the empirical discoveries of the 18th and 19th century this view of *Nature* as an *interrelated system* begins to dominate the philosophical thinking of major scientists.

2.4 The Emergence of Nature as an Interrelated System

> Nature can be thought of as a closed system whose mutual relations do not require the expression of the fact that they are thought about.
>
> A.N. Whitehead, *The Concept of Nature* (1920), 3

Many of the leading philosophers of the 18th and 19th century describe Nature as a system. Although P.L. Moreau de Maupertuis attacks materialism for its inability to explain the formation of organic matter, the title of his book *Système de la Nature* (1768) encapsulates the philosophical thinking of the age. Nature is to be thought of as a system of interrelated components.

The mere thought of Nature as an interrelated system is not a new idea.[21] The organismic worldview thought of Nature as a vast organism. With this view of Nature as an organism, it is natural to think of all phenomena as mutually dependent on each other. Individual components of this world organism functionally contribute to the function of the whole organism: a vast cyclic movement around the stationary centre. This world organism falls under the ideas of *purpose* and *final causes*, as do its individual components. Individual objects possess inherent tendencies – natures – to fulfil their individual purposes. These inherent tendencies provide the cause of their natural motions either towards or away from the centre of the universe.

The mechanistic worldview thought of Nature as a vast clock. Again, it is natural to think of all phenomena as mutually dependent on each other. But the mutual interdependence takes on a new meaning. This clockwork universe does not fall under the idea of purpose but under the idea of mathematical necessity and efficient causation. The dissolution of the notion of causation into mathematical relations lies at the root of the Laplacean identification of causation and determinism (Chap. 5). The constituents of the mechanistic world are still related to each other, but the relation is captured in the laws of nature. The solar system consists of a central body (the sun) and a certain number of orbiting planets. The sun and the planets are the constituents of the system. All these bodies exercise a mutual attraction on each other. The orbits reveal a regular pattern, which is expressed in Kepler's laws. The interaction between the constituents of the solar system is such, that from variations in the orbit of one body, the existence of another body can be inferred. This is how Neptune was discovered. Slight discrepancies in the observed orbit of Uranus led Englishman John C. Adams and Frenchman Urbain J.J. Leverrier to predict the existence of a yet undiscovered

[21] Cassirer, *Das Erkenntnisproblem* Vol. I (³1922/1974), 207, 399; Collingwood, *The Idea of Nature* (1945)

planet in 1845. The gravitational effect of this unknown planet accounted for the slight variations in the orbit of Uranus. In 1846 the German astronomer Johann Galle observed Neptune in its predicted position. The interrelations between the constituents of a mechanical system are of a *mechanical* kind. They are subject to mathematical treatment, to prediction and causal analysis in terms of efficient causes. (An efficient cause accounts for the observable effects in terms of a set of relevant antecedent conditions.) The interrelations between the constituents of an organismic system are of an *organismic* kind. They not are subject to mathematical treatment, to prediction; they are only subject to causal analysis in teleological terms. (A final cause accounts for an observable phenomenon in terms of its function.)

It is important to note, however, that organic interrelations, subject to mathematical treatment, to prediction and causal analysis in terms of efficient causes, exist between organisms. Darwin stressed this phenomenon. Darwin did not think of Nature as a vast organism but as a mechanism.

Darwin belongs to the 19th century when the idea of Nature as a system became prominent amongst leading scientists. Let us remain with the philosophical speculations of the 18th century. Under the impact of the Scientific Revolution, Nature had been transformed from a closed world into an infinite cosmos. The term *cosmos* is increasingly adopted by philosophers and later scientists to capture the idea of an interrelated system. In his *De L'Interpretation de la Nature* (1743), Diderot defines Nature as the concatenation of all phenomena (*enchaînment de tous phénomènes*). It is for him a 'universal system.' For d'Holbach Nature and the universe are an uninterrupted and unfathomable chain of causes and effects. In Nature everything is related by necessary laws. D'Holbach distinguishes Nature as a system in a narrow sense – a connection of elements within a single object or organism – and Nature in a broad sense – a vast system of all the subsystems. In this cosmos, this universal nature, all the subsystems depend on each other and on the cosmic system.[22] In his famous definition of determinism (1820), Laplace took up this view of the cosmos as a vast concatenation of causes and effects, which led him to an identification of determinism and causation (see Chap. 5).

In these early philosophical interpretations of the Scientific Revolution the new view of Nature as an interrelated system is mentioned but not yet systematically worked out. We have to turn to Kant and in particular to Alexander von Humboldt to see it spelt out in its glorious details. In the writings of these two philosophers Nature becomes a *Cosmos*. The adoption of Nature as a cosmos of physical interconnections, expressed in the laws of Nature, reveals the clockwork image as no more than a useful analogy. It is at best an analogous model, which may capture some positive analogies but always at the expense of the negative analogies. All analogous models suffer from this effect: they express the unknown in terms of what is known but they carry no guarantee that the similarities are real.

[22] D'Holbach, *Système de la Nature* (1770), Part I, Chaps. 1–6 ; Leibniz, *Monadology* (1714), §56 equally postulates the universal connectedness of all things.

2.4.1 Immanuel Kant

Immanuel Kant (1724–1804)

To treat Nature as a cosmos of interrelations opens up the possibility of giving a mechanical theory of its development. This is exactly what Kant attempted in his *Allgemeine Naturgeschichte und Theorie des Himmels* (1755). The book is remarkable in its attempt to provide a purely mechanical account of the evolution of the cosmos. From the very beginning of the argument, the idea of Nature as an interrelated system figures prominently in Kant's *Theory of the Heavens.* The Preface speaks of the discovery of the systematic connection in Nature of 'the great chains of creation in the full scope of infinity'.[23] The whole cosmos is the effect of blind mechanical laws of motion. The whole cosmos is a system of subsystems. The Milky Way is a system, as much as the so-called fixed stars. All these systems embedded into larger systems reveal the systematic connection of the whole universe. Kant becomes one of the first thinkers to speculate that the cosmos has a history – a history, which makes it evolve from an original chaos to the systematic order, due to the operation of mechanical laws. Kant speculates that millions upon millions of years will pass by for the present order of the universe to be established. Creation is not the work of a moment.

> There had perhaps flown past a series of millions of years and centuries, before the sphere of ordered Nature (...) attained to the perfection, which is now embodied in it. (...) The sphere of developed Nature is incessantly engaged in extending itself. Creation is not the work of a moment (...). Millions and whole myriads of millions of centuries will flow on, during which always new Worlds and systems of Worlds will be formed, one after another, in the distant regions away from the Centre of nature, and will attain to perfection (...).

Vast amounts of time – mountains of millions of centuries as Kant puts it – will be needed to generate the present order. And the universe will never cease producing new order.

> This infinity and the future succession of time, by which Eternity is unexhausted, will entirely animate the whole range of Space to which God is present, and will gradually put it into that *regular order* which is conformable to the excellence of His plan. (...) The creation is never finished or complete. It has indeed once begun, but it will never cease. It is always busy producing new scenes of nature, new objects,

[23] Kant, *Allgemeine Naturgeschichte* (1755), Vorrede, 227 [author's own translation]; in his Introduction to the *Encyclopaedia*, d'Alembert distinguishes the spirit of philosophical systems, which he rejects, from the systematic spirit, which produces a welcome unification of phenomena; see d'Alembert, *Discours Préliminaire* (1751); Diderot, *L'Interprétation* (1770), §XXI also favours 'recueiller et lier les faits' and is opposed to 'l'esprit rationnel'.

> and new Worlds. (...) It needs nothing less than an Eternity to animate the whole boundless range of the infinite extension of Space with Worlds, without number and without end.[24]

Furthermore, Kant becomes the first scientist to give a systematic account of what the Greeks regarded as *fixed* stars. They were in fact galaxies like our Milky Way. Hence, they were systems in their own right. What Kant means by the expression 'systematic constitution of the cosmos' – *systematische Verfassung des Weltbaus* – is illustrated by reference to our planetary system. The planets and comets in the solar system already constitute a *system* because they orbit a central body. But what makes it truly a system is that its constituents are related to each other in a systematic and uniform fashion, mathematically describable by mechanical laws.[25]

The kind of mechanical account of the evolution of the cosmos – a mechanical system, consisting of subsystems, all interrelated –, which Kant attempts, is revealed in the subtitle of his *Natural History*: *Essay on the Constitution and the Mechanical Origin of the Whole Cosmos, treated according to Newton's Laws*. This puts Kant's attempt squarely in the tradition of classical physics. But there is more in Kant's notion of Nature than the clockwork image would suggest. In his Critical Period, Kant provides a formal definition of Nature.

> By nature, in the empirical sense, we understand the connection of appearances as regards their existence according to necessary rules, that is, according to laws (*Critique* A216, B263; *Prolegomena* §14)

Kant's *Critique of Pure Reason* distinguishes a *material* notion of Nature – as the sum of all appearances (B163) – from a *formal* notion – as the sum of universal laws (B165).[26] It is not the job of the philosopher to furnish the empirical laws of nature. This is the job of the physicist through the study of Nature's appearances. The philosopher's task is to state the fundamental *a priori* laws of Nature:

- In all change of appearances substance is permanent; its quantum in nature is neither increased nor diminished. (*Critique*, 1781/1787 A182, 205; *Metaphysische Anfangsgründe*, 1786, 106)

If we regard substance, on the analogy with the corpuscularian philosophy, as a lump of solid matter, a Principle of Permanence of Substance would not be an acceptable view of Nature. Antoine Lavoisier (1773/74) showed that the quantity of matter does not remain invariant in combustion. A burned-up piece of metal weighs more than the original piece. And after the discoveries of thermodynamics it became fully clear that matter and energy are interchangeable. Kant's *Principle of*

[24] Kant, *Allgemeine Naturgeschichte* (1755), 334–5, English text quoted from Toulmin/Goodfield, *Discovery of Time* (1965), 132–3; for an assessment of Kant's cosmology, see also Falkenburg, *Kants Kosmologie* (2000)

[25] Kant, *Allgemeine Naturgeschichte* (1755), 253, 326–35

[26] See also Kant, *Prolegomena* (1783), §17; *Metaphysische Anfangsgründe* (1786), Vorrede; Cassirer, *Erkenntnisproblem* II (1906), 671; for a discussion of Kant's analogies of experience, see Weizsäcker, *Einheit der Natur* (1971), Chap. IV.2; Mainzer, *Symmetries* (1996), §35.21; Torretti, *Philosophy of Physics* (1999), §3.4

Permanence of Substance does not refer to substance in the traditional sense. It is a conservation principle. Kant states that in nature some quantity remains invariant but it is the job of physics to determine what the invariant quantity is. With this insight Kant became the precursor of the modern invariance view of reality (see below Ch. 2.8).

- All change in matter has an external cause. (*Metaphysische Anfangsgründe*, 1786, 109; *Critique*, 1781 A189)

This, too, proved to be an untenable principle. This discovery of radioactive decay, without any discernible external cause, posed serious philosophical problems regarding the traditional Laplacean notion of causation.

- Principle of Coexistence, in coordance with the Law of Reciprocity or Community. In the first edition of the *Critique* this principle read:

> All substances, so far as they coexist, stand in thoroughgoing community, that is, in mutual interaction (*Critique*, 1781 A211/B256; *Metaphysische Anfangsgründe*, 1786, 110)

The insistence on mutual interaction or the interrelatedness of all parts of the cosmos is, as we have observed, the vital ingredient that takes the philosophical conception of Nature beyond the clockwork image. This is an ontological view of Nature. After the discovery of the relativity and quantum theory, the view of Nature becomes relational and formal: *the real is the invariant*. But even Kant's *Principle of Coexistence* suffered modification. Kant held that 'things are coexistent so far as they exist in one and the same time' (*Critique* B258). With the Special theory of relativity this innocent-looking statement became problematic. Kant, as we shall see, did not share Newton's view that time was *absolute* (independent of all material happenings). He agreed with Newton, as did every other thinker before Einstein, that time was *universal* (the same for all observers). So he would not have considered the simultaneity of events to be a problematic notion. All observers would agree on the periods when things co-existed. But Einstein disagreed: even the simultaneity of events was only relative.

What is important in these considerations is the step from metaphor – the book of Nature is written in the language of mathematics (Galileo), the universe is like a rare clock (Boyle) – to philosophical paradigm of research. Nature is a system of interrelated systems, governed by mechanical laws. For Kant, as the subtitle to his *Theory of the Heavens* indicates, the universe was governed by Newton's deterministic laws. In every particular theory of nature we can only find so much real science as there is mathematics in it.[27] The Kantian notion of Nature becomes irrevocably linked to the necessity of mathematical laws.

The mathematics available at Kant's time was much more suited to an atomistic approach to Nature than to the idea of Nature as a system of interrelations. The differential equations of classical physics could describe the trajectories of the particles to any desirable degree of accuracy. Laplace's demon was at work. The

[27] Kant, *Metaphysische Anfangsgründe* (1786), 14–5

concept of a physical field had not yet been developed. Philosophical speculation was far ahead of concrete developments in science. But it provided the new terminology – physical systems and their interrelations – with which the results of the particular sciences could be integrated into a new conceptual network of what was to be meant by Nature. This in turn would affect the fundamental notions: causation, matter, motion, time and space.

We should note for future reference a certain degree of freedom of the philosophical issues from the concrete science. We have characterized the bond between philosophy and science as a dialectical relationship. There is give and take. Science borrows notions from philosophy. Philosophy takes scientific results as constraints. The philosophical consequences of science are not logical deductions. There is room for manoeuvre. We see this in Kant's case. Although Kant is heavily indebted to Newton for the scientific results, he does not agree with Newton on the interpretations of the fundamental notions. Kant rejects Newton's realist interpretations of time and space (Chap. 4). For Newton both time and space existed in the real physical world irrespective of material happenings. They were physical properties of the universe. For Kant they were objective properties of the structure of the human mind. This idealist view had a powerful influence on 20th century physics. Insofar as Newton had any explicit conception of causation, it would have been a functional view (Chap. 5): the idea that causation was exhausted by the existence of differential equations. Here again, Kant would disagree. Causation was an *a priori* category of the human mind with the help of which phenomena could be ordered into an objective relation.

2.4.2 Alexander von Humboldt

Alexander von Humboldt (1769–1859)

Perhaps the most comprehensive and systematic interpretation of Nature as an interrelated system is to be found in the writings of Alexander von Humboldt. In 1844 von Humboldt published four volumes of a 'sketch of a physical world description' with the simple title *Kosmos.* It represents an encyclopaedic sweep across the whole range of human knowledge from the cosmically large to the microscopically small. Von Humboldt sees scientific research as a synthesis of empirical observation and rational reconstruction (Vol. III, Introduction, 3–6), while the whole of Nature is an interrelated cosmos – 'a mutual interaction of the forces of nature' (Vol. I., Introduction, 27–8). Von Humboldt defines the cosmos, reminiscent of Kant, as the world order. In his description of the highest purpose of a physical description of the world, von Humboldt reveals what he means by world order. It is an anticipation of modern thinking. The highest purpose is the knowledge of unity in the manifold of appearances. It is *unification*.[28] Thus the idea of a cosmos – of an interrelated network of physical laws – offers an immediate advantage to *Homo sapiens*. There is order in nature such that diverse, disparate phenomena can be reduced to some underlying principle, and empirical laws can be derived from fundamental laws. Nature is an interrelated whole such that from the knowledge of some parts we can access to knowledge of other parts. For the purpose of von Humboldt's *Kosmos* is to present the inner concatenation of the general and the particular; it is to seize the permanence of laws in the flux of phenomena (Vol. I., Introduction, 3–5). 'For a reflexive contemplation nature is unity in diversity.' (Vol. I., Introduction, 5; author's own translation) If Nature is a cosmos, an embedding of smaller systems in larger systems, than the discovery of one law points to a higher law, which is yet to be discovered. But although scientific research reveals a universal concatenation, this universal claim is not to be understood as a simple linear order; rather it is like a web (Vol. I., Introduction, 23).

> The more we penetrate into the essence of the forces of nature, the more we reorganize the connection of phenomena, which, if observed individually and superficially, seem to resist every attempt to join them, and the more we render possible the simplicity and terseness of our presentation. (*Kosmos*, Vol. I., Introduction, 21–2; author's own translation)

Von Humboldt concedes that the human mind cannot yet grasp the unity of the cosmic phenomena in all their diversity. His idea of a cosmos is a research programme. However, it is already partially realized insofar as the human mind has already grasped a partial insight into the relative dependence of all phenomena (Vol. I, 57; Vol. III, 7; Vol. IV, 12).

There has always been a strong tendency in Western thinking about Nature to reduce *becoming* to *being*, to explain the change of phenomena in terms of the permanent.[29] But von Humboldt has a thoroughly *dynamic* view of Nature: 'through being a small part of becoming is revealed' (Vol. III, 8). Nature is not a dead aggregate of matter. The natural systems we observe are the result of the 'joint action of the forces in the cosmos, the mutual causation and concatenation of the products of nature' (Vol. I, Introduction, 27–8). Being and becoming cannot be separated – the being we observe is the effect of the process of becoming in the past (Vol. I., 43–4).

We see in this insistence on the dynamic view of Nature the limits of the earlier clockwork image of the cosmos. A clockwork is governed by a very limited set of

[28] von Humboldt, *Kosmos* (1844); for another early statement on *unification* in science and the interaction between branches of science, see Du Bois-Reymond, 'Über die wissenschaftlichen Zustände der Gegenwart' (1882). 'This is why our world exhibits some degree of order rather than complete disorder, why it is a cosmos rather than a chaos,' says Popper, *Quantum Theory and the Schism of Physics* (1982), 178. Modern work on unification in science can be found in Friedman, 'Explanation' (1974), 'Theoretical Explanation' (1981) and *Foundations* (1983); for dissenting voices, see Galison/Stamp *eds.*, *Disunity of Science* (1996)

[29] See Čapek, *Philosophical Impact* (1962); Popper, *Open Universe* (1982); Prigogine, *End of Certainty* (1997), 11, 58

mechanical laws. There is no suggestion of a unity in diversity, no suggestion of a hierarchy of laws. Strange as it may sound, there is no suggestion of a history or an evolutionary development in the clockwork image. In a curious way, a clock is time-symmetric:[30] taking photographs of a clock at different moments in time, does not allow us to deduce a unique direction in the movement of the clock hand. The clock hands could have reached the 'later' stages by going anti-clockwise, rather than clockwise, as conventionally assumed.

Questions of determinism and causation, which are already embedded in the view of Nature as a clockwork, naturally pervade the conceptual network of those who embrace this image. The usefulness of such concepts is only questioned in science, as we shall see, when the experimental evidence raises doubts about the adequacy of certain notions. This happened with the advent of quantum theory. But questions of determinism, of being and becoming, also lie embedded in the conceptual network, in which the Special theory of relativity is grounded. Many prominent physicists took the Special theory of relativity to have demonstrated the Kantian view that the passage of time was a human illusion. Physical reality is static (see Chap. 4). For this reason, von Humboldt's view of Nature as an *interrelated* and *dynamic* cosmos – a cosmos with an evolutionary history describable by purely mechanical law, which Kant tried to chronicle – is an important contribution, the true significance of which could not be appreciated before the advent of the relativity theory and the renewed debate about being and becoming. When von Humboldt published his monumental *Kosmos*, in 1844, the natural sciences finally entered a phase of empirical discovery, which would end the philosophical speculation about the structure of Nature. The notion of Nature as an interrelated system took on empirical significance. Four developments were of particular importance in this step from philosophical speculation to empirical hypothesis:

- Darwin's evolutionary theory
- The discovery of atomic structure
- The laws of thermodynamics
- The field concept.

2.5 The End of Philosophical Speculations About Nature as a System

Before concrete empirical discoveries were made to confirm and establish the new view of Nature as an interrelated system, we observe a number of publications by well-known scientists on systematic ways of uniting the scientific knowledge of the day into coherent encyclopaedic systems. These attempts had philosophical predecessors, as in Diderot's *Encyclopédie ou Dictionaire Raisonné des Sciences, des Arts et des Métiers*, in 28 volumes (1751–72), and von Humboldt's *Kosmos* (1844). We can see in these attempts to unify scientific knowledge a reflection of the growing awareness that Nature is not simply a disconnected assemblage of phenomena. A good example is A.M. Ampère's *Essai sur la Philosophie des Sciences* (1834). As

[30] Cf. Popper, *Quantum Theory* (1982), 88

its subtitle – *Exposition Analytique d'une Classification Naturelle de Toutes les Connaissances Humaines* – indicates, it is an attempt to provide a natural classification of all human knowledge. Similar to social scientists, like Auguste Comte and Adam Smith, Ampère believes that a correspondence exists between the stages of the evolution of knowledge and historical epochs or successive developmental stages of childhood. There exists a hierarchical classification of all knowledge: 'the study of Man must not come before the study of the world and of nature' (*l'étude de l'homme ne doit venir qu'après celle du monde et de la nature*). And all knowledge acquisition proceeds in an inductive fashion: first, collection of observed facts, then the deduction of general laws and cause-effect relationships.

On the other side of the English Channel, J.F. William Herschel, the son of William Herschel – discoverer of Uranus – made similar observations. His *A Preliminary Discourse on the Study of Natural Philosophy* (1830) treats of the 'Subdivision of Physics into Distinct Bodies and their Mutual Relations' (Part II) and of 'the Classification of Natural Objects and Phenomena' (Part II, Chap. V). In the inductive spirit of the age, natural history is presented 'as an assemblage of phenomena to be explained' (1830, 221). Nevertheless, W. Herschel reveals an awareness, indicative of the new view of Nature, that science is more than 'a collection of facts and objects presented by nature.' For he writes:

> Modern chemistry has established that the universe consists of distinct, separate, indivisible atoms, making up larger bodies. Modern [chemical] discoveries destroy the idea of an eternal self-existent matter, by giving to each of its atoms the essential characters, at once, of a manufactured article and a subordinate agent. (*Preliminary Discourse*, 1830, Part I, §28, 37–8)

It would not take long before the atom was also conceived as a system. In due course, concrete empirical discoveries confirmed philosophical speculations about Nature as an interrelated system. These discoveries then led to philosophical challenges to the adequacy of the established fundamental notions.

2.5.1 Charles Darwin

Charles Darwin (1809–1882)

In the *Origin of Species* (1859) Darwin did not make the interrelatedness of Nature a central theme. But this view of Nature, inspired by biology, is an important consequence of the evolutionary theory of species. It could not have been a theme of the pre-Darwinian view of Nature. The *Great Chain of Being*, the dominant paradigm of attempts to explain the diversity of species prior to Darwin, was a sort of biological atomism. Each species was created by a Deity on a particular rung on a hierarchical ladder, which presented the *Chain of Being*. The existence of each being in its predetermined location on the ladder did not depend for its existence on the existence of any other being on lower or higher rungs. The existence of species depended on the will of God. The close resemblance in appearance of the ape to humans did not worry the believers in the *Great Chain of Being*. Although ape and man seemed quite close, God had ensured the gulf between them and guaranteed their separation by placing them on different rungs. Man was placed on a higher rung than the ape. So man was closer to God at the top of the ladder than the ape.

All this changed with the advent of Darwin's evolutionary theory. Darwin's account was at first misunderstood as claiming that humans had descended directly from the ape. Rather, Darwinism postulates that they share a common ancestor. Secondly, the reassuring sense of hierarchy was destroyed, for species evolved from earlier species by a process of natural selection. Thirdly, there was no longer any sense of unidirectional progression towards 'higher' forms of life. Rather, species adapted to their natural habitat. Adaptation can lead to a loss of complex organs, like the eye. The idea of interrelatedness is already built into Darwin's formula: it is a *theory of descent with modification*. Or, in modern parlance:

Evolution = Natural Selection + Random Mutations.

Since present-day species are the *dynamic* result of past changes (in the environment and the genetic make-up of earlier members of a species), evolutionary forces affect species at any moment in time. Species are dependent on the environmental niche, in which they struggle for survival against competitors. Any ecological niche provides food resources for many different species. There is intense competition for the food resources from both members of the same species and members of other species. Both seagulls and seals eat fish. And seagulls compete for fish amongst themselves. According to Darwin's original explanation the scarcity of food and the abundance of competitors for the limited food supplies bring forth a continuous struggle for existence, in which those with a slight advantage tend to survive and those with a slight disadvantage tend to perish. There is thus an *interdependence* of the characteristics of a species and the particular environment, in which it seeks to survive. Ability to gather food and feed the young, and avoidance of predicators through camouflage or other means are examples of such characteristics. The stronger the lion, the faster the antelope. There is interdependence in the biological realm, as Darwin noted in the first edition of *The Origin of Species* (1859):

> (...) the structure of every organic being is related, in the most essential yet often hidden manner, to that of all other organic beings, with which it comes into competition for food or residence, or from which it has to escape, or on which it preys. (Darwin, *Origin*, 1968, 127)

The interrelatedness also exists in a diachronic sense. Species evolve from each other. They have similar body plans and share many features. *Homologies* – the same organ, like the heart, appears in many different creatures under a variety of forms – can be explained by relating the histories of descent to a common ancestor. *Analogies* – organs, which have the *same* function in otherwise unrelated animals – can be explained as responses to similar environmental pressures. It is well known that Darwin's evolutionary theory had a tremendous impact on the view of Nature and the self-image of humans. With Darwin's explanation of the emergence of living creatures, humans seemed to lose their privileged place in the natural world as closer to God than brute beasts. For Darwin subjected all animate matter to the operation of evolutionary forces. The Principle of Descent with Modification, not the Act of Creation, provides the explanatory core of life. This was one of the reasons why Darwin's account was attacked. The other reason was Darwin's materialism. Even complex organs like the eye and the human brain emerge through natural selection. Materialism existed of course before Darwin. Paul Thiry d'Holbach characterized humans as biological machines. They were a product of nature like any other living automat. He even speculated that humans emerged on earth in the course of time.[31] Darwin's materialism is no longer expressed in terms of a simple mechanism. Darwin partakes of the evolution in the view of Nature, which we have observed in the physical sciences. In fact the Darwinian idea of evolution is not compatible with a simple mechanistic view of Nature.[32] The simplicity of the Great Chain of Being is analogous to the simplicity of the clockwork universe: in both images the components are simply juxtaposed. With the idea of interrelatedness, which is also built into Darwin's evolutionary view, the juxtaposition is replaced by the idea of a multi-layered interaction between the components of the system. Theoretically, this opens the route to deduction and unification. Darwin also diverts biological thinking from the path of determinism. It was still prevalent in the physical sciences during his lifetime. Evolution is a stochastic process. Very soon, physics would follow suit.

2.5.2 The Discovery of Atomic Structure

Eventually, this new view of Nature as a system would appear as incompatible with classical atomism. Just as it is mistaken to regard Nature as a mere collection of biological entities, so it is mistaken to regard it as a mere collection of atoms, governed by the laws of motion. The ultimate constituents of Nature are not the point particles of Newtonianism. The atoms themselves are to be conceived as systems. This was confirmed with the discovery of radioactivity and the detection of the electron inside the atom. J.J. Thomson, who discovered the electron in 1897, conceived of the atom as a sphere of positive electricity with negatively charged electrons embedded in it, arranged in concentric rings. This is often called the *plum-pudding model* of the atom. But the pudding model of the atom did not consider the existence of a nucleus. It was quickly shown to be incompatible with new evidence about the scattering of atoms. The system-like nature of the atom was finally established when E. Rutherford and his co-workers discovered the nucleus

[31] D'Holbach, *Système* (1770), Chap. VI; La Mettrie's *L'Homme Machine* (1748) stands in the same tradition, as does, after Darwin's *Origin*, Huxley's essay on Animal Automatism (1874); see Mayr, 'Darwin's Influence' (2000), 67–71

[32] See Collingwood, *Idea of Nature* (1945), 14–5; T.H. Huxley, Darwin's able defender, explicitly embraces the view of Nature as a complex system; see his 'Lectures on Evolution' (1876), 46–7

inside the atom and established its properties (1911). The Bohr-Rutherford model of the hydrogen atom (1913) was inspired by an analogy with our planetary system.[33]

Jean Perrin established the reality of the atom, which had previously been doubted by scientists. For Perrin, the reality of the atom had far-reaching consequences for the view of matter and Nature. The point about the atomic hypothesis, as J. Perrin observed[34], is twofold: *firstly*, matter is porous and discontinuous – it has a granular structure – and *secondly* radioactivity shows that atoms and therefore matter are not immutable. The study of matter is greatly simplified by the atomic hypothesis since

> The whole universe, in all its extraordinary complexity, may have been built up by the coming together of elementary units fashioned after a small number of types.

Thanks to the atomic hypothesis any 'material system' can be decomposed into smaller units. Perrin implicitly indicates that classical atomism is under threat from these developments. The granular structure of matter meant that the world of atoms could not be adequately represented by Newton's mechanical laws. The porous and discontinuous nature of matter forced Max Planck, in 1900, to introduce his famous quantum hypothesis. With some regret Planck admitted that Nature made random jumps after all – albeit quantum jumps. If this admittance posed worries for the classical conception of causation, they were in no way alleviated by the acceptance of the law of chance as the 'general law of atomic disintegration'. The deterministic evolution of mechanical systems along unique trajectories, one of the main assumptions of the clockwork universe, was under threat. As we shall see in Chap. III, Max Planck was the first to state that the chance-like character of radioactive decay introduced a probabilistic element into the physical worldview, which would have significant implications for our understanding of causation and determinism. Although radioactive decay is a probabilistic event, it is nevertheless governed by a statistical law. This law was first formulated by E. Rutherford and F. Soddy in 1901.

2.5.3 Thermodynamics

A significant feature of this new view of Nature as an interrelated system is that the *interrelatedness* becomes subject to mathematical analysis. The laws of thermodynamics, especially the discovery of the first law – conservation of energy – did much to convince scientists of the universal interrelatedness of the whole cosmos.[35] The

[33] For further discussion and literature see Heilbron/Kuhn, 'Genesis' (1969); Heilbron, *Historical Studies* (1981); Weinberg, *Discovery* (1992); Weinert, 'Theories, Models and Constraints' (1999) and 'The Construction of Atom Models' (2000)

[34] Perrin, *Atoms* (1916), 159, 186–9, 11, 8, 193–4; 'Mouvement Brownien' (1909), 5–114

[35] The discovery of the laws of thermodynamics is itself a perfect illustration of the tight connection between science and philosophy; see Elkana, *The Discovery of the Conservation of Energy* (1974); Prigogine/Stengers, *Nouvelle Alliance* (1979), Chaps. V–VII; de Regt, 'Philosophy' (1996); Sklar, *Physics and Chance* (1993) and *Philosophy of Physics* (1992), Chap. 3 provide a good discussion of the philosophical issues involved in thermodynamics and statistical mechanics.

law of the conservation of energy states that the total amount of energy in the universe remains constant. But there are many forms of energy – electrical, chemical, and mechanical – and these forms of energy can be transformed into each other. In this process of transformation, some energy can be used to do useful work but some energy will be lost. Energy is the ability to do work. But while all work, W, can be transformed into heat (energy, E), not all heat (energy) can be transformed into useful work. Some energy, ΔU, will be consumed during the work process. ($E = W + \Delta U; W = E - \Delta U$). Take a mechanical device, like a car. The engine will do useful work (it will run) but some of the energy put into the engine (in the form of petrol) to do the work will be needed to run the engine and cannot be transformed into useful work. In this way 19th science discovered numerous illustrations of the law of the conservation of energy. Hermann von Helmholtz was the co-discoverer of the Principle of the Conservation of Energy. He mentions, as one example, movements under gravitational influence, in which the principle of the conservation of energy was known to apply.[36] In this case, the energy appears in the form of angular momentum. For instance, the closer a planet approaches the sun, the faster it has to travel in its orbit. The principle is also involved in the rotational spin of the earth and the distance of the moon. Due to tidal forces, the earth's daily rotation slows down; as a consequence the moon recedes from the earth by approximately 4 cm per year. It is immediately obvious that the first law of thermodynamics must regard Nature as a system of processes between which a balance of energy holds.

Hermann von Helmholtz (1821–1894)

> If, now, a certain quantity of mechanical work is lost, there is obtained, as experiments made with the object of determining this point show, an equivalent quantity of heat, or, instead of this, of chemical force; and, conversely, when heat is lost, we gain an equivalent quantity of chemical or mechanical force; and, again, when chemical force disappears, an equivalent of heat or work; so that in all these interchanges between various inorganic natural forces working force may indeed disappear in one form, but then it reappears in exactly equivalent quantity in some other form; it is thus neither increased nor diminished, but always remains in exactly the same quantity. We shall subsequently see that the same law holds good also for processes in organic nature, so far as the facts have been tested.[37]

The balance of energy is not restricted to local phenomena: the whole of the universe is involved.

---

[36] von Helmholtz, *Über die Erhaltung der Kraft* (1847), Chap. III

[37] von Helmholtz, 'Conservation of Force' (1862–3), 315–6; see also *Über die Erhaltung der Kraft* (1847), Chap. I, 8

> It follows thence *that the total quantity of all the forces* capable of work *in the whole universe remains eternal and unchanged throughout all their changes.*[38]

The order of life on earth – a process, which requires an input of energy – is enabled by an expenditure of energy – a loss of order – in another part of the solar system. The gradual burning of energy in our local star, the sun, keeps life on earth. The law of the conservation of energy thus displays, according to von Helmholtz, 'a grand connection between the processes of the universe.' In fact, von Helmholtz stresses the heuristic value of the 'the law of the conservation of forces':

> (...) its actual signification in the general conception of the processes of nature is expressed in the grand connection which it establishes between the entire process of the universe, through all distances of place or time. The universe appears, according to this law, to be endowed with a store of energy, which, through all the varied changes in natural processes, can neither be increased nor diminished ... [39]

Nature truly becomes the *cosmos*, as von Humboldt had speculated. Nature is 'a regularly-ordered whole – a kosmos'.[40] In the principle of the conservation of energy, the physical sciences had finally found the physical principle, which justified the 17th century idea of the universe as a self-regulating, energy-sufficient cosmos, which dethroned the energy-deficient universe of the scholastic tradition. So the laws of thermodynamics did much to prepare a new view of Nature. What is decisive in this context is that new empirical discoveries are the driving force behind these changes. (Already von Humboldt had used the empirical discoveries since Copernicus to inspire his philosophical speculations.) What may be called a *philosophical* view of Nature (as an interrelated system) was already a common currency by the 1850s. And scientists are often influenced in their views of fundamental notions by the prevalent philosophical presuppositions.[41] But the new empirical discoveries turned the *philosophical* worldview into a *physical* worldview. In the law of the conservation of energy physicists had empirical evidence of the 'interaction of the forces of nature'.[42] Thus they had evidence that the implicit assumption in the corpuscular philosophy – the universe consists of a collection of atoms, governed by mechanical laws – was mistaken. Rather, the cosmos is a gigantic system, consisting of myriads of subsystems, held together by the balance of energy. The law of entropy – a mathematical statement about the increase of disorder in a closed system – added a significant twist to this view. It is true that the total energy of the universe will never be exhausted (first law). But it is also true that energy may not indefinitely be available to do useful work. At

---

[38] von Helmholtz, 'Conservation of Force' (1862–3), 316; italics in original

[39] von Helmholtz, 'Aim and Progress of Physical Science' (1869), 333

[40] von Helmholtz, 'On the Conservation of Force' (1862–3), 279

[41] This claim will be further substantiated in Chap. 4 on the block universe and in Chap. 5 on determinism and causation.

[42] In 1854, von Helmholtz gave a lecture entitled 'On the Interaction of Natural Forces'; von Helmholtz (1854), 137–74

some distance point in the history of the universe, no energy will be left to restore life and order. The total energy will be totally dissipated. All life will come to an end. Von Helmholtz was the first to describe this final state of maximum entropy as the end of the history of the universe. It has become known as the Heat Death.[43]

Interlude A. *Causation, Determinism* and *Time.* Was this law of the conservation of energy still compatible with the classical views of Nature, of space and time, of matter and motion, of causation and determinism? At first it seemed so. Thermodynamics is an account of macro-phenomena. It must be related to the underlying explanatory theory of statistical mechanics. The paradigm of statistical mechanics is still the Newtonian point particle. The molecular-kinetic theory of gases was constructed on the basis of a billiard-ball model. The invisible gas molecules behaved *like* billiard balls. There was an *analogy* between gas molecules and billiard balls. Their individual trajectories are not known to ordinary human observers so that the physicist can only calculate statistical averages. Nevertheless, for the Laplacean demon even the trajectories of the gas molecules would be perfectly deterministic. And they would be time-reversible. Just as the hands of a clock are time-reversible – it does not matter to the clock mechanism whether it works clockwise or anti-clockwise – so the trajectories of the gas molecules are: the Laplacean demon can uniquely determine the trajectory of a given gas molecule, irrespective of whether it is traced in a forward or backward direction.

For a while such a view was still seen as compatible with a belief in a universal principle of determinism. The cosmos may be a system of subsystems but there was as yet no evidence of the role of chance and contingency in the physical universe.[44] Anticipating the interpretation of the Special theory of relativity, the world could be seen as a map with the trajectories of *all* particles already traced out. The undeniable irreversible processes, observable throughout the universe, could be explained away as a mere macro-phenomenon. At the micro-level the universe would still be reversible and deterministic. The irreversibility would only characterize the statistical ensembles of molecules, and was possibly caused by special initial conditions at the beginning of the universe.[45] At the atomic level, the molecules follow deterministic trajectories. It is just that the limitation of the human intellect was unable to grasp the precise trajectories of the millions of molecules making up a liquid. But essentially Laplace was correct: to the superhuman intelligence of the demon, 'the motions of the largest bodies as well as of the lightest atoms in the world' followed perfectly deterministic trajectories. Radioactive decay and other indeterministic processes in the atomic realm had not yet been discovered.

[43] von Helmholtz, 'The Interaction of Natural Forces' (1854), 171; see Barrow/Tipler, *The Anthropic Cosmological Principle* (1986), Chap. 3.7

[44] However, the idea of the universe as a system harbours the possibility of contingency and chance more readily than the cogwheels of the universe as a clockwork.

[45] Popper, 'Arrow of Time' (1956), 'Time's Arrow' (1965; Layzer, 'Arrow of Time' (1975); we should be aware that today there is evidence for chaos in the micro-world, see Gaspard *et al.*, 'Experimental Evidence' (1998); Dürr/Spohn, 'Browning Motion' (1998)

The message of thermodynamics had not yet been absorbed. Ludwig Boltzmann originally thought that thermodynamics could be reduced to the time-reversible laws of classical mechanics. And he proposed that the law of entropy could serve as a basis for the arrow of time. But after several famous objections (due to Loschmidt and Zermelo), Boltzmann had to concede that the Second Law is of a statistical nature. It states, in Boltzmann's new interpretation, that the increase of disorder in the universe occurs with overwhelming probability. But this leaves room, statistically speaking, for the spontaneous return of a physical system to a more ordered state. Once the molecules have escaped from a perfume bottle, the Second Law does not forbid them from returning to the safe order of the bottle in a spontaneous concerted action. It is just very unlikely in the history of the entire universe. And this injection of probabilistic thinking puts a damper on the hope to return to the deterministic and time-reversible assumptions of classical statistical mechanics. However, presuppositions are a powerful potion. They helped to sustain the classic belief in a deterministic universe. So there was as yet no reason to abandon the classic belief in a deterministic universe. Three examples of the belief in causal determinism in a view of Nature as a cosmos:

1. Thomas Young, in his *Lectures on Natural Philosophy* (1845) considers that the law of causation: 'That like causes produce like effects, or that in similar circumstances similar consequences ensue' to be the most general and most important law of nature.[46]
2. In his original lecture *Über die Erhaltung der Kraft* (1847) von Helmholtz had assumed that the aim of theoretical science was to discover 'final unchanging causes of natural processes', although he concedes that not all phenomena may be reducible to the principle of necessary causation. In an addendum, dated 1881, he regrets the Kantian flavour in this statement and equates causation with lawfulness: 'the principle of causation is nothing but the presupposition of lawfulness in all natural phenomena.'[47] As we shall see later (Chap. 5) this is a functional view of causation, which has been popular with scientists ever since Laplace. Einstein held that 'the differential equation is the only form, which alone satisfies the need for causation of the modern physicist'.[48] As differential equations are also ideal tools for the predictability of the trajectories of physical systems, which is often regarded as one of the important features of determinism, we can see here how easy it is to identify causation and determinism.

The notions of causation and determinism seem to be interchangeable. Following Laplace's identification of causation and determinism, there is in the minds of 19th century scientists no difference between these principles. Claude Bernard, the French physiologist, makes the connection very clear:

[46] Young, *Natural Philosophy* (1845), 11; Huxley, 'Lectures on Evolution' (1876), 47–8 gives eloquent expression to the general belief that 'the chain of natural causation is never broken.'

[47] Von Helmholtz, *Über die Erhaltung der Kraft* (1847/1996), 4, 53; Bunge, *Causality* (^{3}1979), 252

[48] Einstein, 'Newtons Mechanik' (1927), 23 author's own translation

3. For Claude Bernard belief in the absolute principle of *determinism* – the necessary and absolute connection of things – is identical with belief in science. And determinism is identical with causation:

> In fact, the absolute principle of experimental science is a *necessary* (...) known *determinism* in the conditions of phenomena – such that given a natural phenomenon, whatever it may be, an experimentalist could never accept that there was a variation in the manifestation of this phenomenon without there having occurred at the same time new conditions in its manifestation. Furthermore, he has *a priori* certainty that these variations are determined by rigorous and mathematical relations. Experience only shows us the form of these phenomena. But the relation of a phenomenon to a prescribed cause is necessary and independent of experience. It is necessarily mathematical and absolute.[49]

James C. Maxwell (1831–1879)

We may call this the Laplacean view of *causal determinism*, according to which the present state of the universe will be the cause of a later state, just as the present state of the universe was the effect of an earlier state. Laplace conceived of this view under the assumption that the cosmic system is simply a vast mechanism of cogwheels, the gigantic rare clock of Boyle's analogy. But this view can be saved even under the assumption that the forces of Nature, both in physical and biological systems, are as interdependent as the cosmos view of Nature stipulates. James C. Maxwell shows us how. Maxwell defines a material system as a configuration of relative positions, which is affected by both internal and external relations. Knowledge of the internal configuration of the elements of the system gives rise to determinism and predictability.

> A knowledge of the configuration of the system at a given instant implies a knowledge of the position of every point of the system with respect to every other point at that instant.[50]

But Maxwell enters an important reservation concerning the traditional view that 'like causes produce like effects':

> This is only true when small variations in the initial conditions produce only small variations in the final state of the system.[51]

But *if* both internal and external relations can change the configuration, in the sense of changing the initial conditions, then the deterministic hopes of the classical period begin to wane. And if Nature is a cosmos, even the ideal paradigm

[49] Bernard, *Introduction à l'étude de la médecine expérimentale* (1865), 89; italics in original; author's own translation; there are many other references to determinism in this text; see *Introduction* (1865), 69, 73, 87–9, 243–4

[50] Maxwell, *Matter and Motion* (1877), §4

[51] Maxwell, *Matter and Motion* (1877), §19

of Laplacean determinism – the solar system – will in the long run fail to be predictable.[52] Small variations in the initial conditions may produce *vast* variations in the final state of the system – the so-called *Butterfly effect* of chaos theory. Even small variations in the initial conditions may only produce final conditions with a certain probability – Perrin's granular structure of matter, which became the subject of quantum theory. With such experimental discoveries, the possibility of a separation of causation and determinism becomes a distinct possibility, although this step was only finally taken with the advent of quantum theory (Chap. 5). We can anticipate the development by considering a material system in Maxwell's sense, subject to internal and external disturbances.

Let us disregard the Laplacean demon and consider human observers. An astute observer would be able to predict with a certain amount of accuracy the phases of the moon or the succession of low and high tides. The observer may even be mathematically gifted enough to write down equations, which would allow the calculation of the phases of the moon and the hours of the tides with a fair amount of accuracy and reliability. But such an observer may not understand *why* these regularities occur or may have a mistaken causal understanding of these phenomena. The Greeks were able to make fairly precise predictions of the orbits of the planets. These calculations were based on the mistaken model of geocentrism. In this situation we have determinism without causal understanding.

Can we have causal understanding without Laplacean predictability? If you take a sledgehammer and smash a large slab to pieces, you will see that we can. Or consider the drooping eye-lid phenomenon: it is probably an age-related weakness. But not everybody gets droopy eyelids. Its observation in a certain patient will permit to infer the presence of certain age-related factors in the patient so afflicted, but knowledge of the presence of these factors in a certain person will not permit the prediction that this patient will suffer from *ptosis* in the future. Knowledge of causal factors does not necessarily give rise to the precise predictability of Laplacean determinism.

So we can have good predictions without causal understanding, and causal understanding without good predictions. Determinism was such a powerful philosophical presupposition that we will encounter it not only in the views of Nature, but also in interpretation of the Special theory of relativity and quantum theory. We will also later distinguish several versions of determinism.

Let us assume that a limited sense of determinism, not requiring knowledge of causal factors, can be upheld in the cosmos-view of Nature. How does the notion of *time* fit into this view? The assumption of determinism in the classical worldview meant that time was not considered to be a fundamental aspect of the

[52] Rees, *Just Six Numbers* (2000), 175–9; Prigogine/Stengers, *Nouvelle Alliance* (1979), Chap. IX, §2. Classical theory – Newtonianism – can be used to give a causal account of the tides or of the orbit of the planets, at the present time. But to derive from this causal account the perfect predictability assumes that the system will remain stable, that there will be no change in the initial conditions. Then the future trajectory of the system is predictable – but under the assumption of determinism.

physical universe. 'In science time was considered a mere geometrical parameter.'[53] The equations of motion contained a temporal parameter, the symbol *t*. Galileo introduced *t* into the equations of physics. It meant clock time: how much time elapses for a physical process to unfold. The parameter *t* is measurable in hours, minutes and seconds. According to Newton the natural philosopher needs an absolute notion of time to formulate the laws of motion. This clearly expresses the need for precise time reckoning. As an ideal clock, absolute time is not linked to material happenings. 'Newton's distinction between time and concrete becoming was at the root of classical physics.'[54] That means that Newton distinguished between the concrete happenings in the physical world and some ideal invariable temporal framework, against which these happenings could be measured. Einstein's relative time has not changed that situation: Einstein time is clock time. But on a more fundamental level, time is not an important ingredient in a deterministic universe. For the Laplacean demon, the future trajectories of all the particles exist *as if* they had already happened.

It has been commonly assumed that classical physics took Newtonian time as a paradigmatic notion of time. But a commitment to the principles of classical physics does not necessarily commit its proponents to the Newtonian view of absolute time. As we shall see in Chap. 4, Leibniz proposed a *relational* view of time: time is nothing but the order of succession of events in the universe. That is, the universe *is* a clock. And such a staunch defender of Newtonian mechanics as I. Kant proposed an *idealist* view of time: time resides in the mind of observers, and is not inherent to the physical universe. This idealist view of time gained great currency amongst physicists once the Special theory of relativity (1905), with its notion of relative simultaneity, had been formulated. A *relational* view of time was also defended by J.C. Maxwell, who played a major part in the introduction of the idea of physical fields. Maxwell was instrumental in establishing the cosmos-view of Nature. Yet he clearly did not believe in Newton's notion of absolute time:

> We cannot describe the time of an event except by reference to some other event, or the place of a body except by reference to some other body. All our knowledge, both of time and place, is essentially relative.
>
> The position seems to be [he adds in a footnote] that our knowledge is relative, but needs definite space and time as a frame for its coherent expression.[55]

Even in the cosmos-view time remains essentially clock time. However, in the relational view, time becomes dependent on the occurrence of material events. The physicist wants to know *how* time is measured, not what time *is*. Einstein found that how time is measured depends essentially on the reference frame of the observer. Einstein's conclusion that time could not belong to the physical universe only confirmed the subordinate role, which time had played in the classical worldview. Given

---

[53] Prigogine, *End of Certainty* (1997), 58

[54] Čapek, *Philosophical Impact* (1962), Part I, Chap. III

[55] Maxwell, *Matter and Motion* (1877), §§17–18. A relational view of time was also expressed in Young's *Lectures* (1845), 16 when he said that 'we are (...) obliged to estimate the lapse of time by the changes in external objects', like the motion of the stars.

the assumption of determinism, from any time slice in the classical universe, only one trajectory leads to a future time slice. From this basic ontological assumption the predictability of the future trajectory of the universe follows. This means that any configuration of the universe existing at any moment in time – which can be imagined as a time slice on which all simultaneous events take place – 'implies all future configurations of the system, and is implied by all past configurations' (Čapek). In this sense, the future already 'exists' as much as the present and the past, at least for the Laplacean demon. In such a deterministic universe, time only plays a subordinate part. But with the relational view of time, the possibility of a *dynamic*, rather than a static form of determinism is created. The Laplacean view involves a *static* form of determinism: the universe is laid out before the demon's eyes like a map or a filmstrip. It is a block universe. The Maxwellian view involves a *dynamic* form of determinism. The history of the universe is created by the occurrence of events, according to deterministic laws. This is compatible with Kant's evolutionary view of the cosmos and could account for our failure to predict the evolution of the solar system in the long run. Nature may possess a transitional character. It may be temporal in a genuine sense. It may harbour the seeds of genuine novelty. But the lure of the Laplacean demon proved to strong.

Then came the discovery of the 2nd law of thermodynamics: the law of entropy (1850). In non-technical language, the law states that, in a closed system, order either stays constant or decreases. It can never increase without an energy input from outside. The increase in disorder, all other things being equal, points to the ubiquity of irreversible processes, which surround us. Cold liquid never spontaneously heats up again. A disorderly room never spontaneously becomes tidy again. These irreversible processes are intimately connected with the question of the arrow of time.[56] Whether irreversibility is the basis of the arrow of time is a hotly debated, and as yet, undecided topic. For our present concern it suffices to observe that the law of entropy – the tendency of closed systems to decrease order – reinforced the question of becoming and genuine novelty.[57] If irreversibility is not just an epiphenomenon but a genuine feature of the physical world, then time-oriented processes reintroduce the notion of time into the physical universe. The Newtonian-Laplacean tradition of the deterministic clockwork universe favours the static world of being. The Special theory of relativity seemed to lend support to this view (Chap. 4). If the physical universe has essentially a *transitional* nature – if physical events are not just ordered according to a basic *before–after* relationship, a succession of events, but there is a connection between them – then time itself may be a feature of the

[56] There is a vast literature on the arrow of time. See, for instance, Popper, 'Arrow of Time' (1956), 'Time's Arrow' (1965); von Weizsäcker, *Geschichte der Natur* (1970); Lazyer, 'Arrow of Time' (1975); Costa de Beauregard, 'Two Lectures' (1977); Denbigh/Denbigh, *Entropy* (1985), Denbigh, 'Note on Entropy' (1989); Zeh, *Physical Basis* (^{2}1992); Sklar, *Physics and Chance* (1993) and 'Time in Experience' (1995), 217–29; Savitt *ed.*, *Time's Arrow* (1995); Landsberg, 'Irreversibility' (1996); Price, *Time's Arrow* (1996); Savitt, 'Direction of Time' (1996)

[57] Prigogine/Stengers, *Nouvelle Alliance* (1979), Chap. III, §3; Prigogine, *End of Certainty* (1997), Chaps. I, II

physical universe. This was further emphasised by the discovery of the 2nd law of thermodynamics. If the physical world is truly transitional, our view of Nature would be affected by such a discovery. For the moment let us return to some further developments, which affected the view of Nature.

2.5.4 Physical Fields

> The results of the work of Faraday, Maxwell and Hertz led to the development of modern physics, to the creation of new concepts, forming a new picture of reality.
>
> A. Einstein/L. Infeld, *Evolution of Physics* (1938), 125

We have spoken of matter and motion and of time. These notions, and that of space, constituted the four fundamental units of the classical model of physical reality.[58] Even when the clockwork image was replaced by the cosmos view of Nature, atomism and determinism could still be retained, since there were as yet no discoveries, which would enforce a revision of the fundamental notions. James C. Maxwell, in the tradition of Pierre Laplace, Thomas Young, Hermann von Helmholtz and Claude Bernard, expressed his belief in Laplacean determinism. The discovery of the granular structure of matter would change things, as would be the discovery of *physical fields*. We must speak of the field concept. And we must again speak of particles once they have been established as complicated systems in their own right. The physical field enters physics with the work of James C. Maxwell and Michael Faraday. It introduces further conceptual distances between the *cosmos*-view of Nature and the *clockwork*-view. The field concept suggests a different solution to one of the great problems of classical physics: the question of the mediation of action across cosmic distances. Newton's concept of gravitation illustrates the problem very well: how can a central body like the sun exercise gravitational attraction over the planets, when there seems to be no mechanical process in operation. The *clockwork*-view was not logically welded to the concept of action-at-a-distance.[59] But this view was compatible with the idea of unmediated action of forces across cosmic distances. Both Descartes and Newton made major contributions to the *clockwork*-view. Yet Descartes rejected action-at-a-distance, while Newton accepted it, although reluctantly and half-heartedly. Today the question of gravitation has been solved in a complete new theory: Einstein's general theory of relativity. It is a field theory. But the question of action-at-a-distance has not gone away. It has returned, with major implications for the view of Nature, under the heading of *non-locality* in quantum mechanics. This describes empirically well-established non-local correlations between spatially separated particles, between which any ordinary causal influences have been excluded. These correlations pose a considerable challenge to a causal interpretation of the quantum world.

For Maxwell 'the mysterious phenomenon of transmission of force' could not be explained by action-at-a-distance. The discoveries of Ampère, Oersted and Faraday

---

[58] Čapek, *Philosophical Impact* (1962), Pt. I, Chaps. VIII, IX

[59] On this whole question see Hesse, *Forces and Fields* (1961); McMullin, 'Origins' (2000)

about magnetism and electricity suggested that action-at-a-distance was not an acceptable mechanism by which electric and magnetic forces could be transmitted.[60] A new symbolism was required: Faraday's lines of force. They came with a new important tool: the *field* concept. The field concept, with its conception of *physical* lines of forces, introduces an important *alternative* to the usual atomistic view of Nature. Are particles or fields the ultimate physical constituents of Nature? Can particles be explained as disturbances of fields? Or are the particles the ultimate building blocks but 'accompanied' by waves? Such questions have not yet been resolved to the present day. The particle-wave duality is one of the paradoxes of today's view of Nature.[61] But note that this view now comprises particles and fields.

What is a physical field? It is mathematically fully described by Maxwell's equations. It can be photographed (Figs. 2.1b, 2.2b, 2.3b). It can be represented in suitable models (Figs. 2.1a, 2.2a, 2.3a, 2.4). There are electric, magnetic and gravitational fields. And fields interact.

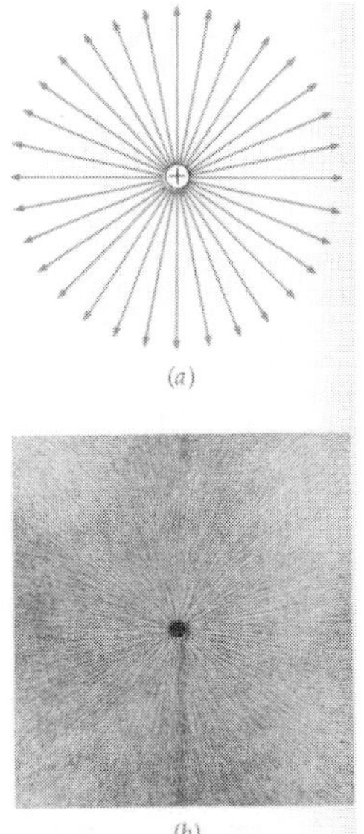

Fig. 2.1. (**a**) Representation of *lines of force* of a single charge; (**b**) a charged object in the centre of bits of thread suspended in oil. Source: Tipler, *Physics* ([2]1982), 592

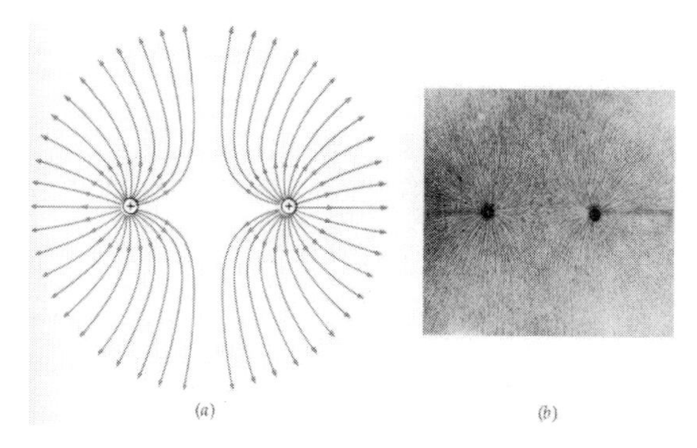

Fig. 2.2. (**a**) Representation of *lines of force* of two positive point charges; (**b**) two equal charges of the same sign, within bits of thread suspended in oil. Source: Tipler, *Physics* ([2]1982), 593

[60] Maxwell, 'On Faraday's Lines of Force' (1873); Hesse, *Forces and Fields* (1961), Chap. VIII

[61] Popper, *Quantum Theory* (1982), Chap. IV; d'Espagnat, *A la Recherche* (1981), *Une Incertaine Réalité* (1985). As we discuss in Chap. III, modern cosmological theories resolve the paradox in favour of waves.

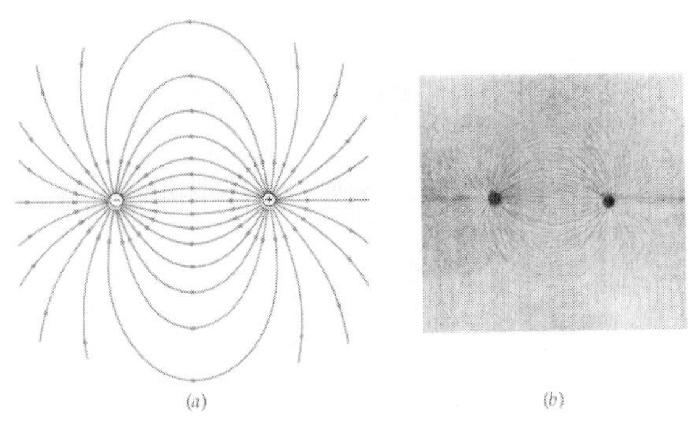

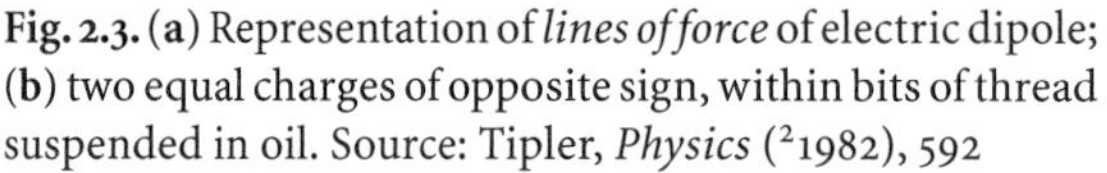
Fig. 2.3. (**a**) Representation of *lines of force* of electric dipole; (**b**) two equal charges of opposite sign, within bits of thread suspended in oil. Source: Tipler, *Physics* (^{2}1982), 592

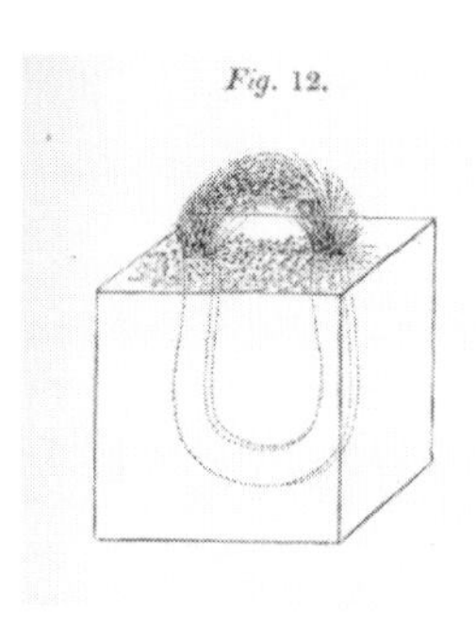

Fig. 2.4. An illustration of Faraday's *lines of force.* Source: Faraday *A Course of Six Lectures* (1860), 37

The *interrelatedness* of electricity and magnetism was one of the great discoveries of the 19th century. An electric current will produce a magnetic field around it (Oersted); a magnet moved through a coil will induce an electric current (Faraday). This meshing between physical fields can be stated in precise mathematical terms. These *quantitative* relationships present an important step into the direction of von Humboldt's ideal of the human mind encompassing 'the unity of the cosmic phenomena in all their diversity'. This aspect of unification is realized in Maxwell's field equations. They mathematically unify the phenomena of electricity and magnetism, and they unify the separate laws, developed previously by Gauss, Faraday and Ampère.

Michael Faraday (1791–1867)

Michael Faraday chose the *interrelatedness* of natural phenomena as the topic of a series of lectures for young audiences. Having demonstrated how the forces of nature – gravitation, electricity, heat – can be used to affect and interfere with natural processes, Faraday concludes with a compliment to philosophers and a general observation. Philosophers, he writes, had been suspecting 'affinity' between forces of Nature for a long time. But the physicist requires experiments to 'prove the universal correlation of the physical forces of matter, and their mutual conversion into one another.'[62] To complete this dialectic we must add that these empirical discoveries may then suggest certain re-adaptations in the fundamental notions.

The electromagnetic field, determined by Maxwell's equations, demonstrated to the physicist, in the words of Heinrich Hertz, the 'principle of the locality of all physical interactions' and 'the mutual material determinateness of all natural

[62] Faraday, *A Course of Six Lectures* (^{2}1860), 148–54

phenomena.'[63] The discovery of the physical field leads to the denial of action-at-a-distance. Thus the *cosmos*-view of Nature incorporates the duality of fields and particles. Fields are as much part of physical reality as particles. But there is another duality: fields encourage the view that Nature is local, particles suggests the view that there are non-local interactions in Nature. One day these dualities may be resolved in a fundamental theory of Nature. We are at a considerable remove from the *clockwork*-view of Nature. We should therefore find that the cosmos-view of Nature affected the concept of physical understanding.

Interlude B. *Models and Physical Understanding.* With the increasing mathematical abstraction, the role of models in the understanding of physical phenomena became ever more important. As Maxwell points out, 'the configuration of material systems may be represented in models, plans or diagrams.'[64] An abstract mathematical theory requires some representation of the physical reality, of which it gives some rational account. This may be achieved by the use of models. A model is a concrete representation of a physical system, which the theory only accounts for in abstract terms. Several types of models can be distinguished, representing different aspects of physical systems (see Chap. 3). What concerns us here is a point about representation. As Maxwell put it:

> The model or diagram is supposed to resemble the material system only in form, not necessarily in any other respect.[65]

Models are ideally suited to represent the cosmos-view of Nature as an interrelated system. Consider a simple scale model of the solar system, the Earth orbiting the Sun.

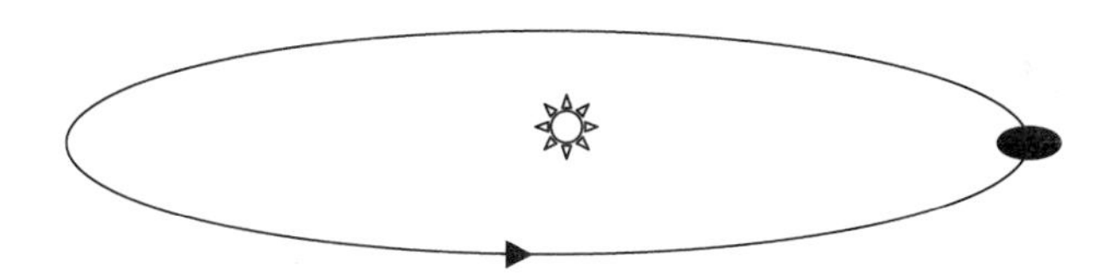

Fig. 2.5. Orbit of Earth around Sun

The model *abstracts* form the existence of the other planets and planetary moons. It *idealizes* the shape of the orbit to a simple ellipsis (with the centre at the origin). And it captures the systematic *interrelation* between the central sun and the shape of the orbit.

On the strength of the analogy with rare clocks, earlier generation of scientists had called for a mechanical explanation of Nature. This limits their need to rather

---

[63] Hertz, *Prinzipien der Mechanik* (1894), Vorwort; Poincaré, *Science* (1902), Pt. IV, Chap. IX also stresses that Nature is to be regarded on the analogies of an organism because of its many interconnections.

[64] Maxwell, *Matter & Motion* (1877), §5

[65] Maxwell, *Matter & Motion* (1877), §5

simple mechanical models, with a *homorphic* or *topographic* representation. The model would be expected, in Maxwell's words, to resemble the material system in spatial and other aspects. The cosmos-view of Nature creates an increased need for different kinds of models, with *homologous* or *algebraic* representations. This is the formal resemblance of the model with the material system, which Maxwell had in mind.

Heinrich Hertz
(1857–1894)

Given such a more formal resemblance, the question of what the model actually tells us about physical reality highlights a representational aspect of knowledge. Heinrich Hertz states explicitly that knowledge is required through the use of models. Hertz was a first-rate scientist and a respectable philosopher. As a *scientist* Hertz empirically demonstrated the existence of electromagnetic waves, predicted by Maxwell's equations. Hertz established the *reality* of the electromagnetic field. With his experiments he provided a confirmation of the philosophical view of the interrelatedness of all phenomena.[66] As a *philosopher*, Hertz wondered about the representational nature of models. There is no direct correspondence between the model and the system modelled. Different models of the same reality are possible. The only requirement imposed on models is that 'the consequences of the models are again the models of the consequences.' That means that the models tells us something about the structure of reality but also that we can derive from the models predictions about the future behaviour of the physical system under consideration. Although the choice of a model does not give us a definite representation of Nature, since different models can be used to represent a physical system, there are nevertheless some criteria, which allow an evaluation of models. These evaluative criteria are *validity* (the conceptual coherence of the model), *truthfulness* (the model relations express the physical relations of the material system) and *usefulness* or *simplicity* (the model must maximize its representation on a minimum of principles). On the strength of these criteria Hertz distinguishes three models in the history of science:

1. The *Newtonian worldview* is based on a system of mechanical principles and the fundamental notions of matter, motion, time and space. But it assumes 'action-at-a-distance' between the material atoms.
2. The *thermodynamic worldview* replaces the concept of force by the concept of energy. It is based on the concepts of energy, mass, space and time.

[66] See Hertz, *Prinzipien der Mechanik* (1894), Introduction by J. Kuczera, 12–36

3. Hertz favours a third picture, which only postulates three independent concepts: time, space and mass. Only local action is admitted. Nature is seen as a system of systems. It shifts the focus of physics from particles to fields.[67]

With the introduction of new models, representation changes its nature. The use of various kinds of models is a recognition that new forms of physical understanding are required for new views of Nature. As Hertz and Maxwell indicate, purely mechanical models of Nature no longer suffice. Relativity and quantum theory lead to a loss of a purely mimetic representation of Nature. A certain direct form of visualization, as scale models capture it, loses its effectiveness. But new more abstract models will take their place. The behaviour of atoms and relativistic objects can still be visualized, but not in a direct mimetic sense. The visualization takes an indirect form. We model atomic and relativistic phenomena *as if* they behaved as the model suggests. The indirect visualization takes a dramatic form in the tracks, which subatomic particles leave in cloud and bubble chambers. The visual tracks make no sense to the untrained eye. They have to be interpreted with the aid of some theory.[68]

What is increasingly important, as Maxwell stressed, is that the models should capture the formal resemblance between the real systems modelled and the representation. The new models are also the harbinger of a new, more formal concept of Nature. This is already anticipated in Kant's distinction between a *formal* and a *material* aspect of Nature. Newton's laws of motion inspired the formal aspect of Nature. With the emergence of the relativity and quantum theory the concept of *invariance* begins to figure prominently in an even more abstract notion of Nature. The Special theory of relativity leads to a four-dimensional reality, the quantum theory to a particle-wave duality and non-locality. A step in the direction of the *invariance-view* of Nature is Einstein's notion of structure laws.

2.6 Fields, Structure Laws and the Decline of the Mechanical Worldview

Maxwell's equations are the fundamental laws, which represent physical fields. According to Einstein, they form a new kind of law – *structure laws* – because they represent the *structure* of the field.[69] The idea of a structure can be related to the emerging view of Nature as an interrelated system. A *structure* in the real world (an ontological structure) consists of the elements and relations, which make up a natural system. For instance, a planetary system has planets as objects and is sustained by, say, Kepler's laws. A molecule is a natural system with a different

[67] Hertz, *Prinzipien der Mechanik* (1894), Introduction; cf. Hund, *Geschichte der physikalischen Begriffe* II (1978), 216–18; Popper, *Quantum Theory* (1982), Chap. IV

[68] Galison, *Image and Logic* (1997)

[69] Einstein/Infeld, *Evolution of Physics* (1938), 143; 236–7, 241–5, 289; Bunge, *Causality* (^{3}1979), 299–301; Friedman, *Foundations* (1983), 32–4; Weinert, 'Laws of Nature' (1993), 'Laws of Nature-Laws of Science' (1995); D'Arcy Thompson, *On Growth and Form* (1942)

structure. Its atoms are arranged in, say, a tetrahedral configuration. These systems themselves can form larger networks, which again consist of objects and interrelated regularities. These regularities are expressed in the laws of science. The laws express structural properties of physical systems. In the laws of science these structural properties are expressed as parameters, which not only state relations between properties (mass, position, charge, density) but also between events (motion, current, magnetism). As physical systems are related in the way the *cosmos*-view of Nature spells out, the laws, which refer to them, are also related. They constitute a network of relationships. This makes unification possible. The 19th century discoveries about the interrelatedness of electric and magnetic phenomena, expressed in precise mathematical laws, bring home this idea of a structure of a field. According to Maxwell, these phenomena are only aspects of an existent *electromagnetic* field. A change in an electric field will produce a magnetic field; a changing magnetic field will induce an electric field. A field, however, is a *local* phenomenon; it dispenses with action-at-a-distance, which was typical of Newtonian mechanism. A further difference is that the field concept is based on the postulate of a finite propagation of the wave. As Maxwell deduced from his equations, electromagnetic waves spread with the velocity of light. Newtonian mechanics has no such inbuilt constraints. Because of the *finite* velocity of light and the *local* character of fields, Maxwell's equations tell the *history* of the field.

> In Maxwell's theory there are no material actors. The mathematical equations of this theory express the laws governing the electromagnetic field. They do not, as in Newton's laws, connect two widely separated events; they do not connect the happenings *here* with the conditions *there*. The field *here* and *now* depends in the *immediate neighbourhood* at a time *just past*. The equations allow us to predict what will happen a little further in space and a little later time, if we know what happens here and now. They allow us to increase our knowledge of the field by small steps. We can deduce what happens here from that which happened far away by the summation of these very small steps. In Newton's theory, on the contrary, only big steps connecting distance events are permissible.[70]

For Einstein the emergence of the physical field concept, with its accompanying structure laws, which describe the history of events in their immediate neighbourhoods, constitutes a break with the mechanical worldview. Structure laws should represent all events in nature. The Special and the General theory of relativity, of which Einstein was the creator, and quantum mechanics all employ structure laws. For Einstein, the field would be the only reality in a future new physics.[71]

This was the future, for the time being the physical worldview had to content itself with the duality of particles and waves. Nevertheless, in Einstein's view the mechanical worldview had already suffered a decline.[72] As we have seen, the decline of the mechanical worldview was precipitated not only by the discovery of physical fields. The discovery of physical fields is itself only part of a groundswell of a new

---

[70] Einstein/Infeld, *Evolution of Physics* (1938), 147; italics in original

[71] Einstein/Infeld, *Evolution of Physics* (1938), 243; Popper, *Quantum Theory* (1982), Chap. IV

[72] Einstein/Infeld, *Evolution of Physics* (1938), 69–122; Čapek, *Philosophical Impact* (1962)

view of Nature: the *cosmos*-view. The other contributory elements were Darwinian evolution, thermodynamics and the discovery of atomic structure.

We have seen that any worldview consists of interwoven strands. The decline of a worldview may affect its strands differently. Indeed, some of the strands of the classical view proved to be tenacious. Determinism survived from the clockwork view into the cosmos-view and from there into Einstein's view of the physical field as the new reality. This included Einstein's view of quantum phenomena and the Special theory of relativity. And Einstein, as we shall see in later chapters, was not alone in clinging to determinism. Determinism even lingers on in quantum mechanics. Other elements of the mechanical worldview, however, proved easier to jettison and were replaced by new elements:

1. The *atomism* of the classical worldview is replaced by the idea of the *system*, of which the field is one physical manifestation. The absoluteness of the classical corpuscle as the ultimate unit of physical reality, with its primary qualities, is no longer tenable. The atom is a system of systems like the cosmos as a whole. The substance of classical particles, their mass, is no longer invariant. According to Einstein's famous equation $E = mc^2$ mass and energy are equivalent. The mass of a particle increases with its velocity. A particle can undergo chemical transformations in which it transforms mass into energy and changes its chemical identity. Furthermore, as the physics of the atom will emphasize, particles have both *wavelike-* and *particle-like* aspects. Both the Special theory of relativity and the quantum theory contributed to turn Kant's *Principle of Permanence of Substance*[73] into a principle of invariance.
2. The notion of causation had a difficult conceptual history in the sciences. On the one hand scientists were strongly influenced by the Lapalacean identification of causation with determinism. On the other hand, experimental evidence at the beginning of the 20th century showed that this identification could not be retained. The reaction to this situation was threefold: some physicists clung to the philosophical presupposition of determinism; others totally abandoned this presupposition and replaced it by the notion of indeterminism; others still separated the notions of determinism and causation. They adopted a probabilistic view of causation in conjunction with indeterminism (Chap. 5).
3. The *fixity* of time and space in the classical worldview is replaced by the idea of the *relativity* of time and space. This is a direct consequence of our inability, according to the Special theory of relativity, to determine in any absolute sense when two events happen simultaneously. The inability to determine the simultaneity of events in an absolute or frame-independent sense has consequences for our view of physical reality. It means that there is no universal *now*. We have already noted that time, in the classical worldview, played a subordinate part. Time was conceived as *clock* time, so that the duration of events could be measured. But clocks only fulfil their purpose if their accuracy can be measured

[73] According to Kant's principle: 'In all change of appearances substance is permanent; its quantum in nature is neither increase nor diminished.' Kant, *Critique* (1781), A 182; the term 'quantum' appears in Kant's original formulation.

against the regularity of some natural process. Today the oscillation of certain atoms forms this natural background. In Newton's time, it was the regularity of planetary motions. Newton suspected, quite rightly as it turned out, that no natural motion in the universe would be totally stable. So he postulated the existence of absolute and universal time. Newton gave the universe a clock. As classical physics also adhered to determinism, time was not seen as a constituent feature of the natural world. If the trajectories of all particles, big or small, are already mapped out from past into future, it is *as if* all events had already happened. The Special theory of relativity seemed to further demote its physical significance. We shall have to consider, however, that although time may be relative, the physical world may not be timeless. But according to the Special theory of relativity, Nature is nothing but a four-dimensional reality.

2.6.1 Four-Dimensional Reality

There is no sense of a universal *now*, independent of the location and motion of observers. According to the Special theory of relativity observers are attached to different reference frames, which have their own spatial and temporal coordinates. Observers in different reference frames judge both the duration and simultaneity of events differently. Then it seems that time cannot be part of physical reality. At least this is how many physicists argued. For if observer O_1, stationed on the platform, cannot agree with observer O_2, travelling on the train, whether the flashes of lightning hit the opposite ends of the train at the same time or not, then the notion of time may have little to do with physical reality. Perhaps it is just a human illusion. Under this scenario, Einstein and others argued, the *dynamic* view of reality in classical physics must be replaced by a *static* view of reality. Physical reality is not one of *becoming* but of *being*. The world of classical particles was seen as moving uni-directionally along a *one-dimensional* time axis and spreading in *three-dimensional* space. The motion of a particle can be represented in diagrams (Fig. 2.6).

This dynamic representation of reality is replaced by a *static* view of reality. In this static view, physical reality just *is*, all events are already laid out from past to eternity. Physical reality is like a *film* with all its frames in place. Time loses even its subsidiary role in the physical world. Because humans cannot view the frames all at once but must read them sequentially, the illusion of the passage of time is created. The adoption of the static view is, according to Einstein, not a matter of choice. The new discoveries, associated with the Special theory of relativity, impose this static picture as the more objective one.

> It appears therefore more natural to think of physical reality as a four-dimensional existence, instead of, as hitherto, the *evolution* of a three-dimensional existence.[74]

In this static four-dimensional world, strict *determinism* reigns. We can think of it as a *map* with all the events already inscribed. There is no strict separation

[74] Einstein, *Relativity* (1920), 150, 144

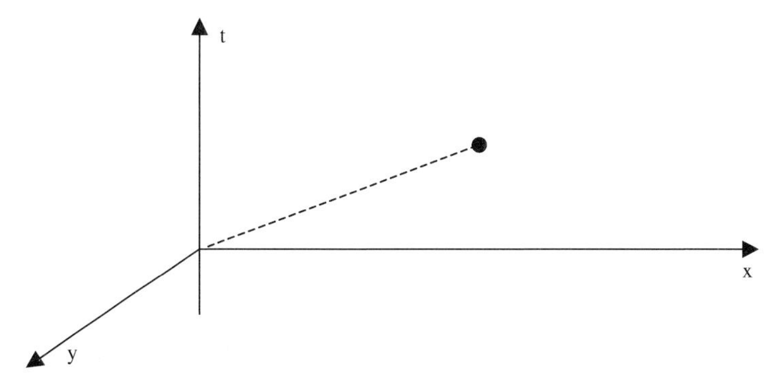

Fig. 2.6. A particle moving with constant speed in the positive x-direction. The third spatial dimension (height) is not shown

between past, present and future. The world just *is*, with all the events placed on the world map. This whole world lies before the eyes of a superhuman intelligence, a Laplacean demon. From any one point in this world map, the Laplacean demon can trace the future and the past. Any world stage implies its future and its past. It is governed by strict laws, so that from the state of the world at one instant any other state of the world at any other instant can be derived.

The classical world of atomic *particles* was located in three-dimensional space and one-dimensional time. The atomic particles themselves, which Boyle regarded as corpuscles, were the ultimate unchanging constituents of nature. Time and space were independent of each other. This is indicated in the classical diagram by the separation of the temporal and spatial axis. But if spatial and in particular temporal measurements now depend on the state of motion of the observer, as the Special theory of relativity teaches us, then this dynamic view will no longer be an adequate representation of physical reality. It must be is replaced by a four-dimensional static world of *events*. It is a *space-time* world. In this four-dimensional static world, space and time are no longer separated. They are forged into a union because each event has its own spatial and temporal measurements. An *event* is a location of an actual or possible happening, which may involve material particles (as in a collision) or simply the crossing of light rays. The location of an event, whether actual or possible[75], is indicated by its position in *space-time. The world of events forms a four-dimensional continuum.*[76] It is called a *four-dimensional continuum* for two reasons:

[75] This means that some events become impossible, as for instance communication with one's past self or communication with aliens who reside in reaches of the universe, which cannot be reached with the fastest signals.

[76] Einstein /Infeld, *Evolution of Physics* (1938), 207. It should be noted that in this book Einstein and Infeld speak of the *time-space continuum* rather than, as it is customary today, of the *space-time continuum.* Echoing Einstein, H. Weyl writes that the physical reality is ' a four-dimensional world': 'It is a four-dimensional continuum, which is neither "time" nor "space". Only the consciousness that passes on in one portion of this world experiences the detached piece which comes to meet it and passes behind it, as *history*,

(a) every event has its own spatial and temporal coordinates, since these depend on the state of motion of the observer; time loses its independent existence **(b)** near every event we can find neighbouring events, which can be reached by a finite number of steps. Two distant events are linked by the intermediate events, which lie between them. There is no action-at-a-distance. This claim was characteristic of the *field* concept.

Structure laws characterize the field. The equations of the theory of relativity are structure laws. Since time has lost its independence and universality, according to the theory of relativity, and there is no action-at-a-distance, the field concept is able to overcome the duality of space and time in the classical theory. "The theory of relativity arises from the field problems."[77] According to field physics, the particle view is no longer tenable. But according to the orthodox quantum view there is a fundamental particle-wave duality.

In this chapter we have already observed that new empirical discoveries lead to a remoulding of our notion of Nature. Hence science may have an impact on our metaphysical views. The most powerful effect of the Special theory of relativity concerns our notion of time. This theory not only demonstrates that temporal measurements become relative to the state of motion of the observer. It also demotes the notion of time from the status of a secondary physical entity in classical views to its disappearance in relativistic views of Nature. Time becomes a *mode* of perception in the idealistic idiom. The apparent denial of the reality of time seems to associate the Special theory of relativity with time-honoured metaphysical debates about idealism and realism. What is of interest here – a recurrent theme in the book – is that purely scientific and evidentially confirmed results are taken to have rather direct philosophical consequences.

2.6.2 Metaphysical Aspects of Relativity

The Special theory of relativity deals with the question of temporal and spatial measurements, which may be made between physical events. It is important to note that the measurements can be recorded by suitable clocks. Human observers are only needed to read the clocks. The theory also tells us that it is inappropriate to imagine only *one* coordinate system for all events in the universe. Rather, each event carries its own coordinate system (see Fig. 4.6, Chap. 4). These coordinate systems provide spatial and temporal reference points. We can imagine them as consisting of spatial and temporal axes through which the spatio-temporal position of events can be uniquely determined. Thus, if we sit in a closed room, we can draw

that is, as a process that is going forward in time and takes place in space.' See Weyl, *Space Time Matter* (1921/1952), 217; italics in original. Similarly, Eddington derives from his study of the theory of relativity a new view of 'the nature of things': 'the real three-dimensional world is obsolete'; 'the four-dimensional world (...) is the real world of physics'. 'All the appearances are accounted for if the real object is four-dimensional, and the observers are merely measuring different three-dimensional appearances or sections...'; Eddington, *Space, Time and Gravitation* (1920), 181

[77] Einstein/Infeld, *Evolution of Physics* (1938), 244

axes on the walls and the floor and thereby uniquely specify where each object in the room is situated. An observer on a station platform and a traveller sitting on a moving train will be attached to two different coordinate systems. If the train stops at the station, the two coordinate systems become momentarily identical. When the train accelerates away from the station, it constitutes a third coordinate system till it reaches a constant velocity again. Thus coordinate systems may be stationary, accelerating or moving at constant velocity, depending on the state of motion of the event. It will always be possible to attached clocks to the coordinate systems. These clocks can be used to measure the duration of events. Let an event E happen on the platform. The platform observer will denote its duration by t_0. This is how long the event took according to his watch. If the same event E is measured by a traveller on a train moving past it with great speed, the traveller's watch will record a *longer* time interval, t_1, for the same event (on the platform). The duration of an event, measured from a fast-moving coordinate system will always appear *longer* than the same event measured from a stationary system.

Let a bar be placed on the platform. The platform observer will measure it and denote its length by x_0. The traveller moving past at great speed will however measure a different length for the bar, x_1. An object measured from a fast-moving coordinate system will always measure a *shorter* length than the stationary system.

Finally, let the bar on the platform be put on scales by the platform observer. This observer will denote its mass by m_0. Now let the train stop. The bar is put on board. The train accelerates to a great speed. The mass of the bar will increase.

These are some of the scientific results. They were taken to have the force of evidence to decide between the metaphysical theses of *idealism* versus *materialism*. The connection between the scientific theory and the metaphysical doctrine lies in the role of the observer. In the Special theory of relativity the observer, it is claimed, becomes the sole bearer of the notions of time and space. This is due to the discovery of the principle of relative simultaneity. We do not need a sophisticated understanding of the doctrines of idealism and materialism to grasp how purely metaphysical views could be drawn from the Special theory of relativity. *Idealism* in the present connection is the view that the mind plays a dominant part in our conception of external reality. The existence of this reality need not be denied but it cannot be understood as it is 'in itself'. Rather the appearance of reality is dependent on the structure of the human mind.[78] This was Kant's view: his system is called transcendental idealism. Kant's idealism adds the view that time and space do not exist in the real material world but are forms of intuition. *Materialism* is a view we have already encountered in Boyle's corpuscular philosophy. The world consists of matter and motion. Matter has primary qualities, which impress on human observers secondary qualities. Sensations of cold and warmth, colour impressions, sounds are all secondary qualities, which exist only in the human mind.

[78] The French physicist Bernard d'Espagnat, *A la Recherche* (1981) and *Incertaine Réalité* (1985) defends a rather similar view when he introduces the concept of *réalité voile* in his discussions of quantum mechanics.

The debate about the metaphysical aspects of relativity was carried out in the pages of *Nature* magazine. Herbert Wildon Carr (1857–1931) was the primary proponent of the view that the theory of relativity had led to a rejection of materialism and forced us to adopt idealism as a general world-view. He observes that the impetus comes from science: a scientific theory has become responsible for a change in way we look at the world.

> It may be obvious at once that the mere rejection of the Newtonian concept of absolute space and time and the substitution of Einstein's space-time is the death-knell of materialism (...). For the concept of relative space-time systems the existence of mind is essential. To use the language of philosophy, mind is an *a priori* condition of the possibility of space-time systems; without it they not only lose meaning, but also lack any basis of existence. The co-ordinations presuppose the activity of the observer and enter into the constitution of his mind.[79]

Carr relies on a classic move of the 'relativist': The measurement of temporal and spatial dimensions becomes dependent on the state of motion of an observer attached to a coordinate system. The finite velocity of the propagation of light deprives observers of any absolute simultaneity of events. What Newton regarded as an unshakeable conviction, namely that time and space belong to the furniture of the material world, are seemingly relegated to the mind of the observer. This spells the end of materialism.

> The concrete unit of scientific reality is not an indivisible particle adversely occupying space and unchanging through time, but a system of reference the active centre of which is an observer co-ordinating his universe.[80]

The role of the mind is enhanced. The Special theory of relativity supports idealism. In the relativist universe, Carr argues, 'substance and cause (...) are definitely transferred from the object to the subject of experience.'[81] Needless to say, such ideas provoked fierce opposition.[82] The discussion, to which Carr dutifully replied, was

[79] Carr, 'Metaphysics and Materialism' (1921), 248; see also Carr, 'The Metaphysical Aspects of Relativity' (1921), 809–11; Carr was the author of *The General Principle of Relativity* (1920). The book received a favourable review in *Nature* **106** (1920), 431–2. Such discussions were commonplace in scientific journals at that time. See for instance V. v. Weizsäcker, 'Empirie und Philosophie' (1917), 669–73; Riezler, 'Krise' (1928); Fleck, 'Krise' (1929); Plessner, 'Problem der Natur' (1930; Jordan, 'Positivistischer Begriff der Wirklichkeit' (1934). It is interesting to note that Philipp Frank could still write 30 years later: 'Everyone reader of magazines and even daily papers knows that the conversion of mass into energy has been used to refute metaphysical materialism and to bolster up metaphysical idealism.' Frank, 'Metaphysical Interpretations of Science, Part II' (1950), 90; for some newer discussions see Dirac, 'Evolution' (1963); Harré, 'Basic Ontology' (1997)

[80] Carr, 'Metaphysics and Materialism' (1921), 248; we find such an overemphasis on the role of the observer in the Special theory of relativity also in Eddington, *Space and Time* (1920)

[81] Carr, 'The Metaphysical Aspects of Relativity' (1921), 810

[82] See Campbell,' Metaphysics and Materialism' (1921); Elliot, 'Relativity and Materialism' (1921); McClure, 'Relativity and Materialism' (1921); Jeffreys, 'Relativity and Materialism' (1921)

somewhat unsatisfactory, because it traded on the ambiguity of the term *materialism*. But it is worth emphasising that physicists *did* draw philosophical conclusions from their discoveries. Einstein and Eddington for instance accepted that *time* had lost its role in the physical universe. Bohr and Heisenberg accepted that the notion of *cause* had lost its role in quantum-mechanical explanations of the atomic realm. And all the major contributors to the two great scientific revolutions of the 20th century – the Special theory of relativity and quantum theory – accepted that the classic notion of *Nature* would have to be revised. Scientific discoveries have important philosophical consequences, of which many physicists were aware. An attempt to assess the metaphysical consequences, if any, of the Special theory of relativity (or any scientific theory for that matter) needs to proceed from the basis of the very assumptions, on which that theories is based. An outstanding example of the kind of epistemological assessment, which can legitimately be drawn from the Special theory of relativity appeared in Volume I of a new scientific journal, published in 1923. It provides us with an important clue as to the new views of nature and reality, which were emerging at this time. The journal was the *Scandinavian Scientific Review*. The author of the assessment was a young Norwegian philosopher – Harald K. Schjelderup.[83] Schjelderup agrees that the Special theory of relativity leads to a rejection of the mechanistic worldview. It eliminates temporal and spatial relations from the static four-dimensional world. *But* it does not lead to idealism. This four-dimensional world exists without the presence of observers. Observers are only required to slice the four-dimensional world into a $3 + 1$ presentation: a world of three spatial dimensions and one temporal dimension. This slicing *depends* on the state of motion of the observer. But the reality of the four-dimensional world is not affected by this slicing act. This world is not relative to the state of motion of observers. Rather, there are certain 'absolute' objects in this four-dimensional world, first introduced by Minkowski (1908). These objects are 'absolute' in the sense that *all* observers will agree on them. They are space-time distances, or *space-time intervals* as they came to be called. The four-dimensional world is one of pure geometric relations. The motion of particles is expressed by their world lines, which describe the trajectory of particles through the four-dimensional space-time. The new physics *geometrises* the physical world. The physical world has a *non-Euclidean* structure. Every event in the world is determined by a set of pure numbers (x_1, x_2, x_3, x_4), which stand, abstractly, for spatial and temporal relations. They are on an equal par. The geometry of space-time[84] – the union of space and time – is fundamentally determined by the behaviour of light signals. If we ignore the existence of matter and energy in the physical universe, as the *Special theory* of relativity did, the light signals, which arrive at and are emitted from every event, form past and future light cones. If we take the existence of matter and energy into account, as does the *General* theory of relativity, then the world lines experience deviations.

[83] Schjelderup, 'The Theory of Relativity and its Bearing upon Epistemology' (1923), 14–65. The editor of *Nature* obviously found it important enough to have it reviewed: 'Relativity and Theory of Knowledge', *Nature* **112** (1923), 377. The review was very positive.

[84] See Liebscher, *Einsteins Relativitätstheorie* (1999)

These indicate the curvature of space-time. The trajectory of particles is described by the *geometry* of their world lines. The interaction of particles is described by the intersection of world lines. The distance between events in space-time is described by the space-time interval, which is the same for *all* observers. The geometry of the theory of relativity is a 'system of pure relations.'[85] This insistence on *abstract relations* is in full agreement with the earlier philosophico-empirical finding that Nature is an interrelated system. Whether Nature is an interrelated system of classical three-dimensional objects, moving along the temporal axis according to the laws of motion, or whether it is a system of geometric relations, does not change the fundamental new insight: the interrelatedness of systems in the natural world.

Which view of Nature, of physical reality, emerges from such a system of pure relations? 'Not a reality of things, but one of *laws* and *relations*.'

> The Relativity theory dissolves all "concrete realities" and substitutes mathematical symbols for them. Not a concrete "matter", which expands in space, is the reality of physics, but the *laws*, which govern larger and smaller areas of the "world", are all that is permanent, and therefore the real "substances".[86]

Carr's conclusion is mistaken: the role of the mind is not enhanced by the theory of relativity and this theory does not support idealism. The observer can always be replaced by ideal clocks and rods, which will register the 'relativization' of temporal and spatial relations. Clocks will slow down or run faster, rods will shrink or expand, depending on their state of motion or the presence or absence of matter. It is true that the theory of relativity makes 'a clean sweep of all naively realistic theories'[87] since it imposes the abandonment of the mechanistic worldview. But it does not embrace idealism. The geometric relations, describing the four-dimensional space-time, are real. And the space-time intervals, which give us the space-time distances between events in the Minkowski world, have the same length for all observers. They do not depend on the state of motion of particular observers or the clocks attached to their reference systems. Nevertheless a number of very eminent physicists embraced an idealist notion of *time* as the result of the discovery of relative simultaneity. Schjelderup's approach to an assessment of the philosophical consequences of the Special theory of relativity was correct: examine whether the assumed consequences follow from the fundamental postulates of the theory. Schjelderup found that idealism does not follow from the relativity theory. Equally, we will argue in Chap. 4 that the Special theory of relativity does not show that the passage of time is a mere human illusion.

2.7 The Demise of the Point Particle: The Wave-Particle Duality

Before we consider the modern abstract notion of Nature, according to which the *invariant is the real* and *the real is the invariant*, we must turn to the contribu-

[85] Schjelderup, 'The Theory of Relativity' (1923), 57

[86] Schjelderup, 'The Theory of Relativity' (1923), 59; italics in original

[87] *Nature* **112** (1923), 377; this quote is taken from the review of Schjelderup's article in *Nature*.

tions, which the quantum theory – the physics of the atom – has made to the notion of Nature. We have seen that Einstein's theory of relativity did not achieve a complete replacement of the particle-view by the field-view, as a new notion of Nature. For at the same time as the Special theory of relativity was launching its attack on the particle-view and the absoluteness and universality of time, the physics of the atom was beginning to suggest further non-classical changes to the notion of Nature: Nature was *discontinuous*, since it apparently allowed quantum jumps. Nature was *probabilistic* rather than deterministic, since atomic events were not uniquely predictable. Nature showed a *Janus* face, since at the atomic level it displayed both *particle-* and *wave-*like characteristics. And Nature was *non-local*, since it displayed strong correlations between widely separated particles.

The Special theory of relativity makes *events* in a four-dimensional space-time the central notion of a space-time view of Nature. The quantum theory of the atom radically transforms the view of particles. All these 20th century changes to the concept of Nature happen on the now well-established and accepted view that Nature is an interrelated system. The Special theory of relativity and the quantum theory are the two most revolutionary innovations of the 20th century. They changed the classical view of Nature beyond recognition.

The quantum theory deals with atomic particles. It views the atom as a system. As it turns out the atom is quite a complicated system, since it consists of a nucleus and a varying number of electrons. The nucleus itself is composed of neutrons and protons. These, in themselves, are made up of quarks. The atom is not 'uncuttable' (as the Greek word 'atom' suggests). It can absorb and emit energy. It can decay and in the process change its chemical properties. But the atom is not simply to be regarded as a particle system. In fact, the atom has a *particle*-nature and a *wave*-nature. It reveals its 'natures' in response to different experimental situations but it has no essence. By an appropriate choice of experimental situations the atom reveals either *particle*-like or *wave*-like characteristics. The physicist must describe it in terms of a *particle-wave duality*.[88] The behaviour of atomic particles challenges our established views of *causation* and *determinism*. Atomic particles even seem to defy the action-at-a-distance prohibition of Einstein's field-view. Instead of obeying the postulate of locality, they seem to follow a rule of *non-locality*. At that time these notions served physicists as a prop to come to grips with the quantum phenomena before them. These early modes of understanding have today given way to new concepts, like *entanglement* and *decoherence*. They suggest new ways of physical understanding (Chap. 5).

---

[88] See Jordan, 'Erfahrungsgrundlagen der Quantentheorie' (1929); Schrödinger, 'Was ist ein Elementarteilchen' (1950); 'Unsere Vorstellung von der Materie' (1952). Planck considered that the 'material point' as the fundamental concept of mechanics is not elementary and may lose its 'sense' when speaking about electrons; see Planck, 'Die physikalische Realität der Lichtquanten' (1927), 529; 'Weltbild der neuen Physik' (1929), 223; Pagels, *The Cosmic Code* (1984); Falkenburg, *Teilchenmetaphysik* (^{2}1995); for work on non-locality, see Maudlin, *Quantum Non-Locality* (^{2}2002)

The classic model to visualize the strange characteristics of atomic particles is the *two-slit experiment* (Fig. 2.7). It illustrates the particle-wave duality and the puzzling non-local behaviour of the system.

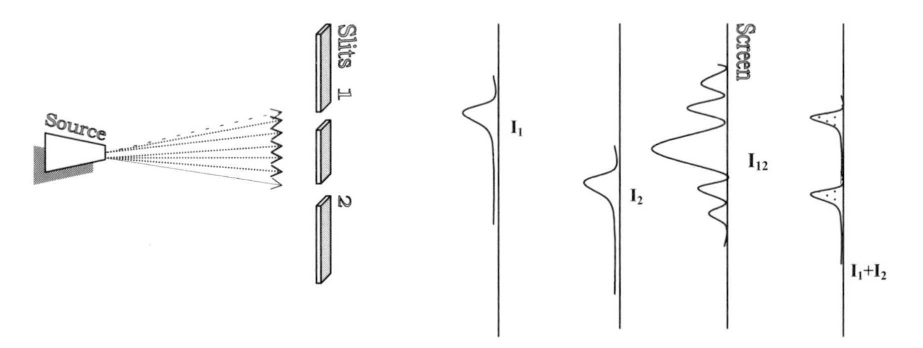

Fig. 2.7. Two-slit gedanken experiment

An atom beam is fired at a screen with two slits, which can be opened and shut by the experimenter. The experiment can induce *particle*-like behaviour in the atoms by simply shutting one of the slits during a particular firing of the atom beam. Let us say that during one run of the experiment $slit_2$ remains shut. The atoms will traverse $slit_1$ and form an intensity pattern on the screen, I_1. If alternatively $slit_1$ is shut the atoms will travel through $slit_2$ and form intensity pattern I_2. There is nothing surprising: if the two screen with intensities I_1 and I_2 are superimposed, like two transparencies, two clear dots will appear $(I_1 + I_2)$. This is what we would expect from two particles. If, for instance, for replaced the atom beam by a sand blast, again keeping one of the shutters closed at a time, the sand grains would form the same intensity patterns. The difference between sand grains and atomic particles appears when the two slits remain open during the experimental runs. The sand grains will behave like particles and form two separated intensity patterns, as we would expect from *particle*-like behaviour. But this is not true of atomic particles, like electrons. The experiment can induce *wave*-like behaviour in the atoms by simply keeping both slits open during a particular firing of the atom beam. If an electron beam is fired at the screen with the two slits open, we will observe the intensity pattern $I_{1,2}$, not I_{1+2}. In other words, we will observe *wave*-like intensity patterns or *coherence*. The troughs and peaks of the two waves, coming from the two slits, will cancel each other out in certain areas of the screen and reinforce each other in other areas of the screen. But the 'particles' enter the slits perfectly separated. How is this coordination of the waves, this formation of peaks and troughs to be explained? How does one particle 'know' what the other is doing? A *non-local* interaction seems to take place, which defies our ordinary sense of causation. A principle of non-locality obtains in the quantum world, in direct violation of the postulate of field physics. In the double-slit experiment this non-locality occurs even when single photons hit the screen – one by one they will form the characteristic interference patterns.

For two-particle systems this phenomenon is increasingly called *entanglement*. It is a central feature of all quantum mechanical experiments. Unfortunately, it cannot be ignored, since the experimental evidence supports it. Atomic particles, for instance, have an observable property called spin. A particle's spin can be designated by a quantum number: $\pm\frac{1}{2}\hbar$. Under suitable experimental conditions, a pair of particles can be prepared in a common source, and then projected in opposite directions towards measuring devices. The experimental arrangement excludes the possibility of any interaction between the particles during their flight. Nevertheless their spin states will display, under certain experimental conditions, perfect anti-correlations. One particle will always have spin $+\frac{1}{2}$ and its partner will have spin $-\frac{1}{2}$. The puzzling question is: how do these anti-correlations arise? No ordinary causal mechanism can be cited. In a later chapter it will emerge that the entanglement between quantum systems is the latest and most fundamental, experimentally well-established evidence of the universal interrelatedness of all objects in the universe. For it turns out that there is no wave-particle duality at the fundamental level.

For the moment let us continue with the early interpretative responses to the situation. The physical behaviour of the atoms seems to display a fundamental *indeterminacy*. It is no longer possible to determine the precise trajectory of each atom on its journey to the screens. Following the classical view of particles the precise determination of the particle's trajectory would require a precise location in time and space. The problem with atoms is that such a precise determination cannot be given *simultaneously*. The atoms can choose a number of trajectories – in the two-slit scenario: go through $slit_1$ or $slit_2$ – but the researcher cannot predict, which one they will choose. The researcher loses a *Laplacean* tool. Consider a beam of atoms sent through a magnet (rather than two slits), as in the Stern-Gerlach experiment (see Fig. 5.13, Chap. 5). The beam consists of bundles of atoms. As it traverses the (inhomogeneous) magnet the beam will be split in two, in the simplest scenario. Its split will be recorded on a screen: two dots will appear, one in the upper part of the screen, one in the lower part of the screen. If we were to concentrate on an individual atom in the beam we would not be able to predict whether it would hit the screen in the upper or lower part. This is its fundamental *indeterminacy*. Heisenberg put it into mathematical relations, which are sometimes known as the *uncertainty* relations. A better term is *indeterminacy* relations, because Heisenberg assumed that this indeterminacy was a fact of Nature, not just a human limitation on what was knowable. It is clear that in the light of the indeterminacy relations, the notions of determinism and causation will have to be modified. Heisenberg and Bohr, amongst others, drew this lesson as *the* philosophical consequence of the quantum theory. The Laplacean ideal of causal determinism fails.

The basic indeterminacy in the atomic realm leads to a fundamental revision in the concept of Nature: *Nature is probabilistic, not* just our knowledge of Nature. In classical physics it was assumed that Nature was deterministic but there were limits to human knowledge. Therefore Laplace equipped his demon with superhuman intelligence. A bottle of perfume contains millions of molecules, say 6×10^{20}. Humans are incapable of predicting the trajectory of each of them, although each molecule's trajectory is perfectly deterministic. For the Laplacean demon all these trajectories are predictable – he could draw a map of all of them. This map would contain all the world lines, which the molecules will ever trace. The Laplacean map would anticipate Einstein's static four-dimensional world. In quantum theory it is the trajectory itself, which becomes probabilistic. This does not mean that pure chance rules. In the two-slit and Stern-Gerlach experiments the particles have a limited number of trajectories available, which they must follow. Some of these trajectories are improbable. Others are more or less probable. These probabilities can be calculated. But they take away the certainty of the Laplacean demon.

The indeterminacies, with which even the Laplacean demon would have to contend, are due to the essentially *probabilistic* behaviour of quantum matter. The indeterminacy makes for uncertainty. Further uncertainties accrue from the *discontinuous* character of Nature. It was a fundamental postulate of Greek science that Nature made no jumps. Nature was continuous. Boyle banned the postulate *Natura non facit saltus* as a principle of explanation from his corpuscular philosophy. Such postulates strived on quasi-metaphysical beliefs to which there was no contrary evidence. Then evidence came to light that Nature did, after all, make jumps. This was realized even before quantum mechanics developed as a proper scientific theory. Radioactivity provided the disturbing evidence. Heavy atoms decay, releasing energy. When an atom decays it changes its chemical nature. But the moment of decay is not predictable. And the cause of the decay is not discernible. Nature makes quantum jumps in discontinuous steps. Energy is released or absorbed in multiples of a known quantity, called the *quantum h*. The amount of quantum can be calculated. This was the great discovery of Max Planck. An atom can only exist in specifiable quantum states: usually a whole number *times* the quantum *h*. An atom can absorb and emit photon energy but only in permissible bundles. When heavy atoms are chemically transformed in radioactive decay, they can only decay into certain permissible states; other states are forbidden.

Every car illustrates this contrast between *continuity* and *discontinuity*. A car has a limited number of gears: on change of gear, the car 'jumps' from one state to another. These are the permissible states: 1–2–3–4. There is no 1½ or 2¾ gear. These are forbidden states. This is the mechanical aspect of *discontinuity*. But the car can move at any velocity value, from zero to a maximum value – there are no forbidden velocity states below the maximum. This is the classical aspect of *continuity*, to which there is no equivalent in quantum mechanics.

If Nature is allowed to make 'quantum jumps', then the classical view of Nature as a causal mechanism is threatened. This was quickly realized by Rutherford and discussed by Planck and Bohr. If heavy atoms just decay radioactively and atoms just absorb and emit photon energy, then the classical ideal of a causal *trace* from the cause to the effect seems violated. *Push-and-pull* notions of causal mechanisms, which fit at least some classical systems, fail spectacularly. As the evidence from, say, the two-slit experiment does not provide us with a delineable trace from slit to screen, the notion of causation may have to be modified.

Interlude C. *The Concept of Nature* and *the Fundamental Notions.* The notion of Nature is intimately bound up with fundamental notions like causation, determinism, time and space. If the notion of Nature undergoes change as a result of new scientific discoveries, it is to be expected that the fundamental notions face the tribunal of evidence too. At the beginning of the 20th century, under the impact of the two great scientific revolutions, Hermann Weyl stated:

> The revolutions, which are brought about in our notions of space and time will of necessity affect the conception of matter too.[89]

But we have already observed that change is differential. The notion of Nature shifted from the *clockwork-* to the *cosmos*-view, yet retained a deterministic commitment for a while. The remaining elements also came under scrutiny with new discoveries. Maxwell, on the background of field physics, adopted a *relational* view of time. Kant adopted an idealist view of time but on the background of the mechanistic worldview. The notion of causation began to totter under the impact of radioactivity and Planck's constant h but Einstein and others blamed the quantum theory, not the Laplacean notion of causation. This is what we mean by the *dialectic* of science and philosophy. This dialectic also affects the notion of physical understanding. Generally, the new view of Nature leads to a modification of visualization. The freeing of human understanding from mechanical models and traceable causal mechanisms meant that the visual and tactile representation of Nature became outdated. Such mechanical representations are often based on *analogue* models, which represent the unknown by the known. But the loss of mimetic visualization also meant a gain in *physical understanding*. New, more abstract models could be envisaged. These more abstract models are better vehicles for the increasing mathematization of Nature. Physical understanding and the notion of Nature cannot be separated either. According to quantum theory Nature is probabilistic, discontinuous, indeterminate, non-local and it displays a fundamental particle-wave duality. Such unorthodox characterizations of Nature call for new forms of physical understanding, aided by new types of models. The dialectic leaves conceptual traces. There are no sudden Kuhn-type conversions, no disruptive shifts in worldviews. There are traditions and there are arguments.[90] The shifts in the views on Nature allows for a rational reconstruction, which is based on the conceptual traces left in the interstice between science and philosophy.

The Special theory of relativity replaces the particle as the fundamental unit by the *event* in space-time. It then abandons the notion of absolute and universal time. The quantum theory deprives the particle of its last vestige of material substance. It is particle or wave, depending on the experimental arrangement. As all scientific theories, the quantum theory faces the question of what it is in Nature

[89] Weyl, *Space-Time-Matter* (1921/1952), 7. At the end of the century Ferris, *The Whole Shebang* (1998), 245 observes that 'science as we know it is built on cause and effect, space and time.'

[90] Weinert, 'Tradition and Argument' (1982); 'Contra Res Sempiternas' (1984); Galison, *Image and Logic* (1997), §9.4 argues for a differential, intercalated progress of experimentation and theory

that endures. Kant had a name for it: the *Principle of Permanence of Substance.* The Special theory of relativity shows that the quantity of mass is not constant: it increases with velocity; and it is equivalent to energy. Einstein had shown that mass and energy are intimately related: $E = mc^2$. And it is precisely in the realm of atomic phenomena that Einstein's famous equation found its most effective area of application. Mass can be transformed into energy and energy into mass. The quantum theory also rejects the permanence of substance: atoms can decay and quantum matter is discontinuous; the indeterminacy relations prevent a precise spatio-temporal localization of an elementary particle in an atomic system. But science cannot be built on shifting sand. If the traditional idea of a permanent substance with its accidental properties has to be withdrawn from circulation, what replaces it? What is *invariant* if it is not material substance? Both in relativity and in quantum theory the *invariant* is expressed in mathematical relations. Due to their reliance on Nature as a system of interrelations, both these theories capture *relational* aspects of Nature in mathematical equations. Thus, the measurement of the duration of events and the lengths of objects is *relational*, since it requires a relation between the co-ordinate system (of an observer, or a clock or rod) and the events to be measured. The measurement of *particle*-like and *wave*-like aspects of atomic units in the two-slit experiment is relational since it requires the relation of a measurement apparatus (and its co-ordinate system) to the event to be measured. But given this apparent dissolution of Nature into relations, some invariance must be found within these relations. It is the *invariant*, which becomes the candidate for the *real*. On the basis of Nature as a cosmos, an *invariance*-view of Nature develops. As Heisenberg expressed it in a characterization of the change, which the notion of substance had undergone as a result of quantum mechanics:

> When therefore we want to speak of the invariant in the flux of phenomena, the invariant is only the mathematical form, but not the substance.[91]

In their insistence on *the invariant as the real* both relativity and quantum theory converge onto a much more abstract characterization of Nature. Such an abstract notion of Nature requires different kinds of models – they will necessarily be more mathematical, since mathematical and geometric relations have taken centre stage.

2.8 Invariance and Reality

> The feature, which suggests reality, is always some kind of invariance of a structure independent of the aspect, the projection.
>
> M. Born, 'Physical Reality' (1953), 149

[91] Heisenberg, 'Atomtheorie und Naturerkenntnis' (1934), 70; author's own translation. See also Heisenberg, 'Philosophische Probleme' (1958), 'Grundlegende Voraussetzungen' (1959). This quote reveals Heisenberg's concern with the *observable* relations in quantum mechanics. The Invariance View goes further: the invariant mathematical form, if confirmed, expresses an ontological *structure*.

In the course of this chapter, we have observed how the image of Nature has undergone not just one but several transformations. All these reconstructions of the notion of Nature were the result of an interplay between philosophical theory and empirical discovery. When the scientific revolution gets underway, the prevailing view is that the *cosmos* is *a vast organism*. Nature is treated as a causal agent. Metaphysical principles like *Nature abhors the vacuum* gloss over the lack of explanatory value. As we have seen, many 17th century scientists and philosophers adopt a nominalist attitude to the notion of Nature. This philosophical work and the physical theories produce a new image of Nature as *a clockwork mechanism*. But soon philosophical ideas begin to portray the *cosmos* as an *interrelated system*. Scientific discoveries in due course confirm this new model. Evolution, thermodynamics, field theories and the structure of atomic matter impose this view of Nature as a system of interrelations. The theory of relativity puts certain *constraints*[92] on the connections between the interrelations in a system. The first constraint is that the propagation velocity of any interactions in the physical world has an upper limit in the velocity of light. This means that it takes time for signals to travel between observers. There is no absolute reference system to which all observers could refer as the standard of space and time. Rather, observers – whether real human observers or ideal clocks and rods – have their own respective coordinate systems or reference frames. All inertial reference systems are on a par with respect to the validity of the laws of motion. This is the principle of relativity[93]. Temporal and spatial measurements, mass and energy determinations are therefore *relational* events: they are *relative* to the reference frames in which they are carried out. The different reference frames attached to different observers are related to each other in an ordered fashion. The most general expression of these ordered relations takes the form of *symmetry*.

The quantum theory also put constraints on the interrelations between the components of a system. The relativistic limit on the velocity of propagation equally applies in the quantum world. With the indeterminacy relations, quantum mechanics imposes a limit on the simultaneous precision of the position *and* momentum, the energy *and* time of quantum systems. If the indeterminacy relations are facts of nature and not just limits of human understanding, then the orbit of an electron around a nucleus will not prescribe a precise, well-defined orbit. Rather, the electron is said to swirl around the nucleus in shells without any precise spatio-temporal orbit. Its orbit is *probabilistic*. There are also limits on the classical notion of a material particle, since a particle may display *wave*-like features in strange *non-local* interactions, as in the two-slit experiment. It is *dualistic*. And if the notion of a material particle is still retained, it absorbs and emits energy in permissible photon bundles. It makes *quantum jump*.

---

[92] The notion of constraint is very important in understanding the nature of scientific theories and models; see Galison, *Experiments* (1987), 'Theory Bound and Unbound' (1995); Weinert, 'Theories, Models and Constraints' (1999)

[93] Einstein, 'Elektrodynamik' (1905); 'Relativititätsprinzip' (1907)

With the dismantling of the classical view of Nature at the hands of the theory of relativity and quantum theory, Nature becomes describable and explainable as a network of mathematical relations. This is just as well, since the classical particle view is no longer tenable and direct visualization has suffered at the hands of relativity and quantum theory. If there is no permanent substance, no absolute time and space the *invariant* must be found in mathematical forms. We have already encountered numerous examples of the mathematical relations, which govern natural phenomena. Many of these mathematical relations, however, vary between different reference frames. It is difficult to see how frame-specific relations can mark out reality. For what appears as a specific relation in one frame does not appear so in another frame. What appears as a *shortened* rod from the point of view of one reference frame will appear as a *longer* rod in another reference frame. The *lifetime* of a particle appears shorter in one reference frame from what it appears in another reference frame. *Reality* cannot depend on particular reference frames. What is *real* must be what remains invariant across different reference frames. The constant is the real, says Max Planck, independent of any 'intellectual individuality'.[94] Let us try to illustrate the idea of frame-dependent reality.

1. Consider an object that is dropped inside a railway carriage. The carriage is initially at rest but is then accelerated to the right. Imagine two observers. One is on the ground: this observer sees the object fall straight down. Another observer is in the accelerating railway carriage: to this observer the object appears to slope downward towards the rear of the carriage; s/he will attribute the trajectory of the object to a pseudo-force. As this pseudo-force is relative to a reference frame, it cannot be 'real'.
2. Einstein used a famous thought experiment to demonstrate that Newtonian gravitational forces could be explained away. A large lab is suspended in empty space with an experimenter inside. As there is no gravitation the experimenter will float freely inside the lab. Let us now image that some 'being' suddenly accelerates the lab uniformly upwards with the help of some rope attached to the lab. The experimenter knows nothing of the 'being' in empty space. S/he will therefore conclude that the lab is suspended, not in empty space, but in a uniform gravitational field. S/he will experience the upward acceleration as a gravitational force. The strange 'being' by contrast will feel that he has to exert a force to overcome the inertia of the lab suspended in space to accelerate it. He will experience the lab as an inertial mass. Einstein concluded from this thought experiment the equivalence of gravitational and inertial mass.[95] It became the foundation stone of the General theory of relativity. According to the theory, gravitation can be 'geometrized'. What Newton considered as a gravitational force, acting at a distance, becomes the local space-time curvature of a mass-energy field. Newton's gravitational force is not 'real'.
3. According to the Special theory of relativity, different observers or clocks, attached to different reference frames, measure different durations for the *same*

[94] Planck, 'Die Einheit des physikalischen Weltbildes' (1908), 49
[95] Einstein, *Relativity* (1920), Chap. XX

events observed from their respective reference frames. They all disagree about what events happen at the *same* time, if they happen to move with constant velocity relative to each other. Time, therefore, is what clocks in these different reference frames record. Different observers disagree about what their clocks tell them. Therefore time is frame-dependent. In a review article on the General Theory of Relativity, Einstein observed that this theory deprived time and space of the last vestige of physical reality.[96]

4. In the view of many proponents of quantum mechanics, the observation of particle or wave properties is dependent on the experimental arrangement. The quantum system displays particle- or wave-like characteristics, depending on the setting of the experimental apparatus. For many proponents of quantum theory this meant that the quantum system, if undisturbed, does not possess any intrinsic properties associated with particles or waves. Rather, in the course of different measurements, mere potentialities jump into certain actualities. This is a challenge to the metaphysician who wonders whether or not it is still possible to speak about the 'individuality' of quantum particles. Even though the physicist may not be charmed by such highbrow investigations, the dependence of quantum mechanical phenomena on the measurement apparatus raises the question of how to characterize quantum systems. Clearly, a classical description is not longer feasible. A much more abstract approach is required. The quantum system and its potential states are located in a mathematical configuration space. The Schrödinger equation describes their dynamic evolution. The actualisation of some of the potential states in experimental situations results in certain measurement readings. This leaves us with the puzzle of how the system can 'transit' from potential to actual values (see Interlude E, Chap. 5). At this stage we may want to turn from frame-dependent to frame-independent realities. In doing so we have the chance of finding invariants. They will help us define the structure of Nature's systems.

Despite all the conceptual upheaval, which the relativity and quantum theory caused in the 20th century, on a very abstract level, a fundamental continuity remained: the idea of *invariance*. The mechanistic worldview regarded time, space and substance as absolute or invariant. These entities provided the backdrop, against which all the changes in the material world could be studied. The modern view of Nature, due to empirical discoveries, no longer regards time, space and substance as invariant entities. It deals with events, processes and world lines. But they give rise to other forms of invariance. Arthur Eddington saw this very clearly in 1920:

> The relativity theory of physics reduces everything to relations; that is to say, it is structure, not material, which counts.[97]

---

[96] Einstein, 'Die Grundlage der allgemeinen Relativitätstheorie' (1916), 86

[97] Eddington, *Space, Time and Gravitation* (1920), 197, 'The Meaning of Matter' (1920), 'Philosophical Aspect' (1920); Schjelderup, 'Theory of Relativity' (1923). While Eddington tended to the view that all we know is 'structure', a similar discussion about structural knowledge of the external world was taking place in philosophy; see Russell, *Analysis of Matter* (1927); Newman, 'Mr. Russell' (1928); Heath, 'Contribution' (1928). This discussion

This assessment of the implications of the new physics for the view of Nature has not changed, as reflected in David Bohm's statement:

> Einstein's basically new step was in the adoption of a *relational* approach to physics. Instead of supposing that the task of physics is the study of an absolute underlying *substance* of the universe (...) he suggested that it is only in the study of *relationships* between various aspects of this universe, relationships that are in principle observable.[98]

Many relations become frame-dependent. Both the relativity and the quantum theory contributed to the dissolution of the particle view. But just as the classical substance underlay many of the observable changes, in the new view of Nature new invariant relationships must underlie the frame-dependence of the relations. In the classical view the substances were the bearers of the changes. In the new view, the invariants must take over this role. The invariants guarantee the transparency between the reference frames. Einstein's principle is relativity, not relativism. The historian of science Gerald Holton reports that Einstein was unhappy with the label 'relativity theory' and in his correspondence referred to it as *Invariantentheorie*.[99]

Instead of a "physics of matter", we get a "physics of principles".[100] What are the new invariants? As the focus of the next chapters is on the block universe and on the question of the passage of time and as well as on the principles of causation and determinism, we should illustrate the *invariants* with respect to these aspects.

Consider *temporal* and *spatial* measurements. Even if temporal and spatial measurements become frame-dependent, the observers who are attached to their different clock-carrying frames, like the respective observer on the platform and the train, can communicate their results to each other. They can even predict what the other observer will measure. The transparency between the reference frames and the mutual predictability of the measurement is due a mathematical relationship, called the *Lorentz transformations*. The Lorentz transformations state the mathematical rules, which allow an observer to translate his/her coordinates into those of a different observer (see Box 4.1, Chap. 4.2). For instance, if the length of the object on the platform is l_0, then the contracted length measured by the train-bound observer will be l. Length l is simply l_0 multiplied by a certain factor. Spatial and temporal relations are governed by the Lorentz transformations between inertial reference frames. The Lorentz transformations allow an objective communication between observer O_1 on the platform and observer O_2 on the train. By employing the Lorentz transformations observer O_1 will be able to calculate which interval observer O_2 will measure for the occurrence of an event; and *vice versa*.

---

has recently been revived in philosophy in a debate about the pros and cons of scientific realism and structural realism, see Worrall, 'Structural Realism' (1989); Ladyman, 'What Is' (1998); French, 'Eddington' (2003); French/Ladyman, 'Remodelling Structural Realism' (2002)

[98] Bohm, *The Special Theory of Relativity* (1965), viii; cf. Barbour, *The End of Time* (1999), 65; Cassirer, 'Zur Einsteinschen Relativitätstheorie' (1921), Chaps. II, III

[99] Holton, *Einstein, History* (2000), 132

[100] Cassirer, 'Zur Einsteinschen Relativitätstheorie' (1921), 62

We see: Temporal and spatial measurements are not invariant *across* different reference frames. They are the result of the different perspectives, or co-ordinate systems, of the observers involved. This *perspectivism* is due to the relegation of spatial and temporal measurements to individual reference frames. It makes no sense to ask for the *real* length of physical objects and the *real* duration of events in a three-dimensional spatial perspective and one-dimensional temporal perspective. A four-dimensional union of space and time has replaced the classical 3+1 view, with its separate spatial and temporal dimensions. There is no interesting physical reality in a three-dimensional world persisting in one-dimensional time. Reality now belongs to the four-dimensional world.

> (...) the four-dimensional world is no mere illustration; it is the real world of physics (...) (...) an observer on the earth sees and measures an oblong block; an observer on another star contemplating the same block finds it to be a cube. Shall we say that the oblong block is the real thing, and that the other observer must correct his measures to make allowance for his motion? All the appearances are accounted for if the real object is four-dimensional, and the observers are merely measuring different three-dimensional appearances or sections; and it seems impossible to doubt that this is the true explanation.[101]

At least the observers can communicate with each other about their respective measurements. That is why the Special theory of relativity leads to perspectivism, not relativism. But they can achieve more than talk about their frame-dependent perspectives. There are frame-independent invariant properties. There is an *invariant* relationship between events – space-time events –, which is the same for all observers. This is the so-called *space-time interval.* It says that the space-time interval, *ds*, for any two events in space-time is the same for all observers, attached to different reference frames.[102] Although the observers disagree about the spatial and temporal distance between two events *separately*, since they carry different coordinate systems, they will agree that the space-time interval between these two events is the same. In the space-time interval, *ds*, we come across an important *invariant* relation: from their different reference systems all observers measure the same space-time interval between events in space-time. Can the invariance of *ds* be made compatible with the idea of the block universe? The idea behind the block universe is that the passage of time is a human illusion. This reflects the demotion of the status of time in the Special theory of relativity. But if at least the space-time interval between two events located in space-time is invariant, then this invariant relation should express a reality in the space-time world, according to this criterion. We shall say more about this invariance in Chap. 4.

Consider now *quantum-mechanical* relations. Just as spatial and temporal measurements become frame-dependent in the Special theory of relativity, certain

---

[101] Eddington, *Space, Time & Gravitation* (1920), 181; see Balashov, 'Relativistic Objects' (1999)

[102] Sklar, *Space, Time and Spacetime* (1974), 260–4; Nahin, *Time Machines* (1993), 310–3; Sexl/Schmidt, *Raum-Zeit-Relativität* (1978), 113–4; Eddington, *Space, Time & Gravitation* (1920), 47–8, 70–1; Chap. 4 below

relations in quantum mechanics are subject to a 'similar' perspectivism.[103] Take an atomic particle, travelling near the speed of light. The particle releases a photon at an angle of 90° from the point of view of its reference frame (Fig. 2.8). At which angle is the photon emitted from the point of view of an observer in a stationary laboratory? The answer is: at 36.9° (Fig. 2.9).

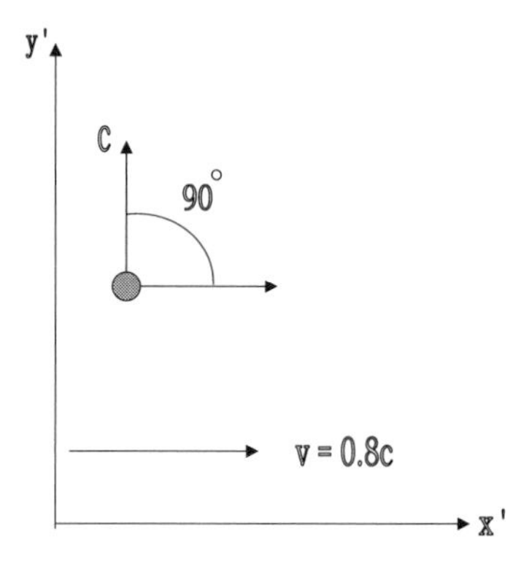

Fig. 2.8. Emission of a photon, at the speed of light, *c*, at right angle from the direction of the motion of the fast-moving particle, which travels at 4/5 of the speed of light

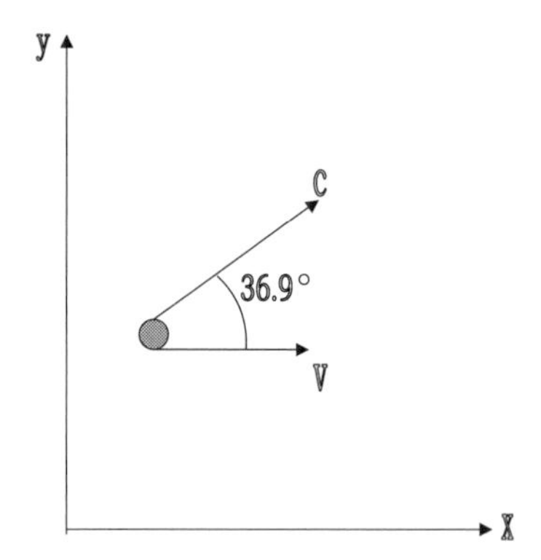

Fig. 2.9. Emission of a photon, as seen by a stationary observer in the laboratory. According to the observer the photon is emitted by the fast-travelling particle at an angle at 36.9°

The measurement of the angle is frame-dependent. We cannot say that the photon is *really* emitted at 90°, just as we cannot say, in the two-slit experiment, that it is really a particle, which travels to the screen. Are there *invariant* relationships in the atomic world? Energy and (angular) momentum are conserved. The laws of

[103] We must be careful not to overstate the analogy. Where there are analogies, there are *dis*analogies. In quantum mechanics we are dealing with a perspectivism of measurement situations. Particle- or wave-, position- or momentum measurements are related by Fourier transforms. The reference frames of the Special theory of relativity are linked by symmetry transformations.

nature are the same in all atomic reference frames. And atomic particles certainly have invariant properties. The spin and charge of the electron, the energy of its ground state and the electron rest mass are examples. There are fundamental constants in the quantum world, like the proton/electron mass ratio, the speed of light c and of course the Planck constant h. These quantities are still relational, since they must be determined from a particular reference system. But they are frame-independent, since their determination in different reference systems produces the same quantities. They must be regarded as belonging to the basic furniture of nature.

The Planck constant h plays a particularly important part in the indeterminacy relations. According to Heisenberg, the indeterminacy relations express mathematical relationships about the structure of the natural world.[104] According to these relations it is not possible to determine *simultaneously* the location and momentum of a particle, nor its energy and the duration, in which it will remain in this energy state. The product of these uncertainties is always greater than the Planck constant. Quantum mechanics can say little about the physical states of *individual* atoms because of the universal validity of the indeterminacy relations. However, given the devaluation of the particle view in modern physics, this retreat from the individual particle is to be expected. The invariant relationships now appear on a higher level. Quantum mechanics expresses formal relationships about the *states* of quantum systems, not individual particles. These systems can take on potential states, which are described in quantum mechanical equations. Mathematics describes the quantum systems in their full potentiality of states. Only some states are actualised in measurements. The result is the probabilistic character of the quantum-mechanical concept of Nature. Thus an electron has a greater probability to be found in its ground state and a certain probability of being in a particular orbital shell. The indeterminacy relations prevent a more precise determination of the electron's whereabouts. We can regard the indeterminacy relations as an example of an invariant relationship in quantum mechanics. We consider them here because they also play an important part in considerations of causation and determinism (Chap. 5). Indeterminacy means that there is no reference frame from which a more precise determination of the location and momentum, the energy and time can be achieved simultaneously.

Invariance expresses the idea that a probability space can be calculated within which the atomic particle could be located with varying degrees of probability. That means that from every reference frame it will be possible to calculate the same probability state of the quantum system. And the atomic constants will be the same

---

[104] Heisenberg, 'Über den anschaulichen Inhalt' (1927); 'Schwankungserscheinungen und Quantenmechanik' (1926–7); 'Die Entwicklung der Quantentheorie 1918–1928' (1929); it should be mentioned here that some, like David Bohm, have contested that the indeterminacy relations express fundamental facts about the structure of the quantum world. For reasons mentioned before, we emphasize the role of the indeterminacy relations. An accurate historical account would have to include the part, which the realization of the non-individuality of quantum particles and the development of quantum statistics played in the move towards an ontology of invariants.

from the point of view of every reference frame. Insofar as we talk about the state of the system with its many potential values, quantum mechanics can make fairly precise statements. They are probabilistic statements about the whole system. These probabilistic statements are lawlike. Invariance means that all observers will agree on the probability space, which characterizes the possible states of the quantum system as a whole.

Invariance attaches to the structure of possibilities of the quantum system, not to concrete quantum objects. Having identified some invariants, we are nudged towards an *invariance view of reality.*

> The appropriate criterion for what is fundamentally real will (...) be what is invariant across all points of view.[105]

What counts as *physically* significant, are the invariant frame-independent structures. As we make transitions from reference-frame to reference-frame, according to mathematical rules like the Lorentz transformations, it is these invariant structures, which are conserved.

Strangely, the invariance view of reality seems to imply that frame-dependent properties are *not* real. Observers in different reference frames will certainly regard the behaviour of their respective clocks and rods as real phenomena. The laboratory observer will entertain little doubt that the angle of forward-radiation is real. Observers in different reference frames seem to infer mutually inconsistent claims about reality. These claims are derived from frame-specific appearances. A proponent of the invariance view is however not forced into a denial that the appearance is real. The colours we perceive are real to us. Yet they are just wavelengths to the physicist. For theoretical purposes, physics is simply not concerned with what happens in particular reference frames. All reference frames ultimately exist on an equal par. The invariance view subtracts the frame-dependence from the reality claims. Physics chases the *physically real* in a frame-independent sense. It seeks to establish the underlying equations, which govern all reference frames. It is the job of symmetry transformations to state how reference frames are related to each other.

What about the problematic nature of *time* in the Special theory of relativity, the troublesome *wave-particle* duality in quantum theory? These are observational and experimental features, which destroyed the mechanistic worldview. The *invariance*-view of Nature accommodates the *relational* and the *formal* aspects of the modern view of Nature. It is a view on which both the relativist and the quantum physicist can agree. And given the unavoidability of reference frames, the link between invariance and reality imposes itself. In the following chapters we will investigate the claims of the scientist-philosopher: whether time must disappear from the physical worldview as merely frame-dependent, and whether the wave-particle duality means that causation must also make its exit from the modern view of Nature.

For the moment let us consider the invariance view in a little more detail. Both the relativity and the quantum theory regard many properties as relational, that is,

---

[105] Hooker, 'Projection' (1991), 493

they are not invariant across different reference frames. But neither theory stops there. The corollary of the relativity principle is that certain properties remain invariant across some or all reference frames. The invariance view would regard the invariants as physically real. It is closely connected with *symmetry principles*. For if reference frames are related to each other by symmetries, then there are elements of structure, which remain invariant. This is just what symmetry principles affirm. Symmetry principles are of fundamental importance in modern science. They state that certain, specifiable changes can be made to reference systems, without affecting the structure of the reference system. Whether we perform an experiment in a reference system in Paris or Tokyo, in the year 1900 or 2000, will not affect the results of the experiment. These results are invariant under spatial, temporal and rotational symmetries. As van Fraassen puts it: 'Symmetries are transformations that leave all relevant structure intact.'[106]

The invariance view of Nature consists of two sub-theses

The invariant is the real. This is a hypothesis about physical reality: what is frame-dependent is apparently real, what is frame-independent *may* be fundamentally real. To claim that the *invariant is the real* is to make an inference from the structure of scientific theories to the structure of the natural world. If we regard reference frames as co-ordinate systems, then reference frames are human inventions. The inference from what our theories tell us about the world to this world itself will always

Box 2.2: Reality as Kickability?

Why can't we be more straightforward and call physically real what is 'kickable' (Popper, *Quantum Theory*, 1982; Hacking, *Representing*, 1983; Deutsch, *Fabric*, 1997, Chap. 4). This criterion serves its purpose. Many objects and entities are 'kickable', even if at a particular moment in time the technology lets us down. At the end of the 19th century many scientists believed that atoms were not real. The technology was not there to manipulate them. The problem with this view is rather that science talks about the reality of phenomena and regularities, which are not kickable, even in principle. The Big Bang is not kickable, nor is the universe as a whole. Although laws of nature can be used to manipulate objects, like molecules and electrons, the laws themselves are not kickable. We cannot kick the Principle of Evolution, the Conservation of Energy or Space-Time. Yet they represent structural regularities, which we want to call real. By contrast, some things are kickable, yet some may not want to regard them as real. The wave-particle duality can be manufactured in experiments, like the double-slit experiment, yet some only regard the waves, others only the particles as real. And of course we can kick relational properties. But we only get perspectival realities.

[106] van Fraassen, *Laws and Symmetry* (1989), 243; see also Rosen, *Symmetry in Science* (1995) Chap. I: Symmetry is immunity to possible change.

be conjectural. But this is not to surrender to scepticism. For some of our conjectures are better confirmed than others. The invariant structures of a well-confirmed theory are therefore good and reliable indicators of what is *real*, irrespective of human intervention. But they are only *conjectural* indicators. They depend on the progress of science. In the Special theory of relativity, Minkowski space-time structure is invariant. According to this criterion, it should be regarded as real. This, however, leaves the claim open to certain interpretations. Does space-time exist independently of material events, as an underlying structure of space-time points? This would make it analogous to Newton's absolute time. Or is space-time itself constituted by spatio-temporal relations between bodies and events? This would make it analogous to Leibnizian relationism. There is no invariant interpretation of space-time. Nor is space-time invariant across different physical theories. In the General theory of relativity, space-time becomes a dynamic entity. In today's quantum cosmology, it only emerges as a semi-classical approximation. Any claim about the reality of space-time must be treated as hypothetical. Firmer evidence, discussed below, is required before the status of space-time itself can be settled. But this is just what this part of the invariance view holds. What remains invariant is however the space-time interval ds.

The real is the invariant. In a physically interesting sense, the real is invariant across different reference frames.[107] It comprises the lawlike regularities, structures and symmetries in the natural world. The *real* will always be frame-independent, since the reality of the structures of the material world cannot depend on the adoption of a particular reference frame. The laws of nature cannot vary randomly from location to location, from time to time, from reference frame to reference frame.

The invariance view of reality asserts that the world consists of structures. These ontological structures are represented in mathematical equations and 4-dimensional space-time diagrams. As a structure consists of objects and vari-

[107] It appears that already Heraclit saw the agreement between sense perceptions of different observers as a criterion of the real; see Schrödinger, *Was ist ein Naturgesetz*? (1987), 61. The thesis that the *invariant is the real* has been a common theme, since Einstein's theory of relativity demonstrated the significance of reference frames and the importance of transformation rules. Dirac considered that the 'important things in the world appear as the invariants (...) of these transformations,' see Dirac, *The Principles of Quantum Mechanics* (^{4}1958), vii, §8. Planck, too, regarded the constant in our worldviews as the real; see *Vorträge* (1975), 49, 66, 77, 291; see also Poincaré, *Science* (1902), Pt. IV, Chap. X; Cassirer, 'Zur Einsteinschen Relativitätstheorie' (1920), 12ff, 28ff, 34ff. The thesis is a common theme in many publications: Born, *Natural Philosophy* (1949), 104ff, 125; Born, 'Physical Reality' (1953), 143–9; Costa de Beauregard, 'Time in Relativity Theory' (1966); Wigner, *Symmetries* (1967), Pt. I; Sklar, *Space, Time and Spacetime* (1974), 359–71; Friedman, *Foundations* (1983); 19, 320–34; Hooker, 'Projection' (1991), 491–511; Norton, 'Philosophy of Space and Time' (1992), §5.4; Maxwell, 'Aim-oriented Empiricism' (1993), 89, 100; Crombie, *Styles of Scientific Thinking* (1994), Vol. I, xii, 6ff; Vol. III, 1763; Cook, *Observational Foundations* (1994), 80f; Morrison, 'Symmetries as Meta-Laws' (1995), 157–88; Feynman, *Six Not So Easy Pieces* (1997), 29–30; Nozick, *Invariances* (2001), Chap. 2; Maudlin, *Quantum Non-locality* (2002), Chap. 2

ous relations between them, the invariance view is in good agreement with the modern emphasis on systems.

Systems are fundamentally interrelated. The invariance view directs our attention to frame-invariant realities. How do we know that a particular property or feature is invariant? As was pointed out above, reference frames are related by symmetries. It is these symmetries, which allow us to distinguish between frame-dependent and frame-independent realities. A convenient distinction is that between *geometric* and *dynamic symmetries*.[108] Rotation, reflection, spatial and temporal translations and space-time symmetries are typical *geometric* symmetries. They take events, things and properties as their objects. Gauge symmetries like the charge-parity symmetry are representatives of the newer *dynamic* symmetries. They take electromagnetic, gravitational, weak and strong interactions between elementary particles as their objects.

Perspectivism and invariance are two faces of symmetries. There are many relational properties, which lay claim to a perspectival reality. But there are also underlying structures, which give us the physically significant realities. At least since the 19th century, with its discoveries of entropy, electro-magnetism, atomic structure and Darwinism, we think of Nature as a system of interrelated subsystems. So we must think of theories as reflecting this interrelatedness. The symmetries play an important part in this interrelatedness. But why should we assume that if the objects and relations are confirmed, the symmetry principles, which govern these structures, are also confirmed? The symmetries tell us more than what the objects and laws tell us. They tell us, which quantifiable changes affect the systems and which do not. So they tell us that the empirical world is one of interrelated subsystems; the interrelation is subject to quantifiable constraints. For if symmetries are immunities to change, this change is not just an epistemological shift in our knowledge claims. The change happens to physical systems. We can track the changes that happen to physical systems through our investigations. Some changes affect the frame-specific parts of the structures, as in time dilation. They leave the system invariant. Others affect what was taken to be the frame-invariant parts, as in broken symmetries. Broken symmetries demonstrate that the immunity is sometimes lifted. A broken symmetry occurs when a change in the reference system changes a regularity expressed in the laws of nature. The most famous example is the violation of charge parity symmetry in weak interactions. This describes an experimentally well-observed situation, in which decay patterns of elementary particles experience exceptions. That is, in which expected decay patterns fail to occur.

The constants of nature[109] are classic examples of the real. The velocity of light, *c*, is invariant across different reference frames in the Special theory of relativity. Both stationary and moving observers will agree on the value they measure for the velocity of light. It does not depend on their framework or on the direction

[108] Wigner, *Symmetries and Reflections* (1967); Morrison, 'Symmetries as Meta-Laws' (1995); Earman, *World Enough* (1989), 173; Rosen, *Symmetry in Science* (1995), 72–6; Mainzer, *Symmetries* (1996), 277, 341–2, 357, 414, 420

[109] Weinert, 'Fundamental Physical Constants' (1998)

in which the measurement is made. As we shall see, the invariance of the velocity of light is a fundamental postulate in the Special theory of relativity. Any argument about the passage of time, as a reality or a human illusion, will have to take this fundamental postulate into consideration. The finite velocity of light also sets a limit on the propagation of causal signals. Light spreads out from any event, forming light cones. Within these light cones, causal chains are irreversible. They are the same for *all* observers.

The Planck constant, h, is invariant across different reference frames in quantum theory. Whether we adopt the framework of a fast-moving photon or that of a stationary observer, the value of h does not change. The Planck constant h plays an important part in considerations of causation and determinism in the philosophical consequences of quantum theory. The constant h stands for the discontinuous and probabilistic character of the modern notion of Nature. Nature appears to make quantum jumps as in radioactive decay. The causal conditions of these quantum jumps are not known. Only the probability of decay for a collection of atomic particles can be stated. This puts a severe strain on the classical ideal of causal determinism. Physicists worried about the philosophical consequences of the quantum theory for the notions of determinism and causation.

In Chaps. 4 and 5 we shall consider the two big questions, related to the two great scientific innovations of the 20th century: are temporal and causal relations real or unreal? But let us first consider how the invariance view of Nature affected the notion of physical understanding. This much more abstract invariance-view of Nature goes hand in hand with an increased mathematization of modern science. It brought the role of models and the problem of *physical understanding* to the fore. How is the mathematical symbolism related to the real world of physics? What do the mathematical symbols mean? How can the mathematical relations be understood in physical terms? As we shall see in the next section, the scientists involved in the great scientific revolutions of the 20th century were keenly aware of the need to clarify the notion of physical understanding.

3

Physical Understanding

> Beyond the knowledge gained from the individual sciences, there remains the task of comprehending. In spite of the fact that the views of philosophy sway from one system to another, we cannot dispense with it unless we are to convert knowledge into meaningless chaos.
>
> H. Weyl, *Space-Time-Matter* (1922), 10

3.1 Understanding and Fundamental Concepts

Concepts like *understanding* and *meaning* are usually associated with a particular view of the Social Sciences. Social life produces and reproduces symbolic meaning. Social scientists need to acquire an understanding of the inherent symbolic meaning in social life. They do this, it is said, by adopting the viewpoint of a passive participant observer. In this view, the role of the social scientist is seen as distinctly different from that of the natural scientist. The object of study of the social scientist is *society*, the network of social interactions. Society does not exist outside the bracket of social interactions. The social sciences deal with the pre-interpreted world of the social participants. The social scientist interprets a social world, which already carries symbolic meaning. The symbolic meaning of the social world is produced and reproduced by the social actors. The study of the social world by social scientists is a matter of human subjects studying other human subjects. It is a matter of symbolic dimensions meeting other symbolic dimensions, a *subject–subject* relation.

The object of study of the natural scientist is *Nature*, the organic or inorganic material world. In this objective sense Nature is not a human product. But, in a symbolic sense, 'Nature' is a creation of human understanding. In their interaction with the material world, humans conceptualise *Nature* in an attempt to understand its functioning. Models, theories and laws are the result. They reflect in symbolic form what successive epochs understood by 'Nature'. The concept of 'Nature' belongs to the category of *fundamental notions* with which humans represent the natural world. Humans also use fundamental concepts to explain how they, as humans, manage to comprehend the world around them. Do to this, humans employ such fundamental concepts as *determinism, indeterminism* and *causation, time* and *space, mass* and *energy, motion* and *rest.* They are fundamental in several senses: a) at a basic level they are needed in every consideration of Nature;[1] b) they are also used to explain how humans acquire knowledge about the material world; and c) they play a central role in everyday discourse. As human knowledge of the lawful processes in the material world has increased tremendously, and sometimes undergone radical change, the fundamental notions have been modified in the light of new empirical discoveries.

Scientific discoveries have had an impact on the fundamental concepts used to describe and explain the natural world. The meaning of the fundamental concepts has departed from their meaning in everyday experience because the scientific worldview has departed from everyday experience. But the everyday experience of the material world is in many ways a reflection, albeit imperfect, of the scientific view of this world. So in many ways the everyday usage of the fundamental concepts is an approximation, albeit rough, to the scientific usage of these concepts.

Natural scientists face a pre-given natural world, not the symbolic, pre-interpreted world of the social scientist. Natural scientists stand in a *subject-object* relation to their object of study. Yet they use symbolic language to make sense of the material world. Questions of understanding and meaning have been familiar to the scientific enterprise throughout its history. Like the ancient Greeks, natural scientists face a complex of often bewildering phenomena. First there is the question of *how* the observable phenomena behave. The observational and experimental data reveal patterns of regularity. Then there is the intriguing question of *why* the phenomena behave in such particular patterns of regularity. In an attempt to answer such questions, the natural scientist aims at *understanding* and *explanation*. Copernicus defended his heliocentric hypothesis (1543) by pointing out that it provided a more coherent understanding of the orbits of the planets than the geocentric hypothesis. In his *Novum Organum* (1620) Francis Bacon proposed 'directions for the interpretation of Nature.' In his interpretation of the phenomenon of heat Bacon gave a clear indication of what later scientists would come to mean by *understanding* in the natural sciences. Bacon reduced the phenomenon of heat to motion. Heat was nothing but motion.[2] The basic move is to go beneath the level of observable phenomena ('heat') to a more fundamental level ('motion'). Bacon's programme had no mathematical precision. This has changed since the Scientific Revolution. But the basic move is still present in today's attempts at understanding. Students of quantum mechanics are familiar with the distinction between the mathematical formalism of the theory and its physical interpretation. The interpretation of quantum mechanics in terms of physical reality poses considerable problems. There are rival interpretations. These often involve suggestions to revise the fundamental

---

[1] This is how Franz Exner, *Vorlesungen* (1919), §§1, 37, 95 in a much-quoted book in the early part of the 20th century characterises fundamental notions; see also Schlick, 'Raum und Zeit' (1917), 21; Schlesinger, *Aspects* (1980), 3

[2] Bacon, *Novum Organum* (1620), Book II, §20

notions with which Nature has been traditionally described. As we have seen, the notion of Nature has borne much of the brunt of these conceptual revisions. However, the question of how the *mathematical* symbols are to be interpreted *physically* is much older than quantum mechanics. A perusal of the relevant literature quickly reveals that the question of the meaning of mathematical symbols and equations and their physical interpretation has accompanied the scientific work of scientists for a long time. Scientists in the 17th century sought to explain the propagation of light and gravitation across empty space by postulating the existence of an ether, which filled space. In a characteristic passage Einstein wrote that the equivalence of gravitational and inertial masses 'had hitherto been recorded in mechanics, but it had not been *interpreted*.'[3] And it was Einstein's interpretation of this equivalence, which paved the way to his theory of general relativity. In the physical interpretations of the mathematical symbols, the fundamental concepts play a significant part. These concepts have acquired a history through the scientific revolutions of the last 400 years. New discoveries about the material world have suggested to scientists that the fundamental concepts have natural limits of applicability. It is only as long as the material world does not throw doubts on the appropriateness of these concepts that they remain unquestioned starting-points of conceptualisation. But when the empirical evidence clashes with the fundamental concepts, scientists awake to the philosophical dimensions of their work. Scientists naturally become concerned with questions of meaning, understanding and interpretation. What do these concepts mean in the work of scientists?

There is a common strand of thought running through the two great scientific revolutions of the 20th century. The new discoveries associated with them required a conceptual revolution. Either completely new concepts were required ('discontinuity', 'uncertainty', 'nonlocality', 'time dilation', 'length contraction'), or a modification of old concepts ('causation', 'Nature', 'time'). Sometimes well-established concepts ('determinism', 'simultaneity', 'absolute time' and 'absolute space') needed to be relinquished. This means that the fundamental concepts used in the natural sciences *depend*, to a certain extent, on the empirical findings. With characteristic clarity, Einstein expressed this dependency succinctly when he said that logically, the fundamental concepts ('time', 'space', 'mass', 'event') were free creations of the human mind, but that they had empirical roots.[4] It is not surpris-

[3] Einstein, *Relativity* (1920), 65; italics in original. See also Einstein, 'Über den Einfluß der Schwerkraft' (1911), 73. Much later Dirac, 'Evolution' (1963), 53 stressed the 'importance of finding physical ideas behind the formalism.'

[4] Einstein, *Relativity* (1920), 141. This discussion appears in Appendix V of this book, which was added in 1952. Einstein shared the conviction that new concepts are required by new experiences with Heisenberg, *Physical Principles* (1930), Chap. IV, §3, 'Deutung' (1955); Bohr, 'Wirkungsquatum' (1929), 483, 485 and Schlick, 'Kausalität' (1931), 145. Note that Einstein, Heisenberg and Bohr fundamentally disagreed about the *interpretation* of quantum mechanics, in particular the usefulness of such fundamental concepts as locality and causation. In philosophy the dialectic between concepts and facts was taken up by Ernst Cassirer, 'Zur Einsteinschen Relativitätstheorie' (1921), 20 who spoke of the 'continuous oscillation between experience and concept.'

ing that some of the leading scientific revolutionaries of the 20th century began to think about the concept of understanding (*Verstehen*) in physics. After all, the behaviour of atoms and fast-moving particles posed considerable challenges to the comprehension of the natural world. Throughout the history of modern science, innovative scientists have been willing to question time-honoured concepts in the face of new experiences. It is the privilege of 20th century physics to have thrown new light on the dialectic between facts and concepts.

Max Planck introduced the quantum revolution. The discovery of the quantum of action (*Wirkungsquantum*), designated by the letter h, introduced discontinuity into modern physics. Planck introduced the letter h in an equation for blackbody radiation to make it compatible with the experimental evidence. As he had arrived at what he called his 'interpolation formula' by way of an educated guess, Planck was faced with the question of the physical meaning of his formula.[5] For Planck understanding took on a particular sense. It was not the 'discovery' of the quantum of action, expressed in the constant h, but the assignment of a physical meaning to this constant which was the 'theoretically most difficult problem.'[6] Planck interpreted the quantum of action - which he had introduced into his distribution law - not as a fictitious entity but as a real physical constant.[7] Experiments quickly showed that this interpretation was correct. It led to a radical rethinking of the physical worldview and, as we shall see, to a rethinking of the fundamental notion of causation.

Throughout his career, Werner Heisenberg was much concerned with questions of understanding and interpretation in the physical sciences. From his early publications on quantum mechanics to his last essays on philosophy, he returned repeatedly to the concept of understanding in physics. Together with Einstein, Planck and Born, he was one of the most philosophical physicists of his generation. It was common practice among many of the founding fathers of quantum mechanics and relativity theory to include philosophical discussions in their technical papers. For Heisenberg understanding in physics meant the reduction of the complexity of phenomena to a few basic and quite general concepts.[8] The possession of such concepts would allow the representation of the underlying unity in a great number of phenomena. But crucially, the discovery of new phenomena would require the revision of concepts, which had served well in the representation of old domains. Heisenberg, like Bohr, stressed that Einstein himself had introduced a revision into the concepts of space and time. The abandonment of concepts like causation, they held, was similarly a consequence of the new discoveries in quantum mechanics.[9] In a more specific sense, then, Heisenberg held that understanding meant the ability

---

[5] Planck, 'Die Entstehung und bisherige Entwicklung der Quantentheorie' (1920), 129

[6] Planck, 'Zur Geschichte der Auffindung des physikalischen Wirkungsquantums' (1933), 27

[7] Planck, 'Die Entstehung und bisherige Entwicklung der Quantentheorie' (1920), 131

[8] *Der Teil und das Ganze* (1973), 46; 'Philosophische Probleme in der Theorie der Elementarteilchen' (1967), 410ff

[9] Heisenberg, 'Erkenntnistheoretische Probleme' (1928), 21–28; 'Die Rolle der Unbestimmtheitsrelationen' (1931), 40–47; 'Prinzipielle Fragen' (1936), 108–18; '50 Jahre Quan-

to detach oneself from old concepts when new domains of experience were under consideration.

> We have understood a group of phenomena when we have found the right concepts for describing these phenomena (...) it is always the simplicity of the concepts in comparison with the great wealth of complicated experimental material, which convinces of their correctness. Usually in a new field many very different experiments can be carried out; and if all these experiments allow a description by the same simple new concepts, these concepts will finally be accepted as the correct ones.[10]

Heisenberg explicitly rejects the identification of understanding with predictability. Ptolemaic astronomy shows very clearly that understanding and predictability are not the same. For Ptolemy, following Aristotle, the earth resided motionless at the very centre of the universe. Ptolemy's complicated meshing of geometric figures to account for the apparent movement of the planets and the apparent rest of the fixed stars provided some fairly accurate predictions. But his geocentric model offered no genuine understanding, let alone explanation. The earth does not occupy the centre of the universe and the planets do not move in circular epicycles, carried on circular deferents, around it. Ptolemy made no claim that his model resembled the real world. It may appear as if by understanding Heisenberg simply meant the mathematical derivation of observable phenomena from fundamental equations. But more is at stake. Mathematical analysis follows the formulation of new concepts. Conceptual comprehension precedes mathematical elaboration.

> Mathematical analysis can be an important help after the correct concepts have been found, since it may then enable the physicists to draw precise conclusions and to compare them with the facts.[11]

This procedure is nicely illustrated in the work of some of the great scientists. In his *Principia* (1687) Newton revises the ordinary notions of space and time, held by the 'common' people. His introduces his notions of absolute space and time before he formulates his equations of motion. Einstein's work shows that concepts like 'the ether', 'preferred reference frame', 'absolute simultaneity' and 'absolute time' needed to be abandoned or revised before the mathematical elaboration could proceed. In 1913 Bohr blamed classical physics rather than Rutherford's nucleus model of the atom for the lack of progress in understanding atomic phenomena.

The physics of the 19th and 20th century paid a price for the introduction of new concepts. An interpretation of the notion of understanding in terms of concepts suitable to new experimental evidence meant a modification of *visualization*. It

---

tentheorie' (1951), 354–60; 'Philosophical Background' (1964); 'Änderung der Denkstruktur' (1969)

[10] Heisenberg, 'The Concept of "Understanding" in Theoretical Physics' (1969), 337; cf. 'Philosophische Probleme in der Theorie der Elementarteilchen' (1967), 411–4. The importance of conceptual understanding prior to the mathematical elaboration of the problem at hand is also stressed by one of the pioneers of loop quantum gravity; see Smolin, *Three Roads* (2001), Chap. 9

[11] Heisenberg, 'The Concept of "Understanding" in Theoretical Physics' (1969), 338

is easier to visualise the atom as a planetary system than to grapple with the abstract idea that it has no definite properties, like position, momentum, orbit. Unfortunately, the evidence does not support the view that the atom is a planetary system. Visualization requires a description of phenomena in terms of causal spatio-temporal relations. Macro-systems can be conveniently visualised. For instance, a planetary scale model will do. But the atom resists such mimetic visualization. The new concepts required for an understanding of the atom imposed severe limits on a direct visual representation. It became a stock phrase in the writings of physicists – whether they dealt with the relativity or the quantum theory – that 'naïve visualizations' had to be abandoned in physical understanding.[12] But even for atoms visualization is not impossible. It will have to be done by appropriate models. The concepts required for the physics of the atom destroyed direct visualization. It was not right to say that the atom *is* a minute planetary system. Only for purposes of illustration could an *analogy* between the solar system and atoms be entertained. A proper representation required what we will later call a structural model.

For Erwin Schrödinger the task of understanding was intimately connected with human ability to construct conceptual models.[13] Such mental constructions assigned underlying *structures* to the observable phenomena. The complexity of the phenomena could be coherently ordered by a *Gestalt*, even though not all aspects of it was subject to observation and experimental checking. Although Schrödinger uses the term *Bild* (picture, image), the primary aim of the conceptual models does not seem to be direct visualization. Rather, some underlying order (*Gestalt*) is to be assigned to the observable phenomena, which renders them understandable. This underlying structure could be expressed in purely mathematical terms. Or an analogy with some familiar structure may be suggested. Perhaps some idealised configuration could represent the underlying order. Schrödinger's notion of conceptual model harbours a complexity, which goes beyond simple mechanical models. It continues the considerations, which Maxwell and Hertz had set in motion at the end of the 19th century.

It is easy to see that some care is needed in this connection. Not every conceptual model will provide understanding. The geocentric worldview of Ptolemy and Aristotle was based on an unrealistic geometric model. It was a useful analogy.

[12] Pauli, 'Relativitätstheorie' (1921), vi; Heisenberg, 'Über den anschaulichen Inhalt' (1927), 172–3; *Physical Principles'* (1930), Chap. I, §2; 'Rolle der Unbestimmtheitsrelationen' (1931), 40–2; Bohr, 'Atomic Theory' (1929), 'Wirkungsquantum' (1929); Popper, *Quantum Theory* (1982), 45, 97

[13] Schrödinger, 'Conceptual Models in Physics' (1928); 'Die Besonderheit des physikalischen Weltbildes' (1947), 37–48. On the demise of visualization in modern physics see Čapek, *Philosophical Impact* (1962), 378–9; Čapek holds (1962, 398) that the end of visualization in modern physics means that the concomitant conceptual revolution is far more radical than the Copernican revolution. But visualization did not end with modern physics. It became more complex; on visualization from a modern perspective see Holton, *Einstein, History* (2000), Chap. IV and Galison, *Image and Logic* (1997). Galison (1997, 22 Fn 30) writes that 'visualization is a contested form of demonstration as well as a contested form of laboratory work' in modern physics.

But it failed to provide understanding. A more careful consideration of the role of *models* is required to decipher their role in understanding.

Let us first observe that such philosophical concerns with the meaning of fundamental concepts and the physical understanding of the mathematical symbolism, used in scientific language, are not confined to the philosopher's ivory tower. In the dialectic between facts and concepts, science harbours an inherent philosophical dimension. Consider the following passage:

> ... one must distinguish sharply between the predictions made by a set of laws and the mental images that the laws convey (what the laws "look like"). I expect convergence only in terms of predictions, but that is all that ultimately counts. The mental images (one absolute time in Newtonian physics versus many time flows in relativistic physics) are not important to the ultimate Nature of *reality*.[14]

In this quote, astrophysicist Kip Thorne embraces an *instrumentalist* thesis about what scientific theories say about the world. Apparently they represent very little of the structure of Nature: 'only predictions count'. Thorne places himself in a long tradition of the realist-instrumentalist debate about the Nature of scientific theories, which stretches back to the Greeks. The publication of the Copernican system (1543) rekindled the debate. The question whether the Copernican system was a faithful representation of reality or a mathematical convenience was finally settled 144 years later when Newton published his *Principia* (1687). The instrumentalist Nature of Thorne's view is immediately apparent from a comparison with Schrödinger's and Heisenberg's views on understanding. In their view the physicist is concerned with the underlying order, expressed in terms of conceptual models or adequate concepts, not just with predictions deriving from the models. For the *realist*, what counts over and above the predictions, is whether the underlying order of the conceptual model corresponds to the order of the real world. Physicist Alan Cook defends such a realist view of scientific theories and the Nature of understanding.

> Physicists make observations upon [the unique world of Nature], observations which of themselves might be no more than curiosities, but it is the aim of physics to put those observations into a rational scheme by which sense may be made of them and of the natural world behind them.[15]

James Cushing, a physicist with an interest in the philosophical foundations of modern physical theories, has given some thought to the question of understanding in the natural sciences. He distinguishes between the aims of *explanation* and *understanding*. Explanation is simply the formal derivation of particular statements from general laws. This is Hempel's *DN* model of scientific explanation. Universal laws entail particular facts. There are alternative philosophical models of explanation.[16] The *causal* model locates explanation in the detection of traceable causal patterns in the physical universe. The *structural* model sees explanation in the assignment

[14] Thorne, *Black Holes* (1995), 85–6; italics in original

[15] Cook, *Observational Foundations* (1994), 96–7

[16] Kitcher/Salmon, *eds.*, *Scientific Explanation* (1989); Salmon, *Scientific Explanation* (1984), Chap. I

of structures to natural systems. Let us ignore the differences between these models of explanation. They share the view that the explanation of some natural system, in biology, chemistry or physics, requires both mathematical precision and a coherent account of a wide range of phenomena. This includes a detailed analysis of structures, mechanisms and regularities. Such coherent accounts often produce precise predictions. In a word, explanations are given by scientific theories.

Sophisticated theories, however, are not always available. Then understanding can be obtained from some suitable model. Heliocentrism is a case in point. Copernicus suggested a planetary model, which assigned a new structure to the solar system. It was based on limited observational data, which dated back to the Greeks. Fundamental problems remained: Is the shape of planetary orbits circular and what keeps the planets in orbit? It took the combined efforts of Kepler, Galilei, Hooke and Newton to transform this model into a coherent theory. A model does not require the precision of a theory. It must at least be empirically adequate; i.e. it must display consistency of the model structure with the empirical data. A model ascribes a structure to a limited range of empirical data.

In Cushing's sense understanding in physics means the assignment of 'picturable physical mechanisms' and 'processes that can be pictured' to the formal, mathematical aspects of physical theories. The formalism of statistical mechanics, for instance, is given an interpretation in the kinetic model of gases.[17] A volume of gas is modelled as a collection of billiard balls. Schrödinger associated understanding with the ability to develop conceptual models, of which mental 'pictures' were only one form. Cushing associates understanding with the availability of picturable models. More precisely, Cushing requires the availability of causal models to provide an understanding of physical processes. This call for visualization harks back to the classical age of physics and its mechanical models. If no representable causal mechanism, no picturable model of quantum relations can be found in the atomic realm, then on Cushing's account, there is no understanding of the atomic realm. There is only a mathematical formalism in the form of the Schrödinger equation.

Such a view may be too stern, as Maxwell already pointed out. Causal models are only one type of scientific model. Schematically, we can say that scientific theories provide explanation whilst models provide understanding. This is because models are subject to fewer constraints than theories. Newton provided an understanding of optical phenomena by a corpuscular model of light.[18] At least this corpuscular model was consistent with a limited range of empirical data. Later Thomas Young, in an early version of the double-slit experiment, showed that the 'atoms' of light

[17] See Cushing, 'Quantum Theory and Explanatory Discourse' (1991), 341; *Philosophical Concepts in Physics* (1998), 338–42; Folse, 'Ontological constraints and understanding quantum phenomena' (1996), 121–36; Beller, 'The Rhetoric of Antirealism' (1996); Salmon, *Causality and Explanation* (1998), 79–91. Contrary to Cushing, Salmon, *Causality* (1998), 9 sees scientific understanding as grounded in scientific explanation. For an up-to-date discussion of computer simulations of statistical mechanics, see Carlson, 'Modeling the Atomic Universe' (1999), 96–7 and Galison, *Image and Logic* (1997), Chap. 8 on Monte Carlo simulations.

[18] See Crombie, *Styles of Scientific Thinking* (1994) Vol. II, 1055

could behave like waves and produce interference fringes. A wave model of light offered an alternative understanding of optical phenomena. It too was consistent with a limited range of empirical data.

To see the point of our distinction between understanding and explanation, consider again the double-slit experiment. We noted that the interference patterns defied our common sense of causation. Yet there have been many attempts in the literature to account for the interference patterns. These models of understanding may be based on different principles:

1. Bohr and Heisenberg exploited the indeterminacy relations and their accompanying uncontrollable 'kicks';
2. Born's interpretation of the wave function gave rise to the idea of probability waves;
3. Feynman proposed his 'sum-over-paths' approach, the modern version of which is Griffiths' 'consistent histories' model;
4. De Broglie and Bohm preferred a 'pilot wave' theory;
5. Deutsch (1997) has recently mooted the idea of shadow photons and the multiverse;
6. The recent discovery of decoherence, which will play a significant role in the subsequent discussion, has provided yet a different solution.

Many different models of understanding have been proposed. No explanation has been reached. Yet each of these models is the source of a possible explanation. Like no other scientific theory, quantum mechanics illustrates the need for models of physical understanding. Even more importantly, as we shall see, it demonstrates that new notions come in tandem with new ways of physical understanding.

As Bohr and Heisenberg emphasised, the nature of the atom imposed restrictions on human attempts at the visualization of the atomic realm. The atom does not behave like a classical particle, whose 'whereabouts' are in principle knowable with great precision. According to Heisenberg we need new concepts to represent new facts. If we are willing to change our concept of causation, then a causal explanation of quantum phenomena may be possible.

Between abstract theory and concrete phenomena, models provide understanding of the material world. Models provide visualizations. But since the decline of mechanical models, in the wake of the retreat of the mechanistic worldview, visualization comes in many different shapes and forms. A popular form of visualization is the *analogue* model. The atom may be modelled on the analogy with the solar system. An electric current may be presented on the analogy with a system of water tubes. Map analogies became prominent in the early literature on Minkowski's geometrical interpretation of Einstein's Special theory of relativity. The analogy aids the visualization but does it aid the understanding? Probably not. This answer is rooted in the nature of analogue models, as discussed shortly. Useful visualizations, which really contribute to physical understanding, must be guided by appropriate concepts. We must possess a good conceptual grasp before a helpful model can be constructed. At one stage the atom was represented as a sphere of positive electricity with electrons swimming in concentric rings around it. There was no nucleus.

This was Thomson's *plum pudding* model of the atom. Then Geiger and Marsden discovered that, under suitable conditions, small particles could be deflected from an atom. Rutherford provided an understanding of this phenomenon of scattering by postulating the existence of an atomic nucleus. This was the *nucleus model* of the atom. This example shows that useful models depend on the provision of empirical data and the existence of appropriate concepts for the physical domain under investigation. There is a need to assign an underlying physical structure (molecular or subatomic) to the detectable phenomena (the behaviour of gases or atoms) in terms of which the measurable processes can be interpreted. Einstein characterised understanding in physics in this way.[19] Physical understanding is achieved when *appropriate* models are available. Appropriate models must be compatible with the evidence. The assignment of the model structure to the empirical data must be reliable and justifiable.

Take the question of *causation* in the atomic domain. A classical notion of causation based on the results of the Newtonian-Laplacean worldview would lead to the denial of causation in quantum mechanics. Yet, as we shall see from empirical discoveries, causal accounts abound in quantum mechanics. Or take the notion of *time*. Its compatibility with relevant empirical evidence must be guaranteed by an appropriate notion of physical time before the adequate modelling of relativistic phenomena can succeed. Thus models play a vital part in physical understanding. However, not all models will do. What, then, is a model?

3.2 Models

There have been many attempts by scientists and philosophers to distinguish various kinds of models and to spell out their role in science.[20] Quite generally, a *model* can be constructed when a limited amount of data is available about some natural or even social system. Typically, the data present the numerical values of a few parameters – like pressure, P, and volume, V, or distance, d, and velocity, v – which can then be related to each other in a model. Models are ideal tools for studying the relationship between parameters, the interrelatedness of Nature. In the model, the parameters need not be numerically specific. They can be indicated by symbols like V, P, d, v and others. What really matters in the model is how they are related to each other. To perform this role models have several important *functions*.

Models concentrate on a few manageable parameters and *abstract* from a number of interfering factors. The interfering factors are bracketed for the purpose of modelling. This operation is called *abstraction*. These interfering factors may be demonstrably negligible, in which case the model will justifiably ignore them. However, closer scrutiny may reveal that the abstracted factors have a non-negligible

[19] Einstein, 'Was ist Relativitätstheorie?' (1919), 127

[20] See Hesse, *Models and Analogies* (1966); Norton, 'Science and Analogy' (1980); Redhead, 'Models' (1980); Cushing, 'Models' (1982); Cartwright, *Laws of Physics* (1983), *Nature's Capacities* (1989); Morgan/Morrison *eds.*, *Models* (1999); Weinert, 'Theories, Models and Constraints' (1999)

influence on the relationship between the parameters, in which case they need to be incorporated in the model.

The real factors operative in the material world may be too complicated to compute, in which case a model needs to introduce mathematical simplifications. The models *idealise* the parameters to make their relationships computable in the models. This is *idealisation*.[21]

Again, more complicated models may be able to reduce the idealisation of the parameters. The inclusion of non-negligible factors and the reduction of idealised parameters are called *factualisation*.

Let us model the solar system with its nine planets. The neglect of their moons or some of their moons is an *abstraction*. If the model assigns circular orbits to the planets, rather than ellipses of various degrees of eccentricity, this is an *idealisation*. If it is possible, via the model, to study how the planets' parameters interact with each other, the model performs a further function. It shows the *interrelatedness* between these parameters. Kepler's third law shows how the orbital period of a planet is mathematically related to its average distance from the sun. This is *systematisation*.

The model's ability to show how parameters may be interrelated and in which way, allows most model to represent natural, social or economic systems. This function is very important in science but also carries an important philosophical message. Nature consists, as we have seen, not of unrelated empirical facts but of *systems* of varying degrees of complexity. A system is constituted by a number of components and the way they interact with each other. This interaction is often expressed in the form of laws of Nature. The solar system is a good illustration of what is meant by a natural system. The awareness that science had to deal with interrelated systems rather than collections of individual facts was of major importance in the history of modern science. It constituted one of the major changes in the concept of Nature.

Humans have only a limited intellectual capacity to handle complex systems. Of all the factors, which are operative in a given system, not all have equal weight and importance. The possibility, which models offer to correlate a finite number of parameters, to abstract from negligible ones and idealise others until they can be computed, is of invaluable service. Humans also have a learning ability, which allows them to render the models more complex, by incorporating more parameters and in a less idealised form. Models can therefore be made to approximate the real systems being modelled.

These are the functions of models: *abstraction*, *idealisation*, *systematisation* and *factualisation*.

There are also various types of models. Most models have *representational* functions. This is important for our gain in physical understanding. They capture

[21] There has been a considerable amount of literature on abstractions, idealizations, factualizations in science: Krajewski, *Correspondence Principle* (1977); Nowak, *Structure of Idealization* (1980); McMullin, 'Idealization' (1985); Brzeziñski *et al.*, *Idealization* (1990); Brzeziñski/Nowak, *Idealization* (1992); Herfel *et al.*, *Theories and Models* (1995); Cartwright, *Dappled World* (1999), 'Models and the Limits of Theory (1999), §9.5; Sklar, *Theory and Truth* (2000), Chap. 3

structural aspects of the natural systems modelled. Typically, models either emphasise the *spatial* ordering of the components in the system – as for instance the spatial distribution of the planets around the sun in the solar system – or place more emphasis on the *mathematical* relationships between the parameters – as for instance in the functional dependence of one parameter on another. When the models emphasise the spatial order, they represent the *topologic structure* of the system modelled. When the mathematical relationship between the parameters comes to the fore, the models represent the *algebraic structure* of the system modelled.[22] In sophisticated models, these two ways of representing may be combined.

In his understanding of 'understanding', Schrödinger referred to conceptual models. But this is insufficient, as models may be mere analogies. Various types of models have been distinguished in the literature – conceptual and practical models, iconic, theoretical and simulation models – but a more detailed distinction is that between *analogue models, scale models, functional models* and *structural models*.[23] This classification can be further improved by distinguishing between *analogue* and *hypothetical* models. *Structural* models can be subdivided into *causal* and *geometric* models. *Conceptual* models must be added to this categorization. Not all models contribute to understanding to an equal extent.

Analogue models represent the unfamiliar or unobservable in terms of the familiar or observable. The model suggests that there is an analogy between certain elements of already known systems and some elements of unknown systems. Analogies are either of a formal or material kind, they are negative or positive. There is a material analogy between, say, the flow of electricity through a parallel electric circuit and a system of water tubes. Both electricity and water flow through a system – the positive analogy – but water molecules are not like electrons – the negative analogy. There are also formal analogies between equations, but the material processes expressed by them may be quite different. For instance, there is a formal analogy between the classical wave equation, concerned with classical waves and the Schrödinger equation, concerned with the behaviour of atomic particles. Usually no distinction is made between analogue models and hypothetical models.

---

[22] This distinction between *topologic* and *algebraic* structures is due to Roman, 'Symmetry in Physics' (1969), 363–69; see Weinert, 'Theories, Models and Constraints' (1999), 313–17. Through a different route, Peter Galison, *Image and Logic* (1997) 19–20 arrives at a similar distinction, i.e. that between the *homomorphic* form of representation (a mimetic preservation of form) and the *homologous* form (a preservation of logical relations between events); see also de Beauregard, *Time* (1987), §1.1.1

[23] This distinction is due to Fürth, 'The Role of Models in Theoretical Physics' (1969), 327–40; on the role of modelling during the time of the Scientific Revolution, see A. Crombie, *Styles* (1994), Vol. II, Pt. IV. In our discussion we reserve no privileged place for computer models or simulation models because computers can simulate all the models discussed in the main text. As structural models, computer simulations have, however, the advantage of interrelating a much larger number of parameters than has hitherto been possible; for a discussion of computer simulation, see Hughes, 'The Ising Model' (1999); on the use of scale models and analogue models in the 16th and 17th centuries see Crombie, *Styles* (1994), Vol. I, Pt. III, §§8,9; Vol. II, Pt. IV, §12

Analogue models are based on formal or material similarity relations. They do not involve idealisations of and abstractions from factors, which we expect to be operative in the system modelled. Early atom models were based on the analogy with the solar system. Before Rutherford proposed his nucleus model of the atom, the Japanese physicist Hantaro Nagaoka proposed a Saturnian model of the atom. But the mere analogy does not assure that the real systems will resemble the analogue model. This is clear from the geocentric worldview (Fig. 3.1). Analogue models are a useful, if limited, step in an attempt to achieve physical understanding. They suggest useful approaches to problem situations. However, we want more from models than just analogies. We want the models to represent structural features of the natural systems being modelled. To achieve real physical understanding we need more sophisticated models.

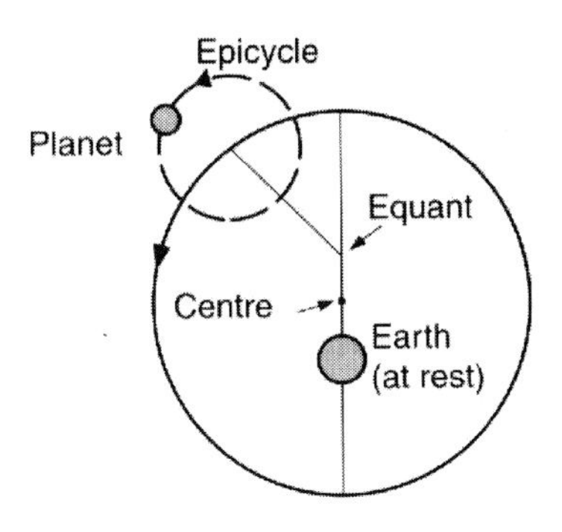

Fig. 3.1. Geocentric account of planetary motion, based on an analogy with geometric circles [Source: Sellers, *Transit of Venus*, 2001, 32; by permission of author and Mega Velda Press]

Hypothetical models – or *as if* models – incorporate idealisations and abstractions. They claim to represent the system modelled *as if* it consisted only of the parameters and relationships stipulated in the model. The economic agent in rational choice theory is a hypothetical model. The Copernican model of the solar system (Fig. 3.2) and representations of energy transitions in atoms (Figs. 3.3, 3.4) represent hypothetical models in the natural sciences. They greatly idealise the parameters – the orbits of the planets and electrons are circular – and abstract from factors such as moons. However, we know that such idealised factors are mathematical simplifications and that abstracted factors are present in the real systems.

Scale models represent real-life objects either in reduced size (toy cars or planetary models) or in enlarged size (a model of the Aids virus or the double helix of the DNA molecule). The human skeleton is a scale model, which can be represented as a real-life size model. Scale models are usually three-dimensional and require a fairly precise knowledge of the operation of the system. Schematic drawings of some engine or experimental apparatus may also be regarded as scale models.

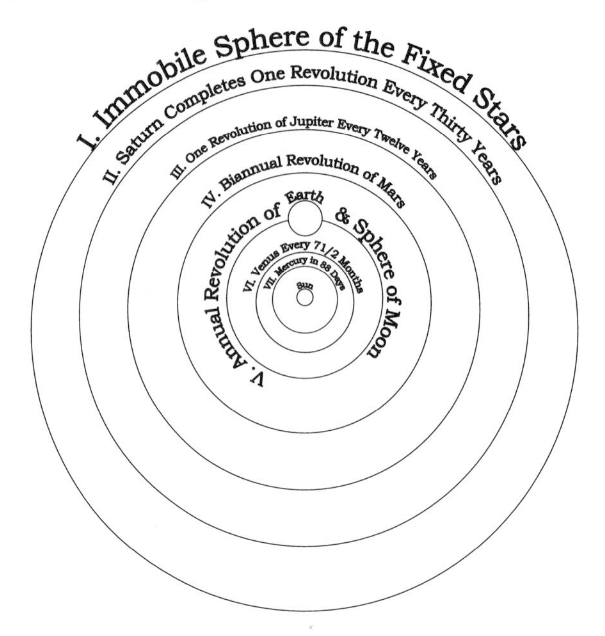

Fig. 3.2. The Copernican Model of the Solar System

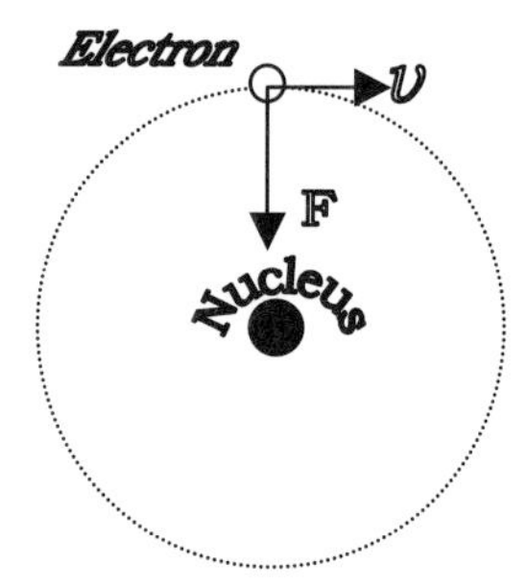

Fig. 3.3. Hypothetical Model of the Hydrogen Atom

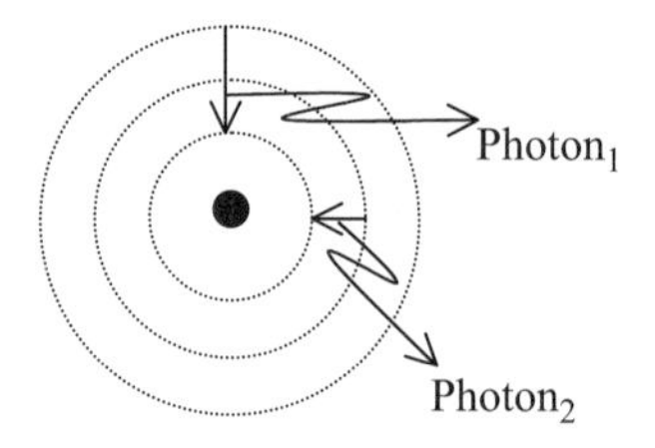

Fig. 3.4. Hypothetical Model of Some Permitted Electron Transitions, accompanied by Photon Emissions, in a Hydrogen Atom

Functional models, as the name suggests, represent the functional dependence between parameters. The Carnot cycle of an ideal gas, relating volume and pressure, or the supply and demand curves in economics, determining the price equilibrium at the intersection of the two curves are illustrations of functional models. There is no need to assign precise values to the symbols, which stand for the parameters. What counts is the nature of the *functional* relationship between some parameters. There is, for instance, an increase of temperature, pressure and density as a function of increasing distance from the surface of the earth towards the core (Fig. 3.5). In these models, the basis of understanding shifts from the topologic to the algebraic structure.

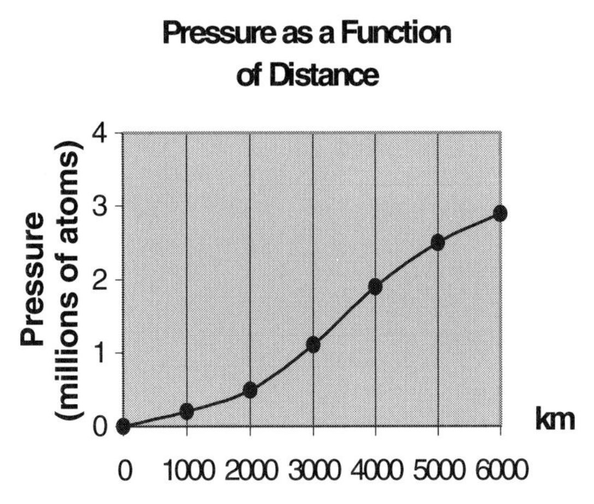

Fig. 3.5. The functional dependence of pressure on the distance between the earth crust and the core

Structural models typically combine algebraic and topologic structures in order to represent how some underlying structure or mechanism can bring about some observable surface phenomenon. Structural models are very useful in the representation of microscopic systems, like the atom. The Rutherford model of the deflexion of a subatomic particle by gold atoms provides a good illustration of this type of model (Fig. 3.6). On the strength of the experimental evidence, Rutherford proposed a *nucleus* model of the atom. The existence of electrons was secondary in this model. Rutherford's central innovation was the postulation of a nucleus inside the atom. This alone could account for the fact that 1 in 8000 projectiles fired at a gold foil was deflected by more than 90°. Famously, Rutherford is reported to have said that this event was as surprising as seeing a 15-inch shell, fired at a piece of tissue paper, rebounding off it.

In Rutherford's model two particles, α and α' (helium atoms, stripped of their electrons), approach a nucleus and are deflected to different degrees, depending on the size of the impact parameter b. The smaller b the larger the scattering angles θ and ϕ. The particle α suffers a deflexion at an angle, θ, which is greater than

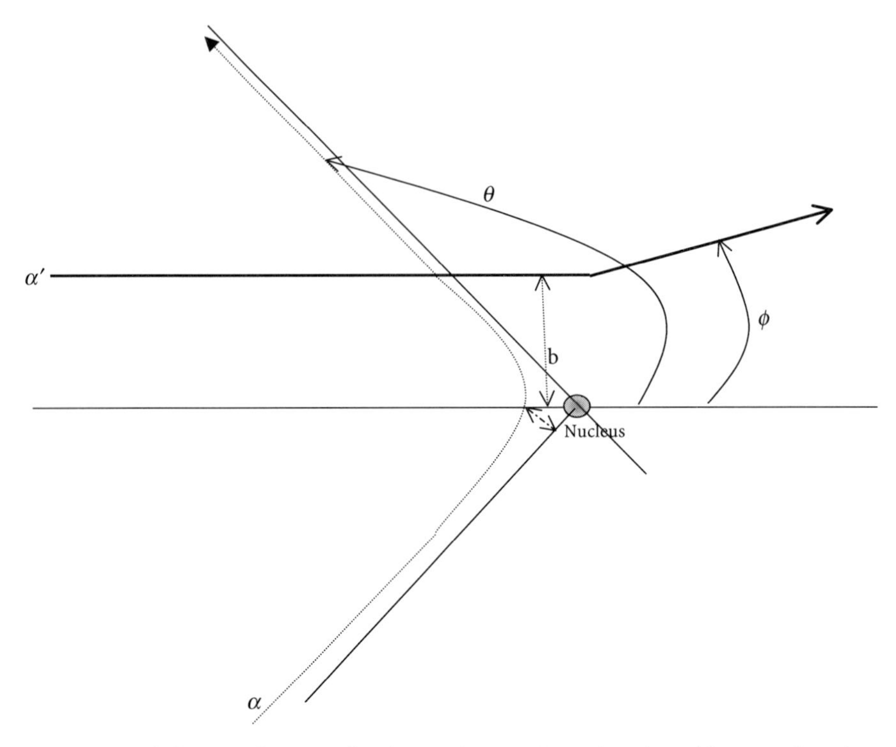

Fig. 3.6. The deflexion of α-particles due to the atomic nucleus in gold atoms (based on Rutherford's original model)

90 degrees. This is a case of large-scale scattering. Through its choice of parameters this model shows how the topologic structure of the encounter of an α-particles with a nucleus will lead to a consideration of its algebraic structure. Rutherford calculated the fraction of particles, which are scattered through various angles. This then leads to various equations, which deliver the quantitative aspect of the model. These two aspects combined make it a structural model. The observable behaviour of the ionised helium atoms (α-particles) as they approach the gold foil – the various degrees of scattering, the most likely degree and the number of scattered particles through given angles – provide the underlying mechanism, which make the observation understandable.

Bohr embraced the nucleus model and developed some postulates, which, although they were *ad hoc*, gave the hydrogen model an algebraic structure. These 'assumptions' permitted Bohr the derivation of some well-known regularity concerning spectral lines and the anticipation of some new, yet undetected spectral lines in the ultraviolet and infrared regions of the spectrum. Bohr also derived the energy levels of the hydrogen atom from his model. The Bohr model (1913), (1915), (1918) represents the first structural model of the hydrogen atom.

When a model combines *topologic* and *algebraic* structures, we have a *structural* model. Variants of structural models are *causal* and *geometric* models.

Causal models often emphasise the operation of continuous, spatio-temporal mechanisms, which link the causal factors to the production of the observable effects. Pasteur's germ model of disease is a typical example. To account for diseases like rabies, Pasteur proposed the existence of micro-organisms, which were held to be causally responsible for this disease. The causal model *traced* a causal chain from the unobserved 'germ' to the observed disease and its symptoms.

Sometimes the existence of a continuous, spatio-temporal mechanism between the cause and its effect is unobservable. This was one of the objections of field physics against gravitation, which seemed to require action-at-a-distance. In the atomic realm such causal tracing also proves to be impossible. No causal mechanism is known today which could account for the 'cooperation' of particles in the double slit experiment, which produce interference effects. Even modern variants of the double slit experiment have failed to find causal, traceable mechanisms to account for the spooky action-at-a-distance. This is true of the famous spin correlations, mentioned above. When no traceable mechanism is available to link the effect to a prior cause, philosophical commitments manifest themselves. Some will say that the philosophical commitment to causation as a traceable mechanism is correct. They are ready to pay the price: something must be wrong with the quantum theory. Others are ready to challenge the philosophical commitment, appealing to the strength of the experimental evidence. The evidence, they say, forces us to change our philosophical commitment to causation, not the scientific theory. If we cannot have deterministic causation, we must be content with probabilistic causation. Then causal models are still possible in quantum theory. In the absence of causal mechanisms, causal models in quantum mechanics will emphasize the existence of a cluster of conditions, between which a conditional dependence exists. In the double slit experiment (Fig. 2.7, Chap. 2.7) there is a conditional dependence between the opening of the two slits and the appearance of interference fringes. In the scattering experiments (Fig. 3.6), there is a conditional dependence between the closeness of the incoming particle to the nucleus and the varying scattering degrees.

Geometric models may just emphasize topologic structures, in which case they are simply a form of hypothetical model (Figs. 3.2, 3.3). But even the most modest geometric model usually comes with a coordinate system. It then becomes a simple form of structural model, because it gives rise to algebraic structures (Fig. 3.7).

The most important use of geometric models, as a form of structural model, occurs in the Special theory of relativity. One of the fundamental postulates of this theory is that no signals can propagate in excess of the speed of light in vacuum. This speed is finite. It takes time for signals to travel across space. We receive signals from intergalactic space, which may have left their source many light years ago. Consider an event E in the present of some observer O. From the present event E light propagates into the future in the shape of a diverging light cone, which forms the future light cone from the point of view of E. At E the observer O also receives light signals from the past. They converge at E and form a past light cone. This cone

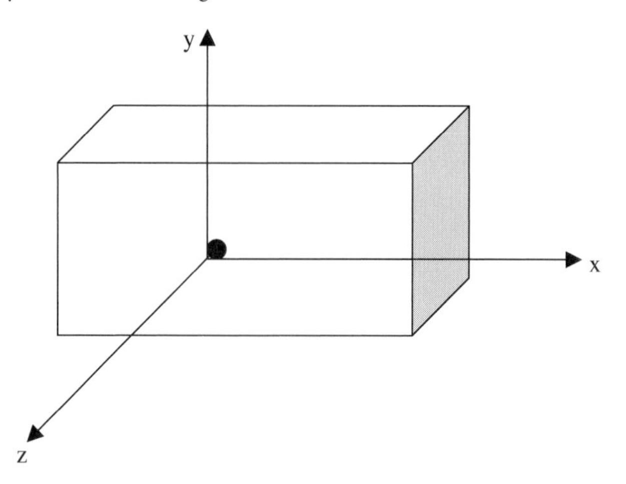

Fig. 3.7. Location of a particle in three-dimensional space. Its location can be precisely determined by attaching it to a coordinate system. Axes *x*, *y*, *z* project its position onto the walls of the space

structure (Fig 3.8) results in the rich algebraic structure of Minkowski space-time, which plays a central role in the conception of the block universe.

Under the proposed classification of models, *conceptual models* can take on a more specific meaning than in Schrödinger's considerations. In the natural sciences, conceptual models permit scientists to introduce thought experiments (*Gedankenexperimente*).[24] In quantum mechanics the most famous example is the double-slit experiment. The double slit experiment served Bohr and Heisenberg to stress the particle-wave duality of the quantum atom. On the strength of this *gedanken* experiment, they also argued that causation had lost its place in the quantum world. The double slit experiment only became a real experiment in the 1960s. In conceptual models, the physically possible operations of natural systems are investigated beyond the limit of what has been observed. The boundary of the physically possible is stretched to the full extent of what the laws of nature will permit. By exploring these boundaries, new insights are gained, new hypotheses are formulated. As a young man Einstein asked himself what the temporal world would look like astride a beam of light. He found that time would stand still. These conceptual explorations ultimately led to the formulation of the Special theory of relativity. Soon afterwards, Einstein again used a thought experiment – the one about the experimenter enclosed in a lab, suspended in empty space – to develop the equivalence of acceleration and gravitation and General theory of relativity.

Now consider the similarity between 'conceptual' experiments in science and in philosophy. Descartes asked himself how much objective knowledge humans could acquire of the natural world, if all our powers of reasoning and observation could be doubted, if both our reason and our perception came under attack from

[24] Norton, 'Thought Experiments' (1996)

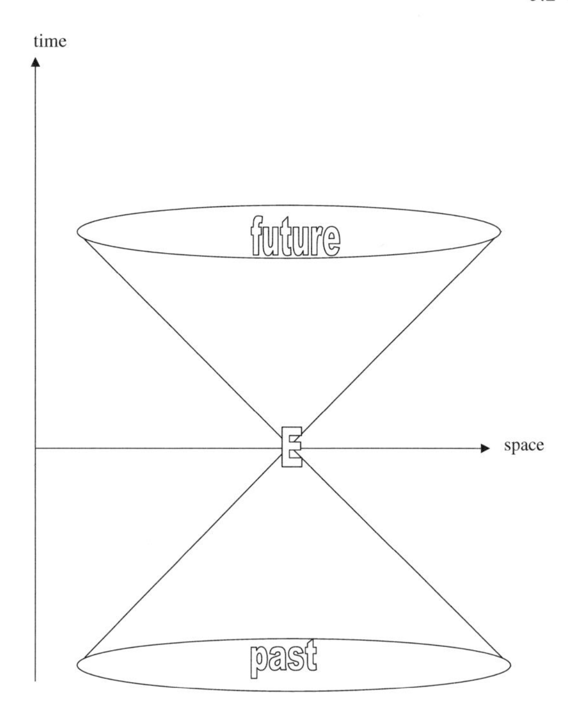

Fig. 3.8. Future and Past Light Cones in Minkowski space-time

the deceptions of a malicious demon. Kant begins his *Critique of Pure Reason* by proposing a Copernican revolution in philosophy. Just as Copernicus had accounted naturally for the apparent retrograde motion of the planets (Fig. 3.1), by exchanging the sun for the earth at the 'centre' of the universe, so Kant hoped to provide a theory of objectivity by making the human mind an active participant in the construction of Nature. *Gedanken* experiments can of course be turned into true experiments, as happened with the double-slit experiment. Conceptual models in philosophy are not testable in this sense. They also lack the mathematical features of idealizations and abstractions. But they are subject to approximations, which measure their adequacy. For instance, we will argue that the *conditional* model of causation is the most adequate model to express causal relations in quantum mechanics. It is also true of conceptual models that they cover a limited amount of data, and often face rival models. But there is no overarching theory, as we so often find in physics, which relates the models falling into its domain. Still the similarities between 'conceptual experimentation' in science and philosophy allow us to include philosophical models in the category of conceptual models. Fundamental notions like Nature, physical understanding, causation, time and space are developed in philosophical models. So philosophy is conceptual model building.

A consideration of the various types of models used in the natural sciences offers us a better grasp of the notion of physical understanding. Genuine understanding requires that models be *representational* of the system under consideration. This requirement excludes analogue models. Although they provide heuristic understanding, even a positive analogy does not guarantee that it has a corresponding relation in the natural world. There may be an analogy between, say, the *camera obscura* and the human eye, but this observation does not suffice to demonstrate how the eye is constructed. The analogue model must give way to a more representational model. For a model to be representational the structure it assigns to the natural system being modelled must correspond to the structure of the real system. The model uses algebraic and topologic structures or both to achieve the representation of the structure of the natural world. Through its representational nature, the heliocentric model advanced understanding, whilst the geocentric model failed. Although the early Copernican system retains the circular orbit of the planets, it assigns a much better topologic structure to the planetary system than the geocentric model. The Copernican model (Fig. 3.2) is a hypothetical model, while the Ptolemaic model (Fig. 3.1) is a mere analogue model. The Copernican model is not very precise in its assignment of algebraic structure, due to its adherence to the circle. The Ptolemaic model is mistaken in the representation of both the topologic and algebraic structure. It does not offer real physical understanding.

This assignment of structure must be guided by appropriate concepts. In the early Copernican system, an important concept is the centrality of the sun. In the development of the heliocentric model by Kepler, Galileo, Newton, Laplace and Kant, it is the notion of a deterministic evolution of the universe. This gives rise to the view of a clockwork universe, in which the concept of absolute time and the principle of causal determinism play essential roles. In the history of atom models the notions of discreteness (Planck's constant) and indeterminism provided essential conceptual guidance.

Physical understanding is a question of degrees. Models are always restricted to a limited set of data. They are governed by their four functions. Models can evolve. As they grow from hypothetical to structural models, the degree of understanding will increase. At crucial moments in the history of science, the fundamental notions exhaust their applicability. They become inappropriate for a further understanding of the phenomena. During such revolutionary periods, scientists and philosophers begin to develop conceptual models, which attempt to recast the fundamental notions. Scientists begin to speak the language of philosophy. Philosophers turn their attention to the revolutionary discoveries. Inherited worldviews become questionable. The delicate balance between facts and concepts is disturbed. Philosophical challenges arise. Feedback loops emerge between science and philosophy.

3.3 Einstein's Problem, Bohr's Challenge and the Feedback Thesis

> One measure of the depth of a physical theory is the extent to which it poses serious challenges to aspects of our worldview that had previously seemed immutable.
>
> B. Greene, *The Elegant Universe* (1999), 386

Prompted by the most recent discoveries in their respective fields, scientists provide new interpretations of the natural world and thereby contribute to its understanding. If the discoveries are of a fundamental nature, conceptual consequences arise. Scientists make philosophical contributions. The heartbeat of science is at its most philosophical rhythm when major conceptual revisions or revolutions are afoot. Then scientists feel the need to extend their reflections beyond the search for mathematical equations. They desire to reach a level of *understanding*, in which some physical meaning is assigned to the mathematical expression of the natural world. But scientific revolutions may also lead scientists to a re-interpretation of the nature of the scientific enterprise. What is interesting in this process, from a philosophical point of view, is that empirical facts filter through to the level of conceptualisation and bring about changes in the way the world is conceptualised. 'Old notions are dissolved by new experiences.'[25] The common territory between science and philosophy lies in this interaction between facts and concepts. In re-interpreting the world *in the light of new experiences* the scientist becomes an active participant in the shaping of human views about the surrounding world. In the face of a 'confusing amount of new evidence' from the atomic realm, Planck called for a new comprehensive physical worldview. This is the role of the *philosopher-scientist* of which scientists are fully aware:

> History has shown that science has played a leading part in the development of human thought.[26]

One of the basic tools of human thought in the process of understanding lies in the provisions of models, based on appropriate concepts. Models are used to gain understanding of the material world. Theories provide explanations. Under suitable conditions this leads to the natural sciences. Conceptual models can also be used to describe and explain the ways humans can acquire knowledge about the natural world around them. Under suitable conditions this leads to philosophy. The interaction between facts and concepts and the interaction between science and philosophy give rise to a fundamental tension, which will reverberate throughout the pages of this study. The more fundamental their concepts are, the more humans cherish them. But when the facts speak against the adequacy of the concepts, something needs to give way. Throughout the history of science, scientists and philosophers have often given up or modified the concepts in favour of the facts. Dissatisfied with the everyday notions of time and space, Newton set forth his notions of absolute time and space. Dissatisfied with the notions of absolute time and space, Einstein set forth his notions of relativistic time and space. In many of his writings, Einstein warned against the fixation on concepts. It would hinder the progress of science. Concepts, which once ordered the phenomena adequately,

---

[25] Born, *Natural Philosophy* (1949), 75

[26] Born, *Natural Philosophy* (1949), 2; cf. Planck, 'Die Stellung der neueren Physik' (1910), 53; Eddington, *Philosophy* (1939), 8; Jeans, *Physics and Philosophy* (1943), 2; de Broglie, *Continu et Discontinu* (1941), 7, 76. According to Heisenberg, '50 Jahre Quantentheorie' (1951), 360 Planck's formula shows that thinking, too, can change the world.

were always in danger of becoming 'thought necessities' (*Denknotwendigkeiten*).[27] Fundamental concepts like Nature, time and causation were dependent on experience. Therefore they were always subject to revision or rejection, depending on their empirical adequacy. Such philosophical reasoning formed the backbone of Einstein's revolution in physics. However, Einstein was not consistent in his epistemological convictions. In his correspondence with Max Born on the interpretation of quantum physics, he declared himself unwilling to give up the notion of causal determinism. In a letter written to Born on April 29, 1924, Einstein refused to abandon the notion of causation in the face of the quantum-mechanical evidence available at that time.

> The idea that an electron subjected to a beam could *freely* choose the moment and direction of its jump is unbearable to me.[28]

Einstein considered that the demand for a causal explanation of the world had a clear sense, even though it could not be strictly realised in practice.[29] Like Planck, Einstein rejected the idea that statistical laws would govern fundamental physical processes. In practice, though, our knowledge may amount to no more than knowledge of statistical regularities. Einstein's refusal to abandon Laplacean determinism is reminiscent of the situation in statistical thermodynamics. Due to our ignorance of how individual molecules move in a collection of particles, we can only formulate statistical regularities about their collective behaviour. But we assume that individual molecules are governed by perfectly deterministic laws. His refusal to abandon the notion of strict causation placed Einstein in a long tradition: the identification of causation with determinism. The originators of this tradition were thinkers like Leibniz, Laplace and d'Holbach. Einstein refused to accept that quantum mechanics had driven a wedge between the traditional identification of causation with determinism.

Einstein's ambiguous attitude towards the fundamental notions – his readiness to reject the Newtonian notions of absolute and universal time and space and his reluctance to abandon the classical notion of causation – points to a general problem. Let us call it *Einstein's problem*. When should fundamental notions be modified, if not abandoned, in the face of experimental or observational evidence? How much authority does the empirical evidence command over the fundamental notions? There is always a precarious balance between the concepts and the facts. Their interaction constitutes the common territory between science and philosophy. But the one does not reduce to the other. We called this interaction the *dialectic* between science and philosophy.

---

[27] See Einstein, 'Ernst Mach' (1916), 102; 'Über die Grundlage der allgemeinen Relativitätstheorie' (1916), 82; *Relativity* (1920), 136–43; 'Space-Time' (1929), 1070; Mittelstaedt, *Philosophische Probleme* (1972), 41

[28] Albert Einstein – Max Born, *Briefwechsel 1916–1955* (1969), 116–7, cf. 44; cf. Figs. 3.3, 3.4 above.

[29] *Briefwechsel* (1969), 44, 210, 269. Einstein's adherence to the principle of causation has attracted much discussion, see Pais, *Bohr* (1991), 119–20 and *Subtle is the Lord* (1982), 464–5.

As we have already noted many scientists did not think of the interaction between science and philosophy as a dialectic. They suggested that it would be better to let the experimental facts do the talking and adapt the fundamental notions to the established facts. The facts, then, would present important *constraints* on the adequacy of the conceptual models.

It is by now a well-established practice in the philosophy of science to couple philosophy with science, to pursue the philosophical concerns in the light of the empirical facts. In his *System of Logic* (1843), John Stuart Mill illustrates this procedure. After the exposition of his four methods of empirical inquiry, Mill proceeds to discuss concrete examples of their use in the history of science. Mill's procedure is typical of much of the effort in the philosophy of science since the seminal work of T.S. Kuhn (1960): back up the conceptual claims by winning support from well-established scientific facts. The danger of this procedure lies in its potential selectivity. The conceptual claims hang on the frame of a few case studies. From the point of view of the present study there is another danger: the philosophical presuppositions present in science are forgotten and ignored. Our study of the notion of Nature has already revealed that the knot between the concepts and the facts, philosophy and science is more complex. Niels Bohr has posed a demanding challenge to philosophy:

> The significance of physical science for philosophy does not merely lie in the steady increase of our experience of inanimate matter, but above all in the opportunity of testing the foundation and scope of some of our most elementary concepts.[30]

The philosopher is invited to move from the facts to the most fundamental concepts. The physicists were aware that 'old notions are discarded by new experiences.' However, it is precisely Einstein's problem to determine to which extent revolutionary scientific discoveries have philosophical consequences. It is the job of the philosopher to evaluate whether or not some of the fundamental notions do bend under the weight of evidence, as the scientists claim they do. When we move from the facts to the concepts, the most natural transition occurs from science to philosophy. The most interesting collaboration emerges. The scientists *question* the fundamental notions, on the strength of the evidence. The scientists claim that empirical evidence can test and refute the fundamental notions. Heisenberg holds that due to empirical discoveries not only the content but also the structure of our thinking can change.[31] New notions are set in place. A dual process is at work. New evidence often does show the inadequacy of the old notions. In this sense new discoveries offer constraints on the conceptual models of the fundamental notions. But the scientist's enthusiasm for the refutation of old notions may be too optimistic. Philosophy does not yield so easily to the verdict of science. The philosophical consequences, which the scientist draws from the new discoveries, may not follow – or may not follow in the way imagined. It is the task of the following chapters to investigate this question. The philosopher works with conceptual models of the fundamental

[30] Bohr, 'Quantum Physics and Philosophy' (1958), 1

[31] Heisenberg, 'Philosophische Probleme' (1967), 421–2, 'Grundlegende Voraussetzungen' (1959), 253

notions. They need to satisfy the constraints imposed by the fundamental scientific discoveries. Investigating and evaluating the philosophical consequences of great scientific discoveries is the philosopher's way of meeting Bohr's challenge.

So the *dialectic* will take the following form. Philosophical presuppositions – i.e. implicit or explicit assumptions like determinism – pervade scientific thinking. New discoveries may highlight the problematic nature of these presuppositions. Many scientists begin to wonder about the adequacy of the old conceptual assumptions. They draw philosophical conclusions from the discoveries. These are the philosophical consequences, which need to be evaluated.

There is further reason why philosophers should take up Bohr's challenge. Why the dialectic between facts and concepts, between science and philosophy, should be approached from the vantage point of some of the most fundamental discoveries in the history of science. In the late 1970 s, Max Jammer[32] wrote a little known paper spelling out the 'philosophical implications' of the new developments in 20th century physics. With quantum mechanics and relativity theory in mind, Jammer formulated his *feedback thesis* (Fig. 3.9). It is, as we can see, an elaborate attempt to come to grips with Einstein's problem and meet Bohr's challenge. It is a conceptual model, which expands on Born's claim that old notions are discarded by new experiences – a claim to which many of the leading scientists of the 20th century subscribed.

In this revision of conceptual foundations lies the interaction between physics and philosophy. But Jammer's feedback thesis is too strong. It tries to meet Bohr's challenge but ignores Einstein's problem. It leaves no freedom to the fundamental concepts to resist the changes, suggested by the empirical facts. It ignores that the fundamental concepts may serve as guidance for alternative models in empirical research. It is true that scientific discoveries have always had philosophical consequences. But concepts are not mere maids to facts. And facts are not simple servants to concepts. There is an dialectic between concepts and facts, science and philosophy, which calls for a revision of Jammer's *feedback thesis*.

Its insistence on the role of philosophical presuppositions make Jammer's thesis partly true. Scientists from Newton to Einstein have presupposed an unanalysed notion of causation, which has an essential Laplacean component. This notion has undergone considerable conceptual changes as a result of empirical evidence from the realm of quantum phenomena. This, in turn, has led to a revision of the notion of causation, but without unanimous agreement amongst physicists and philosophers. Just as scientists operated with a Laplacean notion of causation, so they presupposed the Newtonian notion of absolute and universal time. This was not a unanimous presupposition, since both Kant and Maxwell held alternative views on time within the framework of classical physics. The notion of absolute and universal time was found to be incompatible with the relativistic facts. Minkowski initiated the notion of space-time. Generations of physicists to the present day have been seduced by the idea that space-time led to the denial of temporal change in the

[32] Jammer, 'A Consideration of the Philosophical Implications of the New Physics' (1979), 41–61

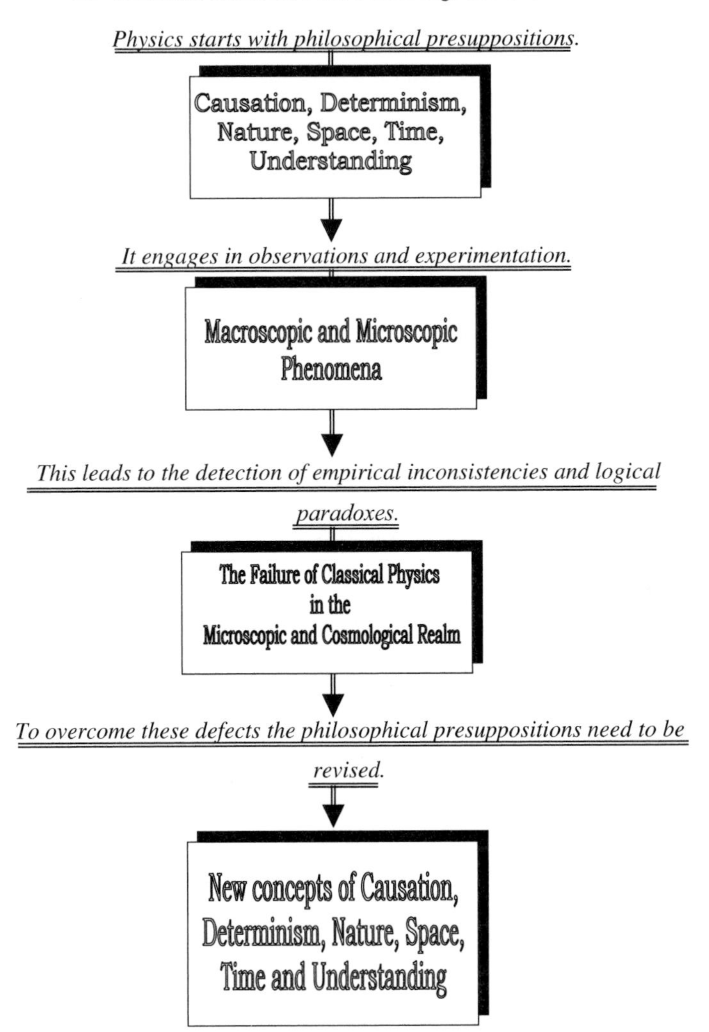

Physics starts with philosophical presuppositions.
Causation, Determinism, Nature, Space, Time, Understanding
It engages in observations and experimentation.
Macroscopic and Microscopic Phenomena
This leads to the detection of empirical inconsistencies and logical paradoxes.
The Failure of Classical Physics in the Microscopic and Cosmological Realm
To overcome these defects the philosophical presuppositions need to be revised.
New concepts of Causation, Determinism, Nature, Space, Time and Understanding

Fig. 3.9. Jammer's Feedback Thesis

physical universe. Becoming was a human illusion. Surprising as it may sound, the 'relativistic physicists' presupposed the Kantian idealist notion of time. The *block universe* was the fundamental reality. Anticipating the solar eclipse of August 1999, Eddington wrote in 1935:

> The shadow of the moon in Cornwall in 1999 is already in the world of inference.[33]

But the *block universe* has never found favour with many philosophers. And some physicists believe in the fundamental temporal nature of the universe. The notion of space-time does not lead as straightforwardly to the denial of temporal change

[33] Eddington, *New Pathways in Science* (1935), 92

as has been assumed by physicists. Philosophical presuppositions must sometimes resist the revision suggested by the empirical facts.

In this book we concentrate on the philosophical presuppositions and consequences of great scientific discoveries, rather than their implications. The notion of presupposition is well known to philosophers and historians of science. In the organismic, the clockwork and cosmos-view of Nature we have already encountered examples of what R.G. Collingwood would call 'absolute presuppositions'.[34] According to Collingwood it cannot be affirmed of such presuppositions that they are true or false, because they serve as anchor points of the prevalent thought systems of particular historical epochs. The historical agents may not even be consciously aware of their presuppositions. They possess, in their minds, logical priority. In this sense Collingwood would probably regard our fundamental notions of Nature, physical understanding, causation and time as further examples of absolute presuppositions. It has often been observed that presuppositions play a pivotal role in human thinking.

> Every epoch has at its disposal a basic system of ultimate general concepts and presuppositions, by virtue of which it masters and unifies the multitude of materials, which experience and observation offer. (...) If we regard the presuppositions of science as arisen [rather than absolute], we recognize them as creations of thought. If we recognize their historical relativity and roots, we open our eyes to their incessant progress and their continuously renewed productivity.[35]

As Cassirer points out, we must be aware that presuppositions will change. Collingwood thought that

> Peope are not ordinarily aware of their absolute presuppositions, and are not, therefore, thus aware of changes in them; such a change, therefore, cannot be a matter of choice.[36]

It is important to realize that presuppositions change. But it is even more important to inquire *how* and *why* they change. Attending to the question of change and why it comes about, leads us to the concentration on philosophical *consequences* of great scientific discoveries rather than their *implications*.

There exist subtle connections between science and philosophy, which have a bearing on the way humans conceptualise the wider cosmos and their access to it. In particular there are important cases in which the scientific discoveries are directly responsible for a change in the way humans conceive of rational schemes for the ordering of their experiences. For this reason it is better to speak of the consequences rather than the implications embedded in great scientific discoveries. *Consequences*, not *implications*, because the scientific community was actively aware of how the current worldviews would be affected by its discoveries and how fundamental philosophical notions, like causation, nature and time, would have to be redefined. If conceptual change occurs, then presuppositions embedded in scientific practice

---

[34] Collingwood, *An Essay* (1940), Chaps. IV, V

[35] Cassirer, *Erkenntnisproblem* I (1922), v–vi; author's own translation

[36] Collingwood, *An Essay* (1940), 48

become the conceptual consequences of the new discoveries. It is appropriate to speak of consequences, because the philosophical outcome of a great scientific discovery may be precisely that hitherto *implicit* assumptions about the nature of reality or research have to be made *explicit* and modified for the sake of further research. The assumptions expressed in the form of fundamental notions are no longer compatible with the evidence.

What about the *presuppositions*? It is true that at any period of time fundamental presuppositions underlie the thought systems of the day. But fundamental concepts are not presupposed in Collingwood's sense. *First*, scientists often explicitly state their presuppositions. In the *Principia* Newton famously defined his notions of time and space before he formulated his laws of motion. Robert Boyle worked out his corpuscular philosophy. Kant and von Humboldt went to great lengths to defend their *cosmos*-view of Nature. *Second*, scientists thematize their presuppositions in the light of new discoveries. Einstein questioned the accepted notion of time. Heisenberg questioned the accepted notion of causation. The assumptions, which scientists make – whether in the form of the old presuppositions or in the form of the new consequences – can guide, constrain and even mislead them. Gerald Holton calls the presuppositions *themata*.[37] Science, then, makes thematic presuppositions in the form of our fundamental notions and others. These thematic presuppositions may remain implicit till new discoveries throw doubt on them. They may be very explicit assumptions of scientific thinking. They are neither true nor false. They may be unverifiable and unfalsifiable. But they are more or less empirically adequate. They are not, as this study will show, incorrigible. They change as a result of new evidence. Such presuppositions are conceptual models. They are not strictly true or false because conceptual models are compatible with different kinds of evidence. Although we cannot strictly falsify such assumptions, they may be empirically

Fig. 3.10. Time Magazine (31 December 1999) declares Einstein *Person of the Century*

[37] Holton, *Thematic Origins* (1973), *The Scientific Imagination* (1978), *Einstein, History* (2000)

inadequate. Models must satisfy such a condition. But it would be a mistake to think that the thematic presuppositions of the old views of Nature become straightforward reformed philosophical consequences of the new discoveries.

Fig. 3.11. German Magazine *Der Spiegel* (13 December 1999) makes Einstein *The Brain of the Century*

The following chapters will spell out the philosophical implications of Einstein's problem. We will try to meet Bohr's challenge and to revise Jammer's feedback thesis. What do fundamental discoveries tell us about the fundamental concepts? How do standard philosophical models of these notions fare in view of their evolution in the natural sciences? Talk of fundamental concepts – *Nature, determinism, indeterminism, causation, time and space, mass and energy* - was very prominent in the physical and philosophical discussions of the great scientists of the past centuries. Everyday discourse not need necessarily employ these concepts explicitly. People tell causal stories without using the notion of causation. People invoke deterministic, if not fatalistic notions without thinking of Newton and Laplace. People are familiar with the image of the clockwork universe without knowing its origin. It is to be expected that the impact of the scientific revolutions of the 20th century on the fundamental notions will eventually percolate down to everyday discourse. This will no doubt happen in distorted form. And it will take time. Eventually, however, the changes in worldviews and the fundamental notions will make themselves manifest in everyday thinking (see Figs. 3.10, 3.11). In a fast-changing world, threatened by environmental disasters and overpopulation, it would be no mean achievement if people began to think of Nature as a *cosmos*, of *time* as a relational concept and of *causation* without determinism and *determinism* without causation. Let us consider what happened to the notions of time, causation and determinism as they faced the verdict of experience over the last 100 years.

Part II

The Scientist Philosopher

4

The Block Universe

> Until we have a firm understanding of the flow of time, or incontrovertible evidence that it is indeed an illusion, then we will not know who we are, or what part we are playing in the great cosmic drama.
>
> P. Davies, *About Time* (1995), 278

4.1 Introduction

Nature is a system of interrelated systems. The laws of nature express the interrelatedness. Different laws of nature express different aspects of how the systems are interrelated. What is real is what remains invariant across different reference frames. On a basic conceptual level, other fundamental notions operate to describe the functioning of Nature. These are the notions of time, space, causation and determinism. As we shall see in the following chapters, these notions have themselves undergone considerable changes as the result of fundamental discoveries, especially in the 20th century. Changes in the notion of *time* are associated with the Special theory of relativity. Changes in the notions of *causation* and *determinism* are associated with quantum theory – the physics of the atom. In this chapter we concentrate on the notion of time, leaving a consideration of determinism and causation for the following chapter. Notions like time, causation and determinism are used at all times in physical theorizing, often as unquestioned presuppositions. This is true also of the Special theory of relativity. While this theory had no reason to question the notions of causation and determinism, it had every reason to question the established notion of time. There are many aspects of the concept of time, which come under scrutiny with the emergence of the Special theory of relativity. Of special relevance for a consideration of the philosophical consequences of great scientific discoveries is the claim, often made on the basis of the discoveries of the Special theory of relativity, that the passage of time is a human illusion. The physical universe is not subject to temporal change. There is no flux, only stasis. It is called the *block universe.* Of particular interest from the point of view of this book is not only that leading physicists adopted this position. Even more interesting is that these physicists, giants like Einstein and Eddington, interpreted the conception of the block universe as an empirical consequence of the adoption of the Special theory of relativity. We will consider the stages of the argument from its earliest inception to later attempts to prove the *unreality* of *time* from the validity of the Special theory of relativity. But the notion of time is complex. According to the Special theory of relativity physical time is clock time. And this theory is a statement about the behaviour of clocks in different coordinate systems. The conception of the block universe, however, embraces features of the notion of time, which go beyond the behaviour of clocks in different coordinate systems. It is best therefore to introduce some of the fundamental features of the notion of time, as they slowly emerged in the general cultural context and the specific context of the physical sciences. It will also be convenient to review some of the philosophical models of time, which are of particular relevance for an assessment of the conception of the block universe. Our aim is threefold: (a) to understand how the conception of the block universe emerged from the Special theory of relativity; (b) to assess whether the unreality of time is an inevitable consequence of the adoption of the Special theory of relativity; (c) to highlight the role of the Kantian, idealist view of time as a philosophical presupposition in the notion of the block universe.

4.1.1 Models of Time

Human preoccupation with the notion of time goes back to antiquity. There are very few areas of human experience, in which questions of time do not play some role. Human fascination with time has been channelled into two main areas of concern.

1. **How is Time Measured?** Very early civilizations, like the Egyptians and Babylonians knew of subdivisions of the year and the month. From very early on certain types of clocks – shadow clocks, water clocks – were in use. However, a significant event in the history of time measurement occurred around 1271: the invention of the *mechanical clock*. This quickly gave rise to much more precise forms of time reckoning and led to modifications of social life. The first clocks lost 1000 seconds per day. Modern atomic clocks are accurate to about one second in a million years. (Typical modern wristwatches lose 10–20 seconds in a year.) The modern notion of *physical* time, which had its origin in Galileo's fall experiments and Newton's definition of mathematical time, is clock time.
2. **What is Time?** This is a much more difficult question to answer. A long line of thinkers especially in philosophy and physics – Plato and Aristotle, Saint Augustine, Galileo, Newton and Leibniz, Kant and Einstein – have made major contributions in an attempt to answer this enigma. Whatever answer is given, it will be more comprehensive than the notion of physical or clock time. The question of *how* time is to be measured can be answered without a clear answer to the question what time *is*. A philosophical inquiry into the nature of time will seek some abstract and formal criteria, which allow at least a tentative answer to the question about the nature of time. According to the findings in

our chapter on Physical Understanding, it will be best to regard these answers as conceptual models. So philosophical inquiry has led to a limited number of conceptual models of time. In the history of queries about the nature of time, we can distinguish between three prominent models.

- *The Idealist View.* In its pure and simple form, an idealist argument about the nature of time arrives at the conclusion that time is nothing but a product of the human mind. Physical reality itself is timeless. The passage of time is a product of human awareness. As we shall see, it is convenient to distinguish a *subjective idealist view* (Saint Augustine) from an *objective idealist view* (Kant). For a discussion of the philosophical consequences of great scientific discoveries, it is ironically the idealist view, which will arrest our attention. For the proponents of the block universe were particularly swayed by the Kantian model of time.
- *The Realist View.* According to this view time and space exist independently of human awareness or human observation. And further that both time and space possess an independent physical reality over and above the material contents of the universe. In his *Principia* Newton endeavoured to provide a more mathematical definition of time and space. Metaphorically speaking, *space* becomes a vessel, a container within which the material objects of the world are placed. *Time* then becomes a river, flowing at a constant rate, against which the duration of material processes in the natural world is measured. Newton makes space and time independent of material objects and events. Einstein's definition of time as clock time and his relativization of time to reference frames stands in direct opposition to Newtonian time. Einstein's Special theory of relativity also quickly led, under the reinterpretation of Hermann Minkowski (1908), to a union of the notions of space and time into a more abstract notion of *space-time*. It will be necessary to review the basic tenets of Newton's mathematical notion of time. But it would be wrong to think that the realist view has suffered a deathblow at the hands of Albert Einstein. Questions about the reality of space-time continue today in a lively debate. The old controversy between Clarke, a defender of Newton's view, and Leibniz-Mach has reappeared today in the guise of the *substantivalism-relationism* debate.[1]
- *The Relational View.* In its most elementary form this view holds that space is constituted by the physical existence of matter and energy in the universe. Time is constituted by the succession of events in the material universe. Time and space possess no independent physical reality. Rather it is the

[1] This debate will not preoccupy us in this chapter, which concentrates on the question of the block universe. The liveliness and technical sophistication of this debate can be witnessed in the following publications: L. Sklar, *Space, Time, Spacetime* (1974); Friedman, *Foundations* (1983); Earman, *World Enough and Space-Time* (1989); Butterfield/Hogarth/Belot *eds. Spacetime* (1996); this volume contains many of the main contributions to this debate published over the last 30 years; see also, Butterfield *ed. Arguments of Time* (1999); Rynasiewicz, 'Absolute versus Relational Space-Time' (1996) and 'Distinction between Absolute and Relative Motion' (2000)

appearance of material processes, which constitute both time and space. Before creation, as Leibniz put it – or before the Big Bang, as the modern cosmologist would phrase it today – there is neither time nor space. Time and space emerge with the appearance of material processes (or change) in the universe. Thus, time and space exist independently of human awareness. Humans form a notion of *space* through the observation of the geometrical order amongst the material objects in the world. And humans form a notion of *time* through the observation of change and the succession of events in the universe.

We should be aware that the relational view commits us to a rather complex notion of time and space. On a very basic level there is the *topological* order of objects with respect to each other: objects lie close to each other or far apart; they are arranged along a vertical or horizontal line; they lie at an angle with respect to each other; they reside on a two-dimensional plane or in a three-dimensional volume. Equally for the *temporal* order, in which objects or events succeed each other. And in which the universe has a history with possible different scenarios: a beginning and an end or a beginning with no end or neither a beginning nor an end. On a very basic level, events are ordered according to a 'before-after'-relation. The extinction of the dinosaurs lies before the construction of the Egyptian pyramids. This topological and temporal order amongst objects and events exists independently of human awareness. Time is a 'before-after'-relation between events; space is a topological relation between objects. The relational view requires that at least some events have occurred and that at least some objects exist in the real world. Thus time and space exist irrespective of human perception and conceptual awareness.

When humans are added to the picture, the notion of time becomes more complex. Humans observe the topological relations between objects and the temporal relations between events. Due to their conceptual abilities, humans construct models of space and time: geometrical models of the local neighbourhood and global properties of the universe; temporal models of the succession of local events and the global arrow of time. With human awareness come calendars and dates, geometries and clocks. Humans employ a fully-fledged notion of time. It will be important to keep these two aspects of time apart. The *before-after* relation is a minimalist notion of time; it is based on natural units of time and does not require human awareness. The *past-present-future* relation is a maximalist notion of time; it is based on a combination of natural and conventional units of time; it requires conceptual awareness.

4.1.2 Natural and Conventional Units of Time

Over the centuries people have developed different ideas about the nature of time. A basic distinction is that between the archaic notion of *cyclic* time and the Judeo-Christian notion of *linear* time. To understand how cultures could develop such different conceptions of time, it is important to distinguish between *natural* and *conventional units* of *time*. *Natural* units of time are based on patterns of recurring regularity in natural phenomena: temporal variations of the climate (winter and summer, the yearly flooding of the Nile) and of plant and animal life (migration of birds, flowering of plants, evolutionary processes); celestial phenomena (the orbits of the planets around the sun). Some basic units of time – like the day and the year – are natural ones. These can be expressed in precise units. For instance, the equatorial rotation period of the earth is 23 hours, 56 minutes and 4.1 seconds – that of Uranus is 17 hours. The tropical (or natural) year, the time occupied by the earth in one revolution around the sun, has a length of 365 days, 5 hours, 48 minutes and 46 seconds.

Conventional units of time lead to socially useful yet arbitrary subdivisions of a longer unit of time into smaller units. The subdivisions of the year into 12 months and the number of days in a particular month are social conventions. The division of the week into 7 days is arbitrary, so is the division of the day into hours, minutes and seconds. The fixing of the beginning of the year on January 1 or the beginning of the day at midnight are social conventions. No natural pattern of regularity corresponds to the conventional units of time.[2]

Our models of time, even today, suffer from a certain amount of underdetermination. Different models of time, like closed or open models of time, may seem equally compatible with the available empirical data. Modern cosmology faces a number of empirically undecided questions: Will the universe expand forever or eventually contract? Is the Big Bang *the* beginning of the universe or only of *our* universe? What is the nature of the Big Bang? A number of competing cosmological models can claim empirical adequacy but at present no model exist, which could account for all the evidence. Cyclic, circular and linear models of time face similar questions.[3]

Although human models of time may forever remain empirically underdetermined, some models will suffer more empirical underdetermination than others.[4] It is clearly important for science to adopt models of time, which are compatible with the empirical evidence. This was the physical motivation behind the transformations, which the notion of physical time has experienced since its inception

---

[2] According to a philosophical school called *conventionalism*, important aspects of the Special theory of relativity, like the choice of the simultaneity relation, are also purely conventional; see Reichenbach, *Philosophy of Space and Time* (1956), 129–35; Grünbaum, 'Space, Time' (1970); Sklar, *Space, Time and Spacetime* (1974), 287–94; Friedman, *Foundation* (1983), 165–76; Norton, 'Philosophy of Space and Time' (1992/1996), 15–8

[3] According to latest astronomical reports, the universe is flat and will expand forever; its expansion may even be accelerating, see *Nature* **404** (27 April 2000).

[4] Despite this aspect of conventionality in our notion of time, it is important to realize that conventional units of time must not fall out of step with natural units of time. If this synchrony is destroyed, the conventional units of time will announce the arrival of natural or social events *at the wrong time*. The calendar could announce Christmas in spring and summer in winter. In the 16th century the *Julian calendar* was falling out of step with the tropical year. The Gregorian calendar was introduced in 1582 to rescue the synchrony.

at the scientific revolution. The notion of *physical* time was first introduced by Galileo, rendered abstract by Newton and revolutionized by Einstein. As the physical sciences provided new insights into the notion of time, the philosophical models faced new constraints. If we accept that the philosophical models must heed the constraints provided by the physical sciences, the question still remains how far these constraints reach into the core of the philosophical models. According to Einstein, Bohr and Heisenberg, philosophical notions should be checked for accuracy against empirical findings. If, however, there is a true dialectic between science and philosophy, the physicists' conclusion that philosophical models of time, causation and Nature must surrender to empirical discoveries may be premature. Even though the scientific revolutions provided new constraints, it remains to be seen whether they also determined the philosophical consequences to be drawn from them. The philosophical consequences spell out our view of the world in the light of the scientific revolutions. But there is a difference between what our view of the world *must* be and what it *could* be as a result of revolutionary discoveries.[5]

To approach this question from the right angle, it will be convenient to briefly review the recent history of the concept of physical time and its entanglement with philosophical notions of time.

4.1.3 Galileo's Physical Time

Galileo Galilei (1564–1642)

In Galileo's fall experiments a new notion of time emerges: *physical* time. Physical time is measurable time. It appears in the equations of motion as the parameter t. This abstract unit t stands for a continuous, linear and measurable scale of temporal units. In the equations of motion t could take any numerical value that could be read off an ideal clock. In Galileo's work physical time becomes *clock* time.[6] The purpose of the introduction of t is to make the duration of a physical event, like the fall of an object from a certain height, measurable. Galileo was not concerned with the question of 'why' bodies moved but 'how' they moved and whether this movement could be described mathematically. However, Galileo was not in possession of an instrument, which could measure the short time intervals of falling bodies. To measure the times of falling bodies, he placed balls on inclined planes and measured the times by indirect means (Fig. 4.1). He reports that in many repetitions of the experiments no time differences were found, 'not

[5] Sklar, 'Time, Reality' (1981), 131 writes: 'While our total world-view must, of course, be consistent with our best available scientific theories, it is a great mistake to read off a metaphysics superficially from the theory's overt appearance, and an even graver mistake to neglect the fact that metaphysical presuppositions have gone into the formulation of the theory, as it is usually framed, in the first instance.'

[6] Elias, *Zeit* (1988), 80, 82ff; Wendorff, *Zeit* (^{3}1985), 205–6; Burtt, *Metaphysical Foundations* (1924), 91ff

even 1/10 of a pulse beat.' He then varied the lengths of the inclined planes and their inclinations and always found that the lengths, L, were related to the squares of the time, t, as:[7]

$$L \sim t^2 .$$

How were the times measured in the experiments? Galileo reports that a bucket of water was used with a hole in its bottom, from which a thin water stream flowed (Fig. 4.2). During each experiment a cup was placed under the water stream and the water was collected. These small water samples were then weighed on precise scales. From the differences in weight, the proportions of the weights were obtained and the proportions of the times.

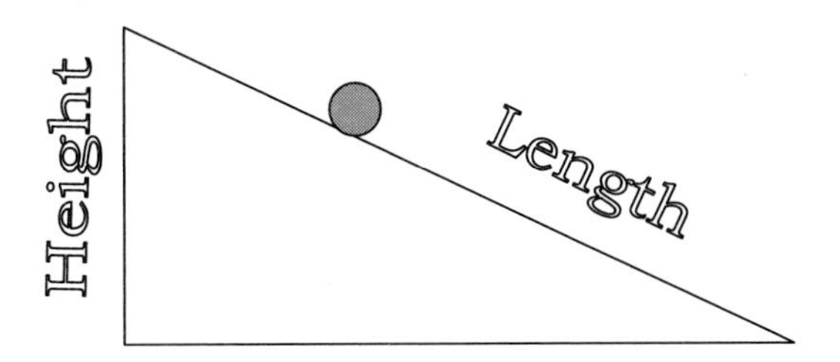

Fig. 4.1. 'Fall' experiments on inclined planes

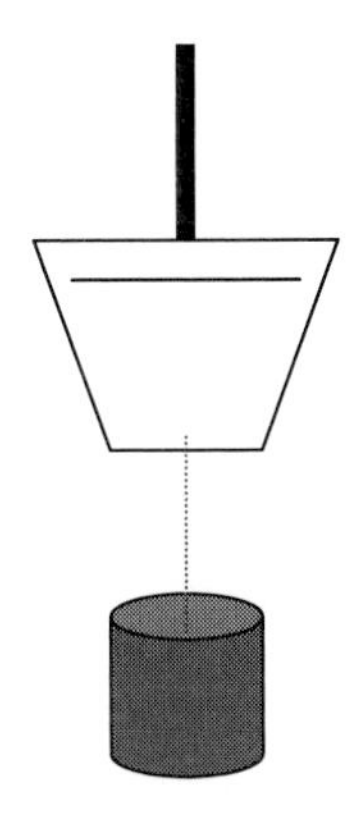

Fig. 4.2. Use of 'water clocks'

Thus, the famous parameter, t, which is so significant in science, had made its appearance in physics. (Note that in Kepler's third law $T^2 = A^3$, T refers to the orbital period of a planet.) It expresses clock time or the measurable time, which elapses between two physical events. In Newton, the notion of time is much more systematic. While Newton's equations of motion also use the parameter t, Newton

[7] Galileo discovered a proportion between distance and time, which today is expressed in the equation: $s = (vt_o) + 1/2at^2$

makes a much more systematic use of the notion of time. Newton incorporates his peculiar notion of *absolute* time as a fundamental notion in his theory of motion. Newton defines this notion in his *Principia* (1687) before the introduction of the laws of motion. It can be interpreted as a mathematical notion of time. It is also important in the history of the notion of physical time because it provided the foil, against which Leibniz formulated his relational view. Einstein, in his discussions of the relativity of time, makes constant references to Newton's notions. The effectiveness of Leibniz's objections to Newton's notion of absolute time is still a point of debate today. The decisive rejection of Newton's notion of time comes with Einstein. There are in fact two aspects to Newton's notion of time, which came under scrutiny. Newton's notion of time is both *absolute* and *universal*. It is *absolute*, because Newton makes time independent of any material processes in the universe. It is *universal*, because for Newton, as for every physicist and philosopher before Einstein, all observers, wherever they are placed in the universe, measure the same time and length intervals. With Einstein, the concept of absolute and universal time is abandoned. Time becomes simply what a clock measures.[8] As clocks behave differently in different physical situations, time becomes relativized.

4.1.4 Newton's Absolute and Universal Time

Isaac Newton (1642–1727)

During his lifetime Newton made various statements on the nature of time and space. This has given rise to various, often hostile interpretations of his notions of *absolute space* and *time*. In particular in his later years, Newton tends to conflate theological considerations with physical and philosophical discussions. In the *General Scholium*, added to the second edition of the *Principia Mathematica* (1713), he declares that 'the discourse (of God) from the appearances of things, does certainly belong to Natural Philosophy.'[9] Although Newton does not identify time and space with Deity, he does make them dependent on the existence of a Deity:

---

[8] Bondi, *Assumption* (1967), 41

[9] *Newton's Mathematical Principles*, Cajori Edition (1960), 546. Interpretations of Newton's notions of *absolute space* and *time* have varied considerably over time. Leibniz and Mach took Newton to have defended some metaphysical ideas about time and space – the *container* and *river* metaphors capture these metaphysical aspects – while commentators in the 20th century, like Toulmin 'Criticism' (1959) and Strong 'Newton's "Mathematical Way"' (1957) have tended to interpret Newton's notions as theoretical concepts, which are required for the formulation of Newton's laws. See also Earman, 'Who's Afraid of Absolute Space?' (1970). Both parties have found textual evidence in Newton's writings. Newton's later writings tend to support the metaphysical interpretations; Newton's earlier writings tend to favour the technical interpretations.

> He is not eternity and infinity, but eternal and infinite; he is not duration and space, but he endures and is present. He endures forever, and is everywhere present; and by existing always and everywhere, he constitutes duration and space.[10]

Further pronouncements in Newton's *Opticks* (1706) that 'space is God's sensorium'[11] led Leibniz to an outright attack on the metaphysical (unempirical) elements in Newton's physical system. Ernst Mach also accused Newton of having fallen under the influence of medieval philosophy and forsworn his empiricist principles.[12]

However, these more metaphysical or theological reflections arose in a different problem situation from the one, in which the notions of *absolute time* and *space* were first introduced. The original problem situation, in which these notions make their first appearance, lies in Newton's *Mathematical Principles of Natural Philosophy* (1687), in which Newton expounds his theory of motion. As Newton avows his commitment to empiricism in the same book,[13] it is reasonable to assume that Newton did not engage in idle speculation when he placed his definitions of absolute time and space at the beginning of his treatise on mechanics. These definitions are introduced even before the laws of motion are formulated.

The best approach to Newton's notions of absolute time and space is therefore to inquire which *conceptual* job they are supposed to fulfil in the overall system of mechanics. After defining such uncommon notions as 'matter', 'motion' and 'force', Newton observes that the 'common people conceive such quantities as "time", "space", "place" under no other notions but from the relation they bear to sensible objects.'[14] That means that everyday notions of space and time are derived from the experience of the co-presence and the succession of material objects and processes in the macro-physical world. To avoid prejudices and errors, which may arise from such conceptions, Newton decides to introduce some convenient distinctions (Table 4.1).

Newton illustratesthese conceptual distinctions by way of an example: a ship is moving westwards on the eastwards moving earth and a sailor is moving east on the ship.

All these movements have *respective* velocities. For instance,

- The place, which the ship occupies on the surface of the earth, is moved towards the east *absolutely* with respect to absolute space.
- The ship is moving westwards with a *relative* velocity, i.e. with respect to the ocean floor.

---

[10] *Newton's Mathematical Principles*, Cajori Edition (1960), 545; see also Toulmin, 'Criticism' (1959), 220–1

[11] Toulmin, 'Criticism' (1959), 221; Leibniz, *Philosophical Writings* (1973), 205, 207; Čapek, *Philosophical Impact* (1962), 40

[12] Mach, *Mechanik* (1976), 217, 222

[13] *Newton's Mathematical Principles*, Cajori Edition (1960), 398–400, 546–7

[14] *Newton's Mathematical Principles*, Cajori Edition (1960), 6

Table 4.1.

Absolute or Mathematical Time	*Duration*, with no relation to anything external. *Absolute time* would exist even without any material objects in the physical universe. According to this view, absolute time could have existed before the creation of the universe by a Deity or the Big Bang. *Image*: the river of time. *Approximation*: pendulum clocks, Jupiter's satellites[15], fixed stars
Relative or Common Time	Some sensible or external measure of duration by means of motion. *Approximation*: clocks and calendars.
Absolute Space	Immovable space with no relation to anything external. *Absolute space* would exist even without any material objects in the world. *Image*: a receptacle or container, within which material objects are placed; the physical properties of the objects, like velocity, momentum and position can be 'measured' with reference to the container walls. *Approximation*: cosmic space.
Relative Space	Some movable dimension or measure of absolute spaces. *Approximation*: Celestial space, determined by its position in respect of the earth.
Absolute Motion	Translation of a body from one absolute place into another (where 'place' is a part of space which a body takes up). *Illustration*: the displacement of a material body, not measured with respect to other material bodies in the universe but with respect to the 'container walls' of absolute space.
Relative Motion	Translation of a material body from one relative place to another. *Illustration*: the displacement of a material body measured with respect to other material bodies.

- The sailor moves eastwards on the ship with a relative velocity, i.e. with respect to the deck.
- The sailor is moved *truly* or *absolutely* with respect to immovable space towards the east.

[15] *Newton's Mathematical Principles*, Cajori Edition (1960), 6; Toulmin, 'Criticism' (1959), 16–7; Strong, 'Newton's Mathematical Way' (1957), 415–25. Note, however, as Bondi, *Assumption* (1967), 49 pointed out (reporting a remark by Max von Laue) that the working of the pendulum clock comprises the whole earth. This can be seen from the fact that the period, T, of a pendulum depends not only on its length but also on the gravitational constant, g, near the surface of the earth, i.e. $T = 2\pi\sqrt{l/g}$

What is the conceptual function of these notions in Newton's theory of mechanics? Unlike Leibniz and Mach, many modern commentators regard them as 'theoretical constructs, whose existence must be presupposed in order to explain the phenomena available at the experimental-observational level.'[16] This is necessary because the observable phenomena may lack the uniformity, required for a reliable notion of time. Newton himself points out that there may be no uniform motion 'whereby time may be accurately measured. All motions may be accelerated and retarded.'[17] It is known for instance that the speed of the earth's rotation varies due to tidal friction and seasonal changes.[18]

Newton was not aware of such rotational inaccuracies of the earth. But he rightly suspected that reliance on physical motions may not be trustworthy. However, 'the flowing of absolute time is not liable to any change.'[19] For Newton this notion of time must be *absolute* – independent of physical events – and *universal* – the time must be the same for all physical observers, whatever their location in the universe and their state of motion.[20] Newton defines the notions of time and space before he formulates his laws of motion, because the laws of motion presuppose temporal and spatial notions. Statements like

> A body on which no net force is acting moves in a straight line at constant speed,

make no sense unless a reference frame is defined, relative to which the movement of the body can be described as uniform and in a straight line.[21] The fixing of a reference frame is already important in everyday life. Image two people standing face-to-face to each other. They receive the order to move to the 'left'; if they do not agree on the directions of 'left' and 'right', they will move in opposite directions. Equally, a straight line drawn on the surface of the earth is 'straight' to

---

[16] Sklar, *Philosophy of Physics* (1992), 23; cf. Jammer, *Concepts of Space* (^{2}1969), Chap. 4; Bondi, *Assumption* (1967), 22–3; Toulmin, 'Criticism' (1959), §III; Earman, 'Who's Afraid' (1970), §III; Stein, 'Newtonian Space-Time' (1970); Burtt, *Metaphysical Foundations* (1924), 248–55. To consider Newton's notions of absolute space and time as theoretical constructs leaves open the question of whether or not they have a referent, i.e. whether anything in the physical universe corresponds to absolute space and time. For a discussion of this question, see Friedman, *Foundations* (1984); Butterfield *et al.* (1996)

[17] *Newton's Mathematical Principles*, Cajori Edition (1960), 8

[18] See Clemence, 'Time and Its Measurement' (1952), 264–6; Sexl/Schmidt, *Raum-Zeit-Relativität* (1978), 28

[19] *Newton's Mathematical Principles*, Cajori Edition (1960), 8; see also Rynasiewicz, 'Absolute versus Relational Space-time' (1996), §3; Čapek, *Philosophical Impact* (1962), 35–8; Costa de Beauregard, 'Time in Relativity Theory' (1966), 417 writes: 'According to Newton's absolute space principle, there must exist an absolute spatial reference frame relative to which all movements can be thought of as taking place.'

[20] As Whitrow, *Natural Philosophy of Time* (1980), 34 (fn) points out the Special theory of relativity changed the meaning of those terms; because of the relativity of simultaneity, 'universal' cannot mean 'with respect to all possible frames of reference.' However, the question whether time or space-time can be absolute – independent of the existence of physical events – is still a matter of debate today (see footnote 1).

[21] See Macauley, 'Motion' (1910–11); Born, *Einstein* (1962), 54–58

the surveyor's eye, confined as the surveyor is to local purposes; but this line is not 'straight' from the point of view of an observer in space. Two lines may be parallel at the equator. But because of the earth's curvature, these two lines will cross at the North Pole.[22]

Thus, the central notions, which appear in Newton's laws of motion – 'state of rest' or 'rectilinear uniform motion' – require reference to absolute standards or, as Newton thought, reference to absolute space and time.[23]

It is not necessary for us to discuss the various thought experiments – the *bucket experiment* and the *two-sphere-experiment* – by which Newton hoped to show that there was at least indirect empirical evidence for the postulation of his absolute notions. It is, however, important to realize that, with Newton, speculations about the 'absolute' nature of space and time were given an empirical twist, since Newton believed that there were observable forces (centrifugal forces in the rotating bucket), which indicated the existence of absolute space.[24]

Discussions about Newton's concepts of absolute space and time continue to the present day. Reactions to Newton's postulations can be put into two categories.

1. The notions of absolute space and time have been branded an unobservable and superfluous metaphysical superstructure.[25] It is true that the later Newton added theological elements to his notions, first introduced in the *Principia*. But by reconstructing Newton's views as theoretical constructs needed for a solution of the dynamical problems of relative motion, some of this criticism can be averted (at least for our purposes). What is much more important is that criticism of Newton's notions led to an important alternative: the *relational view of time* (Leibniz, Mach). This is important because it denies one of the assumptions in Newton; that time and space, in an absolute sense, are independent of existing things.
2. Newton's notions do not just stipulate 'absoluteness', they also imply 'universality': every observer, in every reference frame, measures the same temporal and spatial dimensions, as long as they detect absolute motion. For Newton absolute and universal time and space are privileged reference systems, which alone give the 'true' temporal and spatial extensions. Think back on the sailor on the ship: his 'true' motion cannot be his relative motion with respect to the ship or the ocean floor. His 'true' motion, Newton thought, must be revealed by reference to absolute time and space. Unfortunately for Newton no preferred reference frames have been found and, according to the Special theory of relativity, cannot even exist. Einstein shows that the notion of absolute and universal time makes no sense: (a) Einstein only accepts clock time, ul-

[22] Born, *Einstein* (1962), 54; Thorne, *Black Holes* (1994), 108

[23] See Norton, 'Philosophy of Space and Time' (1992/1996), 5–6

[24] Sklar, *Space, Time, Spacetime* (1974), 165

[25] Reichenbach, quoted by Earman, 'Who's Afraid of Absolute Space' (1970), 287–8; Mach, *Mechanik* (1976), 223; Leibniz, *Philosophical Writings* (1973), 205–6, 227; Toulmin, 'Criticism' (1959), 213; Zwart, *About Time* (1976), 25; Sklar, *Space, Time, Spacetime* (1974), Chap. III

timately based on atomic oscillations; (b) observers even in inertial reference frames do not measure the same time intervals for the same events; (c) they do not measure the same lengths and (d) they do not even agree on when two events happen simultaneously. Furthermore, the strict separation between time and space, familiar from Newton's formulations, classical physics and everyday experience disappears. Instead, we require a notion of *space-time*.[26] As we shall see in a later section, the emergence of the notion of relative simultaneity of events and the interpretation of the Special theory of relativity in terms of *space-time*, lay the foundations for the view of the block universe and the unreality of time.

4.1.5 The Relational View of Time

Gottfried Wilhelm Leibniz (1646–1716)

The relational view of time is due to the German philosopher Gottfried Leibniz and was defended by the Austrian physicist Ernst Mach. To put it succinctly, we can say that it shares with Newton's views the idea that time and space are universal, but it denies that they are absolute. Time and space are universal in the sense that all observers in all inertial (and non-inertial) reference frames agree on the same spatial intervals (length) and the same temporal intervals (duration) between two events. But the relational view rejects Newton's strong realism about space and time. Space and time are not absolute in the sense that they require no reference to the external material world. As Leibniz says:

> time is nothing apart from temporal things; instants apart from things are nothing; time only consists in the successive order of things.[27]

> Space and time, in themselves and outside the world, are imaginary. Space comprehends all places, just as time comprehends all durations; but places in space and durations in time, unless occupied, are as ideal as space and time.[28]

[26] See Friedman, *Foundations* (1983); Earman, *World Enough and Space-Time* (1989); Sklar, *Space, Time, Spacetime* (1974); Butterfield/Hogarth/Belot *eds.*, *Spacetime* (1996). A popular exposition of these concepts and ideas can be found in Gamow, *Mr Tompkins* (1965), 15. Note that in Newton's own formulations, space and time are absolute, separately existing entities. But Newton's theory can be formulated in terms of space-time structure. In a neo-Newtonian space-time structure, time is still an absolute entity but space no longer is.

[27] Leibniz, *Philosophical Writings* (1973), 212

[28] Quoted in Benjamin, 'Ideas of Time in the History of Philosophy' (1966), 20

With respect to the absoluteness of time and space – its ontological status – a comparison of the realist and the relational view shows the following differences[29]:

Table 4.2.

	Newton	**Leibniz**
Space	All things are placed *in* space as to the order of their situation. The universe *has* a receptacle.	Space is the order of co-existing things. The universe *is* a receptacle.
Time	All things are placed *in* time as to order of succession. The universe *has* a clock.	Time is the order of the succession of events or the order of things in relation to their successive positions. The universe *is* a clock.

Thus, for Leibniz there is no place or time prior to creation: instants apart from things are nothing and outside created things there would be no space, i.e. no order of situations.[30] Such a view has important implications for the human determination of space and time. Time is an abstraction at which we arrive through the observation of changes in the physical world. Space is determined through a consideration of a body K in relation to other bodies A, B, C.[31]

Ernst Mach (1838–1916)

The relational view makes an important distinction between *physical* time and *human* time. In terms of the earlier distinction between natural and conventional units of time, Leibniz explains the acquisition of human time as an abstraction from the physical 'before-after'-relation of events. The physical 'before-after'relation marks physical time. The conceptual 'past-present-future'-relation marks human time.

The relational view does not affect a reconstruction of Newton's notion of absolute time as mathematical time as much as Newton's metaphysical musings, which make time and space appear as some ontologically independent entity. We must realize, however, that the relational view, as defended by Leibniz and Mach, is based on certain assumptions, which may themselves be questionable.

---

[29] See Leibniz, *Philosophical Writings* (1973), 211, 212, 220, 237; Newton, *Principia* (1687), 8; Whitrow, *Natural Philosophy* (1980), §1.10; Rynasiewicz, 'Absolute and Relational Space-Time (1996), 285–6; Langevin, *La Physique* (1926), Chap. V gives the following characterization: 'l'espace est l'ensemble des événements simultanés – le temps est l'ensemble des événements qui se succèdent en un même point.'

[30] Leibniz, *Philosophical Writings* (1973), 212, 237; as we shall see later, Saint Augustine married a relational with an idealist view of time, which also places the beginning of time at the moment of creation.

[31] Mach, *Mechanik* (1976), 217, 224; Leibniz, *Philosophical Writings* (1973), 260–1; Zwart, *About Time* (1976), 26–33; Smolin, *Three Roads* (2001)

- Mach's view is based on an extreme form of empiricism, which will only accept what is given to us by experience. Mach's position may be regarded as a form of Instrumentalism. It regards all theoretical accounts as convenient orderings of the empirical data, with no claim as to the physical reality or unreality of the postulated theoretical components. The empirical data are ordered according to one overriding principle: simplicity.[32] Mach's instrumentalist philosophy exerted a heavy influence on Einstein's early thinking.[33] Einstein's notion of clock time stands in the tradition of Galileo's physical time.
- Leibniz's objections are based on the metaphysical principle of *sufficient reason*. This principle states that nothing happens without there being a sufficient reason why it should be thus and not otherwise.[34] As we shall see in the next chapter, the French astronomer Pierre Laplace interpreted the Leibnizian principle as the axiom of the universal causal concatenation of all events and thereby established an influential identification of causation and determinism, which was cast into doubt by the experimental findings of quantum mechanics. For Leibniz, the conceptual job of this principle is to keep out notions like absolute space and time. For if there were *absolute space*, then there should be a reason (apart from God's pure will) why bodies are placed in it in a particular spatial configuration rather than another. If absolute space were a 'container', then we should be able to find a physical cause why the bodies in it are not further to the north or the south in terms of an imaginary coordinate system, painted on the walls of the container. But according to Leibniz, it is impossible to state such a physical cause or for God to have a reason. Equally for time.[35] If there were *absolute time*, then it would make sense to ask why God did not create the world a year or two later or earlier. But there is no discernible reason why God should have moved the succession of real events a few notches up or down some absolute but imaginary time scale, consisting of empty and occupied instants. So once again, absolute time cannot exist.[36]

Despite such 'dubious' metaphysical underpinnings, of which it would have to be divested, the relational view of time makes some important observations. *Firstly*, it helps to distinguish, more clearly than the idealist view (to be discussed later), between *human* and *physical* time. *Secondly*, it identifies physical time with the order of *succession* of physical events in the world. This emphasis on physical change as the basis of time has a natural affinity with the notion of the transience of time. Where there is material change, there is time, irrespective of the existence

[32] Mach, *Mechanik* (1976), 226, 467; cf. Earman, 'Who's Afraid of Absolute Space' (1970), 298; Sklar, *Space, Time, Spacetime* (1974), Chap. III

[33] See Holton, 'Metaphor' (1965), 38–52; *Thematic Origins* (1973), Chap. II

[34] Leibniz, *Philosophical Writings* (1973), 207, 211, 221; Earman, 'Who's Afraid of Absolute Space' (1970), 311

[35] Leibniz, *Philosophical Writings* (1973), 212

[36] For a discussion of Leibniz's principle of sufficient reason see Newton-Smith, *Structure of Time* (1980), 104ff; Sklar, *Space, Time, Spacetime* (1974), Chap. III; Barbour, 'Relational Concepts of Space and Time' (1982/1996), 141–164; Maudlin, 'Buckets of Water and Waves of Space' (1993/1996), 263–84

of observers. Leibniz speaks of the *order* of successive events.[37] What precisely is this order? A proper answer to this question can be given, once we have worked out Minkowski's influential notion of space-time.

The relational view stands in stark contrast to the philosophical conclusions drawn from the Special theory of relativity. For that theory seemed to suggest that the passage of time was a human illusion. The physical universe was a timeless block universe. So it was the idealist view of time, as we shall see, which carried the day with the majority of physicists. Let us now turn to the Special theory of relativity and the idealist view of time in order to see how their 'alliance' came about.

4.2 The Special Theory of Relativity and the Idea of the Block Universe

4.2.1 The Special Theory of Relativity (1905) – Some Results

Albert Einstein (1879–1955)

We have discussed changing conceptions of time and the input of physics and philosophy to our notions of time. Throughout these discussions we have assumed that time remained the same for all observers. This assumption is not only in agreement with our ordinary understanding of time; it was given a theoretical underpinning in Newton's notion of absolute time. According to Newton, there is an absolute temporal reference frame, relative to which all movements can be thought of as taking place. Equally, there is an absolute spatial reference frame, which Newton calls absolute space.

As we have seen Newton's laws of motion make no sense without a specification of the temporal and spatial

[37] Leibniz, *Philosophical Writings* (1973), 237. The relational view seems to be based on the existence on *actual* events. Then the issue arises whether there could not be *possible* or even *empty* moments of time and space. The question of *possible* orders of coexistence and *possible* orders of successions naturally arises when it is assumed that Leibniz only has the coexistence and succession of *actual* events in mind. But there is textual evidence that he has not; see Leibniz, *Philosophical Writings* (1973), 220–1, 235, 237 and Benjamin, 'Ideas of Time in the History of Philosophy' (1966), 20. For further discussion see Sklar, *Philosophy of Physics* (1992), 21–2 and *Space, Time and Spacetime* (1974), 168–73, 222–3; van Fraassen, *Time* (1970), 99; Friedman, *Foundations* (1983), 63 erroneously restricts Leibnizian relationism to actual events. Another problem frequently raised with respect to the relational view is the possibility of *empty* moments of time. For a discussion of this issue, see Newton-Smith, *The Structure of Time* (1980) and 'Space, Time and Space-Time' (1988), 22–35; Teichman, 'Time and Change' (1993). The relational theory is experiencing renewed interest, see Barbour, *End of Time* (1999), 'Development of Machian Themes' (1999); Belot, 'Rehabilitating Relationism' (1999); Pooley/Brown, 'Relationism Rehabilitated?' (2002)

dimensions to which they refer. Newton requires some reference frames to have a special status: they are *inertial* reference frames, in which the classical laws of motion hold and which are dependent on the fundamental assumptions of absolute time and space. According to Newton there are preferred reference frames, which are those in which the laws of motion hold. These laws do not hold in accelerated systems.

The most puzzling aspect of Einstein's Special theory of relativity is that three important features of space and time, which we normally take for granted, are cast into doubt and are made dependent on the velocity of the motion of the reference system, in which the observers find themselves.

1. The *length* of time observers measure for an event E varies according to whether the observers are stationary or are in constant movement with respect to the event to be measured. Generally, with respect to a stationary E, the duration of E is always shorter for an equally stationary observer than for the moving observer. From the point of view of the moving observer event E seems to take longer. This phenomenon is called *time dilation*. It is equivalent to the statement that moving clocks slow down by a factor of $\sqrt{1-(v^2/c^2)}$ or approximately $1/2(v/c)^2$, as seen from the system at rest (and vice versa).[38] The duration of an event, which takes place in the same reference system as the clock that measures it, is called *proper time*.
2. The *lengths* of objects in the x-direction (the direction of movement) as measured by observers in different reference frames again vary according to whether the observers are stationary or in movement with respect to the length to be measured. Generally, with respect to a stationary object, the length of an object in the x-direction is longer for an equally stationary observer than for the moving observer. From the point of view of the moving observer, the object seems to shrink. This phenomenon is called *length contraction*. It is equivalent to the statement that moving objects to the stationary observer appear to shorten by a factor of $\sqrt{1-(v^2/c^2)}$ and vice versa.[39] The length of an object, which is placed in the same reference system as the measuring tape that measures it, is called *proper length*.
3. Finally, moving and stationary observers do not agree on the *simultaneity* of two events. Two events that are simultaneous (appear to happen at the same time) in one reference frame are not simultaneous in another reference frame moving relative to the first with constant velocity. To speak of time in the physical sense, is, according to Einstein, to make statements about simultaneous events.[40] To say that event E happens at time t is to say that E is simultaneous with a certain position of the clock hand on the clock face. This clock is stationary at the

[38] Einstein, 'Zur Elektrodynamik bewegter Körper' (1905), 36

[39] Einstein, 'Zur Elektrodynamik bewegter Körper' (1905), 35. Note that in 1960 calculations showed that this is not how the objects appear to the observer: objects appear to rotate by a certain angle; see Weisskopf, 'The Visual Appearance of Rapidly Rotating Objects' (1960); Terrell, 'Invisibility of Lorentz Contraction' (1959)

[40] Einstein, 'Zur Elektrodynamik bewegter Körper' (1905), 27–9

location of the event E (otherwise a moving clock would appear to slow down from the point of view of E); and this stationary clock at E is synchronized with another stationary clock (just as our everyday clocks are synchronized with, say, the BBC pips, which in turn are synchronized with UTC – Universal Time Coordinated). This idea of *relative* simultaneity (as opposed to Newton's absolute simultaneity) became the decisive feature of the Special theory of relativity in the assertion of a block universe.

To understand these curious consequences of the Special theory of relativity for the conception of time, we have to recall the two fundamental principles, on which the whole of Einstein's theory is based.

a) The generalization of the *Galilean principle of relativity*. According to Galileo and Newton, all inertial reference frames, i.e. those that are either at rest or in uniform motion with respect to each other, are equivalent with respect to the validity of the mechanical laws. Einstein generalizes this principle to include all the laws of nature: in two reference frames, moving uniformly with respect to each other, all the laws of nature are exactly identical.[41] Another way of saying this is to say: the laws, according to which the states of physical systems change, are independent of the systems, moving relative to each other at uniform velocity, to which the changes are referred.[42] If, for instance, we have a stationary observer and a moving rod, which the observer wants to measure, then it does not matter which of the two systems we regard as stationary and in movement respectively. From the point of view of the stationary observer, the moving rod appears contracted in the x-direction (the direction of movement) by an amount of $\sqrt{1-(v/c)^2}$. From the point of view of the moving rod, the observer seems to rush past and appears contracted by the same amount. Absolute uniform motion cannot be detected.
b) The other important principle relates to the *constancy* of the *velocity* of *light* in vacuum (ca. 3×10^8 m/s). It is the same in all reference frames moving uniformly to each other. It is *neither* dependent on the velocity of the emitting body *nor* on the direction, in which the light ray is emitted. To fully appreciate the significance of this result[43], imagine a train, moving with a constant velocity ($v = 30$ m/s) in the positive x-direction. A marksman, M, is positioned on the roof of the train. M fires two *bullets*: one in the direction of the moving train, another in the opposite direction. The bullets are fired with $v' = 800$ m/s. This scene is observed by an observer, O, at rest on the embankment of the rail track. For M the bullet will have the same velocity in both directions. But O will calculate different velocities for the bullets, as they are fired in the direction of the train or in the opposite direction. In the direction of the train, O will calculate: 30 m/s + 800 m/s = 830 m/s; in the opposite direction, O will calculate 800 m/s – 30 m/s = 770 m/s. Thus, O must take into account the velocity of the

[41] Einstein/Infeld, *Evolution of Physics* (1938), 176
[42] Einstein, 'Zur Elektrodynamik bewegter Körper' (1905), 29, 51
[43] Russell, *ABC of Relativity* (1925), Chap. 3 provides further vivid illustrations

reference frame, within which the event is observed and O's calculation produces a combined velocity of the event within its reference frame. O's calculations are an illustration of the classical addition of velocities theorem: $V = v \pm v'$.
The observer on the embankment will calculate different final velocities (830 km/s and 770 km/s) for the bullets. The marksman now exchanges his rifle for a torch and shines a *light* in both directions. If we apply the addition of velocity theorem, which worked in the case of the bullets, we arrive at a contradiction with the postulate of the constancy of light. If the value of c in vacuum really is constant, then this must be the limit velocity of any physical signal. Hence the addition of velocities theorem of classical physics must be wrong for it would yield superluminary velocities, in contradiction to the constancy of c. The old velocity theorem $w = v + v'$ must be replaced by a new theorem:

$$w = \frac{v + v'}{1 + \frac{v \cdot v'}{c^2}} .$$

It yields the old theorem in the limit when v and v' are much smaller than c.

With these postulates in place it is possible to consider how the everyday notions of time and space and their sophistication in Newtonian mechanics yielded the notion of *time dilation*, *length contraction* and the *non-simultaneity* of events in reference frame, moving relative to each other.

As the notion of *relative simultaneity* became the cornerstone of the argument in favour of the block universe, we will only consider the transition, brought about by the Special theory of relativity, from *absolute* simultaneity in Newton's worldview to *relative* simultaneity in Einstein's worldview. Consider the lack of synchronization of moving clocks.[44] How can we synchronize two clocks separated in space? We could place observer C at a point midway between the two clocks situated at A and B. C sends a light signal to A and B, where observers set the clocks to a prearranged time when they receive the signal. The signal is reflected back to C with the same velocity (Fig. 4.3).

The two clocks are in synchrony according to our definition if the time it takes the signal to cover the distance C–A–C is the same as the time it takes the signal to cover the distance C–B–C: $t_A - t_B = t_B - t_C$. That is, they measure the same time interval for the transmission of the light signal.[45] We can now define the simultaneity

[44] Einstein, 'Zur Elektrodynamik bewegter Körper' (1905), 28, 30; Tipler, *Physics* (1982), 941; Born, *Einstein's Theory of Relativity* (1962), 225–32; Gamow, *Mr. Tompkins* (1965), 13–4; Feynman, *Six Not So Easy Pieces* (1997), Chap. III. Peter Galison, 'Einstein's Clocks' (2003) argues that at the time of Einstein's search for a solution of the distant simultaneity problem, a material culture existed in Central Europe, which made the coordination of clocks, as for instance required by rail travel, of great practical interest.

[45] There may be an apparently simpler method: move clocks, which have been compared and synchronized in one location, to different positions. This method does not work because moving clocks run slowly according to the time dilation result of the Special theory of relativity.

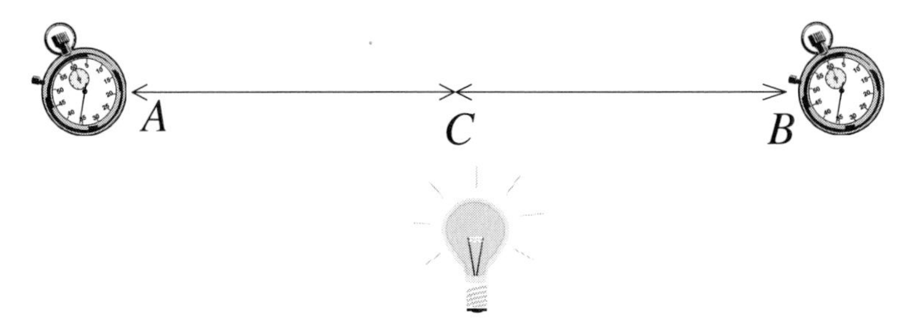

Fig. 4.3. Synchronization of clocks

of two events in a particular reference frame. Two events in a reference frame are simultaneous, if the light signals from two distant events reach an observer halfway between the events at the same time.

But two *events*, which are simultaneous in *one* reference frame, S, will not be simultaneous in *another* reference frame S'. Consider a moving train, with observers A', B' and C' situated at the front, back and in the middle of the train respectively. The train is attached to reference frame S'. It moves past a platform, reference frame S, with observers A, B and C positioned at the front, back and middle respectively. Let the event be that the train is struck by lightning at the front and back (A', B') (Fig 4.4).

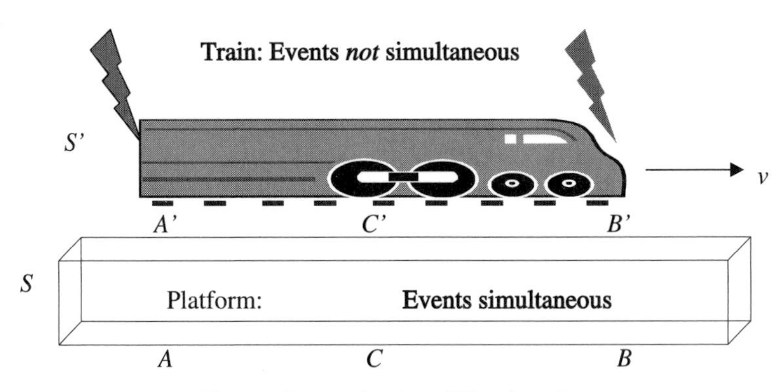

Fig. 4.4. Determination of Simultaneity

How do different observers judge the 'simultaneity' of the event? Platform observer C sees these two events at the same time, hence in S they are simultaneous. Are these events simultaneous for C' in reference frame S'? No! C' sees the flash from the front of the train before he sees the flash from the back of the train. C' travels forward to meet the flash from the front but moves away from the flash coming from the back of the train. It will take longer for the flash from the back to reach C' than it will take the flash from the front. C' will therefore conclude that the events are not simultaneous. For C' the front of the train was struck before the back of the train, since light has a constant velocity.

4.2.2 The Special Theory of Relativity and Models of Time

The results of the Special theory of relativity constitute important constraints, which any philosophical model of time, if it is to be adequate, should fit. This requirement is in line with Bohr's challenge to philosophy. So far we have introduced the *realist* and the *relational* view of time. How do these models of time fare with respect to the Special theory of relativity?

- The *absolute* view of time, in the Newtonian sense, finds itself in direct contradiction with the results of the Special theory of relativity. According to the principle of relativity, absolute uniform motion cannot be detected. There is no preferred reference frame according to which proper time and proper length are the 'correct' temporal and spatial measurements. Nor is there something like absolute simultaneity. The synchronization of clocks is also relative to the frame, in which events are measured. Time, according to Einstein, is frame-dependent.
- According to the *relational* view, time is the order of succession of events. It shares with Newton the view that all observers measure the same spatial and temporal intervals. With the advent of the Special theory of relativity, this implication must be dropped. All measurements of temporal and spatial intervals are relative to the reference frames, in which these statements are made. It appears at first that the relational view of time should fit in well with relative simultaneity, proper time and proper length.[46] It makes time dependent on the order of the succession of events, as long as we add that the succession, duration and simultaneity of event must be related to the reference frames under consideration.

At this point it is important to appreciate that this relativity of reference frames does not make temporal and spatial measurements hopelessly relative. The Special theory of relativity is able to predict by how much such measurements differ between different reference frames. The transformation theorems of the theory, the Lorentz transformations, allow a transition between different reference frames.

Although the relational view seems to be a natural philosophical view to complement Einstein's clock time, this is not how physicists judged the connection between the relativity theory and the philosophy of time. For the physicists the relativity theory led to a *static, timeless* conception of the universe. The idea of the *block universe* depends crucially on the relationship of the Special theory of relativity with idealist views on time.

4.2.3 The Special Theory of Relativity – Some Early Reactions

Having formulated his *Special theory of relativity*, Einstein had developed the central notion, which was to serve as a basis for the argument of the block universe. This was the notion of *relative simultaneity*. But if the *block universe* really is a philosophical consequence of the Special theory of relativity, Einstein was slow to

[46] Zwart, *About Time* (1976), 28–33, 161–2; Friedman, *Foundations* (1983) argues against Leibnizian relationism.

draw it. So were the first exponents of the new theory. This situation is very different from the philosophical excitement, which surrounded the second revolution at the beginning of the 20th century. The discovery of atomic phenomena and the slow emergence of a *theory* of quantum mechanics, as we shall see in Chap. 5, led to an instant awareness on the part of the participating scientists that fundamental philosophical issues were at stake. The quantum phenomena seemed to threaten the classical notion of *causation*. Physicists were quick to notice the threat and to offer remedies in various forms. The first publications on the Special theory of relativity, by contrast, did not refer to the block universe as a philosophical consequence. Technical articles on relativity began to appear in *The Philosophical Magazine* in April 1907. In July 1914, Max Planck published an article in this magazine, which makes a step towards an *idealist* interpretation of the notion of time as a result of the Einsteinian revolution in physics: the determination of time, Planck writes, has become dependent on the motion of the observer.[47] In *The Physical Review* articles on Einstein's relativity theory began to appear in July 1910 (Volume **XXXI**). In 1913, this journal published an article by R.D. Carmichael[48], which, in its title, held out the promise to discuss its philosophical aspects. But while the article gives a clear exposition of the postulates of the theory of relativity and explains its physical consequences: relative simultaneity, length contraction and time dilation, it does not take the step from the motion-dependence of time determination to the block universe. Its only timid philosophical suggestions concern 'the Philosophical Controversy concerning the One and the Many'. The British journal *Nature* opened its pages to the relativity theory in 1913 (Volume **90**). At first this took the form of numerous reviews of new books on the theory and of brief discussion notes on articles, which had appeared in other scientific journals. In 1916 Arthur Eddington began a series of articles on the General theory of relativity, in which he embraced the idea of a block universe. This hesitant beginning burst into a climax in 1921 when *Nature* devoted a large part of Volume **106** to a discussion of relativity and its various aspects. Einstein opens the discussion with a brief outline and the section ends with the publication of a bibliography on relativity. It is an indication of the revolutionary impact of the relativity theory that in 1910, J. Laub published the first bibliography, in which more than 130 publications on the Special theory of relativity are listed.[49] It is clear from an examination of the scientific journals of the time that Einstein's theory found almost immediately numerous adherents.[50]

---

[47] Planck, 'New Paths of Physical Knowledge' (1914), 65. This is a translation of a speech, held in Berlin in 1913 under the title: 'Neue Bahnen der physikalischen Erkenntnis' (1913), 74

[48] Carmichael, 'On The Theory of Relativity: Philosophical Aspects' (1913). A more interesting connection between the theory of relativity and materialism, made by H.W. Carr in the pages of the journal *Nature* (1920) was discussed in the chapter on Nature.

[49] Laub, 'Über die experimentellen Grundlagen des Relativitätsprinzips' (1910), 405–411

[50] This empirical fact throws some doubt on Kuhn's often-quoted statement that a new theory only gains a foothold in the scientific community, when the proponents of the old paradigm have died out. It is true that the theory of relativity had some persistent opponents, like O. Lodge. But many of the great, established scientists, men like A. Eddington, M. von Laue,

There were of course opponents like Oliver Lodge and satirists like Leo Gilbert. O. Lodge did not question Einstein's theory as a set of mathematical equations. He wanted to reject relativity as a philosophy.[51] The world was not merely a 'being but truly a becoming', but this was to be understood in a pre-relativistic sense. Lodge was a defender of the classical aether theory and regarded time and space as unchangeable. The Minkowski universe presented to him a' cold abstraction of the space-time manifold', which was repugnant to common sense. Consequently, Lodge preferred the Lorentz-Fitzgerald mechanical contraction hypothesis to 'complicating time and space.'

The Special theory of relativity also inspired some to satire. Leo Gilbert calls the relativity principle 'the latest fashionable folly in science'.[52] In his preface the author states that 'this book deals with one of the most interesting errors of mankind.' It derides, largely on the strength of common-sense arguments, the 'forgery', which the Special theory of relativity allegedly commits with the notion of time. Gilbert's book received a rather positive review in *Nature*. The anonymous reviewer predicts that the book will increase rather than lessen the general interest in the theory of relativity. The review also gives a flavour of a common-sense revolt against the counterintuitive consequence of the relativity theory for our notion of time.

> That the more extravagant conclusions resulting from the extreme adaptations of the principle should be held up to ridicule is quite wholesome, as it reveals the weak points in the argument and prevents the unwary from carrying it too far. (...) Our notions of time and space become almost interchangeable, and the "present moment" becomes meaningless without considerable restriction so soon as relative motion is involved.
>
> Leo Gilbert burlesques these innovations with much humour and ingenuity, and will no doubt largely prevent them being taken too seriously. Since Einstein himself has practically abandoned the principle of the apparent constancy of the velocity of light in all circumstances, and even his mathematical methods have failed to deal with accelerated motion, there is little left of the imposing mathematical superstructure, and what "craze" there was has given way before a sober appreciation of an interesting speculation on its merits.

The reviewer was out of touch with the spirit of the scientific community. It is always dangerous to appeal to common sense in the judgement of scientific matters. This does not mean that scientific judgements should be accepted uncritically. Especially

M. Planck, A. Sommerfeld enthusiastically embraced the new ideas. The bibliography at the end of this study demonstrates by the sheer volume of the published articles the extraordinary interest in the new theory. A majority of writers endorsed the relativity theory.

[51] Lodge in *Nature* **106** (1921), 795; **93** (1914); **104** (1920), 543; **106** (1920), 325–6, 357–8; **107** (1921), 716–9, 748–51, 784–5, 814–5; **110** (1922), 446; **114** (1924), 318–21; in the German-speaking world H. Driesch defended common sense against relativity. For instance Driesch, *Relativitätstheorie und Weltanschauung* (^{2}1930), 81–2 declares, reminiscent of Kant, that *one* time and *Euclidean* space are the *noli me tangere* of thinking.

[52] Gilbert, *Das Relativitätsprinzip* (1914). This satire earned the author a favourable review in *Nature* **93** (March 19, 1914), 56–57, from which some passages are quoted.

when scientific judgements are used to make philosophical pronouncements, care is required. There is no doubt, however, that physicists eventually formulated some philosophical consequences of the Special theory of relativity, which gave renewed prominence to idealist views of time. Historically, physicists arrived at the conviction that the passage of time was a human illusion in *two* steps. For Einstein and some of the most eminent proponents of his theory, the *first* step was the argument from the validity of the Special theory of relativity (relative simultaneity) to the acceptance of the *block universe* as a philosophical consequence of relativity. This step depended on Minkowski's formulation of the concept of *space-time*, in a speech in Cologne in 1908. This concept, too, quickly established itself and most publications on the relativity theory make use of it. After some hesitation, Einstein eventually saw its merits. The *second* step leads from the idea of a block universe to the endorsement of an *idealist view of time*. Einstein did not make this step. However, as we shall see, it is perfectly possible to defend an idealist view of time without endorsing the block universe. But it is not possible to embrace the block universe without accepting an idealist view of time.

To assess the impact of idealist views of time on the physicists of the 20th century we shall retrace these two steps. First, then, the step from relative simultaneity to the block universe. A proper discussion of this requires an understanding of Minkowskian space-time. Although Einstein acknowledged the importance of the notion of space-time, in his philosophical discussions, he made very little technical use of it. Yet Einstein embraced the idea of the block universe, at least in his early writings. Einstein's thoughts offer us a non-technical introduction into the idea of the block universe. Once we have grasped the idea, we turn to idealist views on time. It will be useful to review the philosophical ideas of the two most prominent proponents of an idealist view on time in the history of occidental thinking: *Saint Augustine* and *Immanuel Kant*. But it is really the Minkowskian notion of space-time that *seems* to support the view of the universe as a block universe. Idealist views of time follow on its heel. The Minkowskian notion serves as a platform for an assessment the *Philosophy of Being*. But it is also claimed that the space-time concept is compatible with a *Philosophy of Becoming*. Both philosophies have been heralded as the philosophical consequence of the Special theory of relativity under the Minkowski interpretation. Both philosophies must be evaluated within this framework.

4.2.4 The First Step: Einstein and the Idea of the Block Universe

In his famous article of 1905, which established the Special theory of relativity, Einstein said nothing about the idealist view of time or the block universe. But with *relative simultaneity* the paper establishes the result, which is central to the idea. In 1908, Minkowski announced the union of space and time in his concept of *space-time*. After some initial hesitation, Einstein eventually accepted the fruitfulness of this notion. Having shaking the traditional understanding of the uniqueness of time measurements for all observers – an everyday conception, which Newton had only tried to render more precise – he had freed himself from the grip of philosophical preconceptions about time. But it took some time before Einstein committed himself in writing to the idea of the block universe. In his paper on the General Theory of Relativity (1916), he writes 'that the requirement of general covariance deprived the notion of space and time of the last remnants of physical reality.'[53]

Einstein had a quite different attitude towards the notion of *causation* in quantum mechanics. He argued persistently that physical science could not abandon the notion of causation. The notion of time, by contrast, was dispensable. We called this *Einstein's problem*: to which extent do scientific discoveries have philosophical consequences? It is not just Einstein's problem. It is a general problem residing in the interstice between science and philosophy. On the one hand, philosophical presuppositions must remain open to revision as a result of new discoveries (Bohr's challenge). On the other hand, philosophy must evaluate how far philosophical concepts need to yield to empirical evidence.

In 1916 Einstein was philosophically ready to relegate the notion of time to the set of physically dispensable notions. Ironically, at about the same time, he expressed his allegiance to the *classical* notion of causation (Chap. 5). The more *conservative* his response became to the budding quantum theory, the more *radical* his views on time became. His *shaky* allegiance to the block universe appeared a few years later. It came without argument and without idealist citations. His source is Minkowski, not Kant. Other physicists – apart from von Laue, there were Eddington, Weyl and later Gödel – did embrace Kant. For those who endorse the block universe, idealist implications are difficult to keep at bay. Einstein did not realize this. As his fame grew and he authored popular books and encyclopaedia articles on the theory of relativity, Einstein never endorsed an idealist view of time. For instance in his article on 'Space-Time' for the 14th edition of the *Encyclopaedia Britannica*, Einstein makes the distinction between *subjective* time – the way each individual experiences the succession of events (as we shall see, this is Saint Augustine's psychological time) – and *objective* time – the time order of external events. But the assumption of a uniform time order for all observers had come under a cloud. The assumption that the simultaneity of external events had absolute meaning for events separated in space was demolished by the Special theory of relativity. As we have seen, one of its postulates is the constancy of the velocity of light (in vacuum). On the basis of this postulate,

---

[53] The requirement of general *covariance* means that the general laws of nature must be expressed in equations, which are valid for *all* co-ordinate systems. Einstein, 'Zur Grundlage der allgemeinen Relativitätstheorie' (1916), §3; for more detailed discussions see Norton, 'Philosophy of Space and Time' (1992), §5.4.3; Friedman, *Foundations* (1983), Chap. II.2; Costa de Beauregard, 'Arguments for a Philosophy of Being' (1966), 429. Some readers may be tempted to object that the discussion of the block universe cannot be restricted to the Special theory of relativity. My excuse is again that the problem of the block universe emerged in the minds of physicists as soon as the principle of relative simultaneity had been established. According to the General theory of relativity, what the Special theory says about the world is a valid approximation at local space-time points. The discussion in §4.3 of the present chapter indicates the role of time in the General theory and discusses how time is treated in quantum theories of gravity.

> No absolute meaning can be assigned to the conception of the simultaneity of events that occur at points separated by a distance in space (...). If no coordinate system (inertial system) is used as a basis of reference there is no sense in asserting that events at different points in space occur simultaneously. It is a consequence of this that space and time are welded together into a uniform four-dimensional continuum.[54]

From the illusion that the meaning of simultaneity is self-evident, the impression arose that this four-dimensional continuum could be broken down – in the manner of classical physics – into 'the three-dimensional continuum of space and the one-dimensional continuum of time.' The theory of relativity destroyed this illusion. And it showed that the concept of time was not fundamental for the description of the physical world. The phrase 'time of an event' has no meaning, until it is related to a reference-frame in which the event occurs. Every reference-frame has its own particular time.[55] And so the breakdown of the four-dimensional continuum into three-dimensional space and one-dimensional time is the work of the observer, attached to a particular reference frame. According to Einstein's often-repeated phrase of that period, *physics becomes a sort of statics in a four-dimensional continuum.* Or, as he also put it:

> From a "happening" in three-dimensional space, physics becomes, as it were, an "existence" in the four-dimensional "world".
>
> Since there exists in this four-dimensional structure no longer any sections, which represent "now" objectively, the concepts of happening and becoming are indeed not completely suspended, but yet complicated. It appears therefore more natural to think of physical reality as a four-dimensional existence, instead of, as hitherto, the evolution of a three-dimensional existence.[56]

This is Einstein's expression of the idea of the *block universe*.[57] The distinction between past, present and future is suspended. There is only a timeless block of events, eternally present. Yet happening and becoming are not completely suspended! It was of course the worry about time, which inclined Einstein towards a static view of the universe. One consequence of the Special theory of relativity was that time was not universal. Every reference frame carries its own clock and observers attached to these reference frames may disagree about temporal measurements and the simultaneity of events. How will this affect the representation of physical reality?

Classical physics, before the advent of relativity threw the notion of universal time into doubt, preferred a *dynamic* representation of physical reality. In such

---

[54] Einstein, 'Space-time' (1929), 1073. Čapek, *Philosophical Impact* (1962), 334–6 quotes Whitehead as saying that there is no such thing as 'nature at an instant'. The elimination of the idea of a cosmic now represents, according to Čapek, 'one of the most serious threats to the classical Laplacean world scheme.' See also Čapek, 'Time-Space (1983)

[55] Einstein, *Relativity* (1920), Chap. IX (26)

[56] Einstein, *Relativity* (1920), Appendix II (122), Appendix V (150)

[57] See Einstein, *Relativity* (1920), Chaps. XVII, XXVII; 'Brief Outline' (1921), 783; 'Space-time' (1929), 1072; Einstein/Infeld, *Evolution of Physics* (1938), 199–208; Nahin, *Time Machine* (1993), 101–12, 137f, 208; Christenson, 'Special Relativity' (1981)

a representation the position of a material particle changes with time. The time and space axes are clearly separated and the motion of the particle is represented as a line in a two-dimensional diagram (Fig. 4.5).

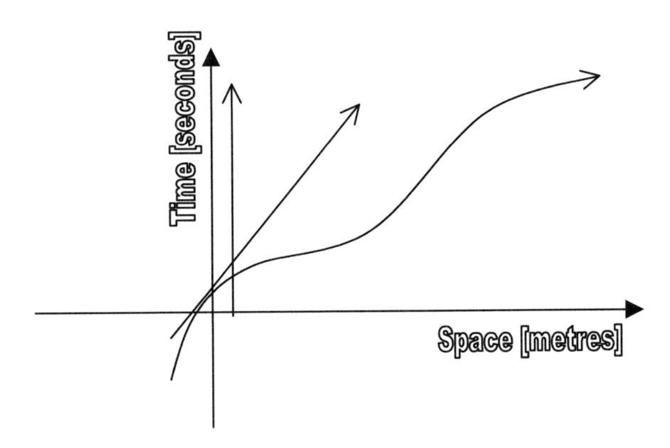

Fig. 4.5. Representation of a stationary particle (*vertical line*), a uniformly moving particle (*straight, inclined line*) and an accelerated particle (*curved line*)

The assumption is that the time axis indicates the same time for all observers. And it is precisely this assumption, which the Special theory of relativity has shown to be mistaken. As the *dynamic* representation requires the objective splitting of space and time axes, it has become inappropriate. With the disappearance of absolute and universal time from physics, such a 'division into time and space has no objective meaning since time is no longer absolute'.[58] It may therefore be more objective to consider a *static* representation of reality. For the *static* representation we need a *space-time diagram*. These diagrams were first introduced by Minkowski to represent the famous union of space and time. Here are the core ideas: *First*, as every reference frame has its own space and time coordinates, time and space can no longer be separated as in classical physics. Results of the measurements of temporal and spatial intervals are not the same for all observers. *Second*, there is an absolute limit beyond which a material particle cannot travel. This limit is represented by the constancy of the speed of light, *c*. *Third*, from every event light signals propagate at a constant speed. To every *event* correspond 4 definite numbers. Three space components – x_1, x_2, x_3 – and a time components – t (also labelled x_4). 'Therefore: The world of events forms a *four-dimensional continuum*.'[59] Every observer assigns different spatial and temporal coordinates to an event, depending on the state of motion. But this is not the whole story. As we shall see later, Minkowski also showed that there was some absolute measure involved, on which observers could agree. This will be called the *space-time interval*, ***I***. For the moment let us ask: How can an event like the motion of a particle in such a space-time continuum be represented?

---

[58] Einstein/Infeld, *The Evolution of Physics* (1938), 108; see also Frank, 'Relativitätstheorie' (1910)

[59] Einstein/Infeld, *The Evolution of Physics* (1938), 107; italics in original

We place the event into a four-dimensional space-time continuum and consider how it would appear from the point of view of observers who are in different states of motion with respect to each other. We can construct the diagrams in several steps.[60] As a *first* step let us ignore the existence of a limit, as it is defined by the world lines of the light signals. The resulting diagram looks like a conventional diagram (Fig. 4.5) but there is a significant difference: the unit on the time axis is still *seconds* [s] but the unit on the space axis is *light seconds* [Ls] or the distance that light travels in 1 s (300 000 km). And we speak of the lines as *world lines* – Fig. 4.6:

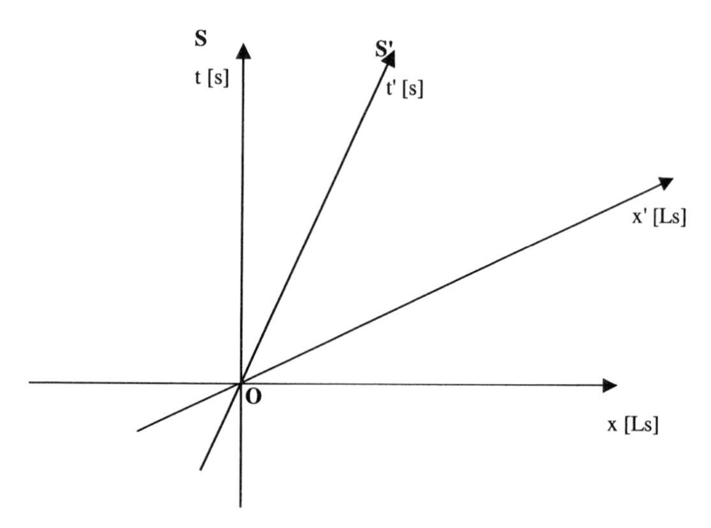

Fig. 4.6. The coordinates of two inertial systems *S*, *S′* that are in relative motion to each other. The rise of time axis t' indicates the velocity of this inertial system with respect to *S*. The steeper the world line, the smaller the velocity of the inertial system. The angle between the two time axes (t, t') must be the same as that between the two spatial axes (x, x')

World lines are curves in space-time diagrams. They describe the motion of particles or the propagation of light signals. In Fig. 4.6, the world lines of two particles are indicated. In the first inertial system, S, the particle is at rest. The second particle is attached to the inertial system, *S′*, which has a constant velocity with respect to S. (According to the principle of relativity, the particle attached to *S′* could also be regarded as being at rest. Then the particle attached to *S* would be in constant motion with respect to *S′*.) We can now perform the *second* step and include, as limit velocities, the world lines of photons. No material particle can travel faster than the speed of light, *c*, according to the Special theory of relativity. That means that no world line of a material particle must incline *below* the world lines of the photons. As the space axis is given in light seconds and the time axis in

[60] See Sexl/Schmidt, *Raum-Zeit-Relativität* (1978), §7.3; Sklar, *Space, Time and Spacetime* (1974), 56–61; Nahin, *Time Machines* (1993), Technical Note 4; Joos, *Theoretical Physics* (1951), Chap. X, §8; Reichenbach, *The Philosophy of Space and Time* (1958), §29

seconds, a light particle, which travels one second in time will cover the distance of one light second (300 000 km). So the world lines of light particles must form an angle of 45° with respect to the time axis. That means that in system S the world line of light signals, which are emitted from the origin, O, ($t = 0, x = 0$) diverge at an angle of 45° into the future. They form a *future light cone*. Signals also arrive from the past at the origin. They converge onto the origin at an angle of 45°, constituting the *past light cone* at O. How are the world lines of light signals to be represented in S'? According to the Special theory of relativity, the velocity of light, c, is the same for all observers, independent of their state of motion and independent on the direction of the light source. That means that for observers in S', the rise of the world lines of light signals must be the same as for observers in S. A good analogy – an *analogue* model – helps to visualise and understand this situation.

> So, let us suppose that we are aboard an ocean liner and we want to have a graphical record of our journey. The simplest thing to do is to take the map and at each hour, say, put a point at the proper latitude and longitude where the boat is, and write next to it the time. It is much more revealing, however, if we make a three-dimensional model and put the dot not on the map but above the intersection of the proper longitude and latitude. The perpendicular distance from map to point should be proportional to the time elapsed from the departure. If we do this each hour and finally connect all these points with a thin wire, the wire will contain in a graphical form all the information about the journey.
>
> This information shows not only the location of the boat, at a given time, but much more. If the wire is straight, it tells us that the boat has travelled on a straight line with constant speed. If the wire lies in one plane (normal to the map) but describing a curve in this plane, the boat was travelling on a straight line though the speed did change during the course, and so on.
>
> Now if we take a particle instead of the boat and note mentally in a four-dimensional space its position (x, y, z) at time t for each instant of time, we get a similar plot. Each point of the plot specifies an event, and the resulting curve is called the world line of the particle; it describes the history of this particle. The four-dimensional continuum in which the plotting takes place is called space-time. The immense importance of this concept is as follows: suppose that several different observers, each using a different inertial system of reference, are observing the motion of a particle, and each is asked to construct a space-time diagram of the motion. According to the Special theory of relativity, each observer will construct exactly the same curve in space-time for the history of the particle. The different states of motion of the observers (since they use different inertial systems of reference) will manifest themselves by the fact that the coordinate axes in space-time, x, y, z, t (which localize an event in space-time relative to the inertial system of reference used by the observer), will be different for different observers. The relation between these axes is given by [the Lorentz transformation]. Thus we can pass from one set of axes in space-time to another set by the Lorentz transformation.[61]

[61] Balzacs, 'Relativity' (1929), 98; see Box 4.1 below for the Lorentz transformations

We can easily insert the world lines of the light signals into Fig. 4.6 and complete the representation. This is shown in Fig. 4.7.

In such a *static* representation,there is no place for an objective *becoming* or an objective *now*. The physical world just *is*. In the first phase of his career Einstein contended himself to affirm the objectivity of the *static* picture and the block universe. A further question did not occupy him: How do the impression of *change* and *becoming*, of the *flow of time* arise? It did not occupy him yet. For Einstein did eventually entertain doubts about the block universe. For instance, he remarked to Rudolf Carnap, 'that there is something essential about the now.'[62]

Einstein did not realize that to embrace the notion of the block universe is at the same time to agree to an idealist account of the origin of time. For if the physical world just *is*, then the impression of change, of becoming and the flow of time, must be located in the mind of the observer. Idealist philosophies of time from Saint Augustine to Kant drew this conclusion. Weyl and Eddington explicitly embraced the idealist view of time, which is already implicit in the notion of the block universe. Ernst Cassirer and others tried to assess the theory of relativity in the light of Kantian philosophy.[63] Einstein's views on this issue were so elusive that *Nature* published

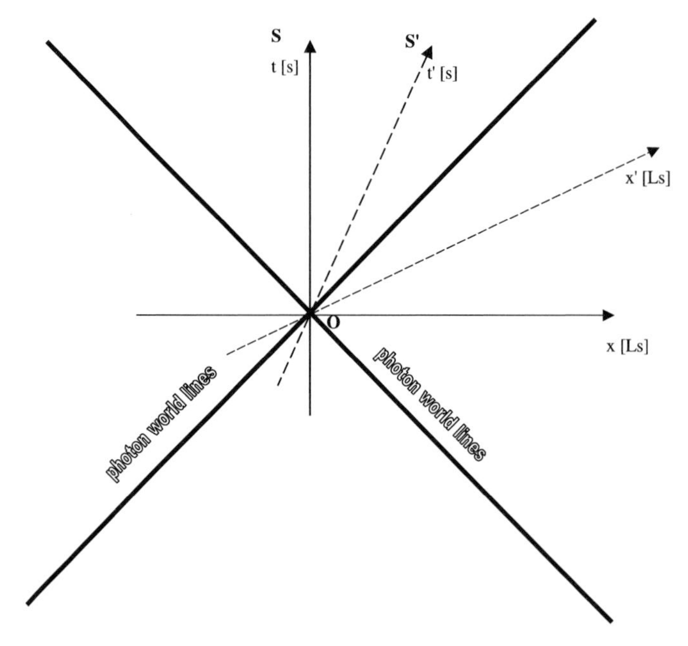

Fig. 4.7. This is the same configuration as in Fig. 4.6. But this time the world lines of the photons, emitted from *O*, are included. They form a diverging future cone and a converging past cone on *O*

---

[62] Davies, *About Time* (1995), 77

[63] Cassirer, *Zur Einsteinschen Relativitätstheorie* (1921); A.C. Elsbach, *Kant und Einstein* (1924); reviewed in *Nature* **114** (1924), 748. P. Carus, *The Principle of Relativity* (1913);

a brief note, entitled 'Einstein and the Philosophies of Kant and Mach'. The note reports a verbal communication, which Einstein made at a conference of the *Société Française de Philosophie* held in Paris in 1922. It confirms that Einstein always eschewed the issue of idealism. Responding to a question from the philosopher Brunschwicg, Einstein is reported to have replied:

> Now there are two opposite points of view: Kant's apriorism, according to which certain concepts pre-exist in our consciousness, and Poincaré's conventionalism. Both agree on this point, that to construct science we need arbitrary concepts; but as to whether these concepts are given *a priori* or are arbitrary conventions, I am unable to say.[64]

The relationship between concepts and facts was essential to Einstein's thinking. In later life he came to believe that humans should use concepts freely but that the adequacy of these concepts should be subjected to the scrutiny of empirical facts. We have already noted Einstein's vacillation in these matters: Newtonian presuppositions of time were relinquished but not the Laplacean notion of causation. His commitment to the block universe was also shaky. After the loss of his friend Besso and shortly before his own death, Einstein expressed the idea of the block universe or the *static* picture of reality in the following words:

> For us believing physicists, the distinction between past, present, and future is only an illusion, even if a stubborn one.[65]

But when he was invited to comment on Gödel's connection between the theory of relativity and the block universe, Einstein again eschewed the issue of idealistic philosophy. Quite 'aside from the relation of the theory of relativity to idealistic philosophy'[66] Einstein considers the question of the direction of time. Without realising it, he injects a dynamic element into the *static* representation of reality and therefore the block universe. Imagine we send a signal from A to B through P. This is an irreversible process. On thermodynamic grounds he asserts that a *time-like* world line from B to A through P in a light cone, takes the form of an arrow

---

reviewed in *Nature* **93** (1914), 187. H.W. Carr, *The General Principle of Relativity* (1920); reviewed in *Nature* **106** (1920), 431–2. It was quite common at this time to discuss philosophy in science journal. For instance, between 1913 (Volume **92**) and 1924 (Volume **114**) the journal *Nature* published numerous reviews of books on the theory of relativity and philosophy. There are many examples of this practice with respect to fundamental issues like causation, Nature and time. A general example is Crew, 'The Debt of Physics to Metaphysics' (1910)

[64] *Nature* **112** (August 18, 1923), 253; italics in original

[65] Hoffmann, *Albert Einstein* (1972), 257–8; historical references to the idea of the block universe are to be found in Nahin, *Time Machines* (1993)

[66] Einstein, 'Reply to Criticism' (1949), 687–88; Planck showed that entropy is a relativistic invariant, a point to which we shall return, see Einstein, 'Über das Relativitätsprinzip' (1917), §15; Heilbron, *Dilemmas of an Upright Man* (2000), 30–1. We will shortly see that Eddington, too, came to doubt the reality of the block universe. According to Eddington, *Physical World* (1929), 92 the problem with the Mindowski view was that it 'leaves the external world without any dynamic quality.'

making B happen *before* P and A *after* P (see Fig. 4.8). This secures the 'one-sided (asymmetrical) character of time (...), i.e., there exists no free choice for the direction of the arrow.'[67] This is true at least if points A, B and P are sufficiently close in cosmological terms. But the asymmetrical character of time is here based on a fundamental *earlier-later* or *before-after* relation between physical events without reference to an observer. There is an event, B, at which the signal is emitted. And there is a later event, A, at which the signal is received. Einstein had claimed that the *static* representation is more objective than the *dynamic* representation. Now there is a subtle shift. He introduces elementary change – the motion of the signal from B to A. It takes time for the signal to reach A. This means that world lines can develop a *history*. In his reply to Gödel, Einstein thus gives an indication that the loss of objectivity of the time axis need not lead to a static world. It is interesting to note that in one of the first books on relativity in the English-speaking world, published by E. Cunningham in 1915, a similar interpretation of the union of space and time is suggested.

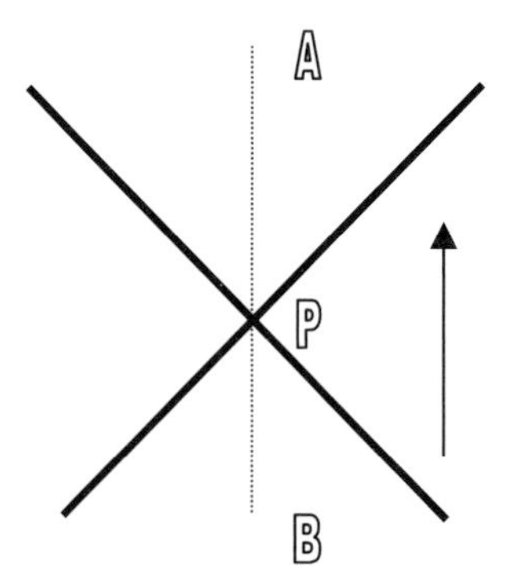

Fig. 4.8. Einstein's consideration of the direction of time in response to Gödel's idealistic interpretation of the special theory of relativity. A *time-like* world line exists between events, which lie within, not outside, the light cone

> The motion of a moving point through all time is represented by a single curve, the points on the curve being ordered to correspond with the succession of events in time, but the interpretation of the curve as representing an ordinary motion is not unique; it depends upon the choice of the direction in the four-dimensional region which is chosen to be the time axis.[68]

[67] Einstein, 'Reply to Criticism', 687; cf. Einstein, *Relativity* (1920), 139–41, where again a distinction is made between a subjective concept of time and time as a sequence of events in space and time. Cf. Prigogine, *The End of Time* (1997), 165.

[68] Cunningham, *Relativity and the Electron Theory* (1915), §60. Robb, *A Theory of Space and Time* (1914) analyses spatial relations in terms of the time relations 'before' and 'after'. A 'before-after'-relation of two instants is an asymmetrical relation. Similarly Reichenbach, *The Philosophy of Space and Time* (1958), 183 holds that the world line at space-time point P represents the *flow* of time at P; later (1958, 270) he adds that world

The block universe may not be *the* philosophical consequence of a four-dimensional view of physics. What does it mean to say, then, that *past*, *present* and *future* are an illusion? What does it mean to say that the objective physical world simply *is*, without *happening* and *becoming*? Does such a *static* world mean that every event that will ever happen is determined to happen by its past? Is it not possible that a world line, as that of Einstein's signal sent from B to A, which is ordered according to the 'before-after'-relation, constitutes a temporal sequence, a slice of time, which exists without human awareness?

Einstein consistently ignored the idealistic implications of the block universe. It is perhaps not surprising that Einstein also grew more suspicious of the view of the physical world as a block universe, once Gödel had claimed that the relativity theory provided 'proof' of the idealist view of time. For Einstein, at least in his later years, was a realist. This means that he was committed to certain philosophical presuppositions. He believed in the reality of a spatio-temporal causal order of the universe. And he believed that a causal order of nature was a deterministically predicable order of nature. This is very clear from his contributions to quantum mechanics. But how can such a causal, asymmetric order be compatible with the idea of the block universe? Einstein was not such a naïve realist to think that the spatio-temporal causal order of a real, external world was reflected, *mirror*-like, in our scientific theories. In fact, he held that the fundamental concepts and laws of a scientific theory were free inventions of the human mind. Furthermore that 'a theory can be tested by experience but there is no way from experience to the setting up of a theory.'[69]

Although Einstein embraced the block universe, as a philosophical consequence of the Special theory of relativity, his allegiance stood on shaky grounds. Becoming, change, happening and time are not completely abandoned.

> (...) in respect of its *rôle* in the equations of physics, though not with regard to its physical significance, time is equivalent to the space co-ordinates (apart from

lines 'exhibit most clearly the singular character of time.' With his geometrization of physics, Minkowski inspired the notion of the block universe, but he himself, if pressed on this point, may not have accepted this notion. Although in his lecture 'Raum und Zeit' (Cologne 1908) he speaks of a four-dimensional physics [(1909/1974), 57] in the same talk he concedes that a 'necessary' time order can be established at every world point [(1909/1974), 61]. Schlick also warned against reading metaphysical speculations into the introduction of time as a fourth dimension in the four-dimensional representation of the world: in the four-dimensional space-time manifold, 'the system of all world lines represents the temporal course of all events of the universe'. Schlick, 'Raum und Zeit in der gegenwärtigen Physik' (1917), 181 [translated by the author]. This idea that the flow of time can be read into the world lines of particles, under certain conditions, has been taken up in modern discussions, see Dieks, 'Special Relativity and the Flow of Time' (1988), 456–60; Prigogine, *End of Time* (1997), Chap. 8.2; Friedman, *Foundations* (1983), 18, 34, where geodesics are introduced as *histories* of particle trajectories.

[69] See Einstein, *Mein Weltbild* (1977), 115, 38–9; Einstein, 'Autobiographical Note' (1949), 89; Born, *Einstein's Theory* (1962), 334. Holton, 'Metaphor' (1965) describes Einstein's pilgrimage from early positivism, influenced by Mach, to rational realism.

> the relations of reality). From this point of view, physics is, as it were, a Euclidean geometry of four dimensions, or more correctly, a statics in a four-dimensional Euclidean continuum.[70]

As a realist Einstein would require that our theories represent, in approximation, the external structures of the universe. And the external structures represent fundamental constraints on the validity of the theories. The *static representation* of physical reality in a four-dimensional continuum may be a more objective representation, in view of the relativity of simultaneity. But Einstein never explicitly argues that the static *picture* entails the block universe. Given his Popperian characterisation of scientific theories as sophisticated conjectures, Einstein would have had to find some arguments to be able to *infer* the block universe from the static *representation* of physical reality. If the static representation of physical reality is the better one, this must be testable against what we know about the physical world. As we know from Chap. 2, the invariants of Nature play an important part in the representation of reality. So we cannot accept the block universe as the philosophical consequence of the Special theory of relativity till we have examined all the available evidence. Max von Laue sounded a note of caution when he pointed out that Minkowski space-time was only 'a symbolic representation of certain analytic relations between four variables.'[71]

We have gained an impression of the block universe and how it may arise from the Special theory of relativity. But how can we show that a *static representation* of physical reality entails a *static world*? In search of an answer to our questions, we must turn to the philosophical views of some of Einstein's most prominent contemporaries. Men like Weyl, Eddington and Jeans were not afraid of embracing the idealist view of time, especially of a Kantian persuasion. They firmly believed in a static physical world of *being*. The world of *becoming* was only a human illusion. They made the step from a *static* representation to a *static* world. If physics demonstrated the *reality* of the block universe, philosophy provided the philosophical foundation to this view in the form of the Kantian philosophy of time and space.

4.2.5 The Second Step: Idealism and Determinism – New Models of Physical Understanding

We have already observed that Max Planck hinted at an idealist interpretation of the notion of time in the Special theory of relativity. Planck observed that 'the principle of the constancy of light had made an absolute determination of time, i.e. one that is independent of the observer, utterly impossible.' In one of the first full-length studies of the Special theory of relativity, Max von Laue made the step to idealism explicit. The relativity principle attributes a particular time axis to each coordinate system. But this does not mean that our knowledge of the external world has become mere opinion. Kant's philosophy can be used to show

---

[70] Einstein, 'A Brief Outline' (1921), 783; italics in original

[71] von Laue, *Das Relativitätsprinzip* (^{2}1913) 51; author's own translation; the first edition appeared in 1911

that although knowledge depends on the observer, this does not render knowledge claims subjective. Einstein's philosophical audacity lay in his destruction of the traditional prejudice that a unique time axis existed for all reference frames.

> However great the transformation is into which he forces our whole thinking it harbours not the slightest epistemological difficulty. For time and space are, in Kant's terminology, pure forms of intuition; a scheme, into which we must order the events, so that they take on objective meaning – in contrast to subjective, highly accidental perceptions. It is therefore one of the conditions of the possibility of the objectivity of experiential facts.[72]

For instance, two astronomers situated on planets in relative uniform motion to each other will attach different time coordinates to their observations. But these time coordinates refer to different systems. And the so-called Lorentz transformations (Box 4.1) allow a translation of one set of observations into another. It is, one may add the analogy, like Celsius and Fahrenheit temperature scales used in different countries. These can be translated into each other.

Box 4.1: The Lorentz Transformations

$$y' = y$$

$$z' = z$$

$$x' = \frac{x - vt}{\sqrt{1 - \left(\frac{v}{c}\right)^2}}$$

$$t' = \frac{t - (vx/c^2)}{\sqrt{1 - \left(\frac{v}{c}\right)^2}}$$

Von Laue's words show very clearly that the physicists did not operate in a philosophical vacuum. Von Laue was the first but by no means the last physicist to appeal to Kantian notions of time to spell out what they took to be the philosophical consequences of the Special theory of relativity. Later Eddington, Gödel, Gold and Weyl, amongst others, were to draw similar conclusions. Von Laue, however, does not appeal to the block universe. He only stipulates *compatibility* between Einstein and Kant. This compatibility takes the form of coexistence between physics and philosophy. Relativistic physics showed, in the words of Hermann Bondi[73], that the concept of a *unique universal time* had to be abandoned. Time is only what a clock tells us. This was not alarming, as philosophy had already provided a conception

[72] von Laue, *Das Relativitätsprinzip* (^{2}1913) 37. Max von Laue later declared that he only arrived at a satisfactory understanding of the theory of relativity when he interpreted it in the light of the Kantian doctrine of space and time; see von Laue, 'Erkenntnistheorie und Relativitätstheorie' (1960), 61

[73] Bondi, *Assumption* (1967), 41

of time, which seemed to be compatible with the consequences of relative simultaneity. But Eddington, Weyl and others go further than von Laue: from *relative simultaneity* to an *idealist view of time* and the *block universe.*

Let us briefly pause to summarise our findings.

Einstein	**Von Laue**
Einstein takes it that the relativity of simultaneity implies a philosophy of being. Relative simultaneity makes the static picture of physical reality more objective. The physical world can be represented as a block universe. But Einstein shies away from an endorsement of an idealist view of time. With his insistence on the irreversibility of time, he even hints at a dynamic view of time. Nevertheless, as we have argued, acceptance of the idea of a block universe entails an idealistic view of time. It also entails determinism.	Von Laue acknowledges the Einsteinian discovery of the frame-dependence of time coordinates. But the relativity of simultaneity does not, for him, carry any implication of a philosophy of being. Rather, space-time diagrams are symbolic representations. However, the frame-dependence of time coordinates implies an objective idealist view of time. Acceptance of an idealist view of time does not imply a block universe.

Epistemologically speaking, we can use our conceptual building blocks to construct different views: (a) the Special theory of relativity and idealist view of time, without the block universe (von Laue); world lines have a history and constitute the flow of time; (b) more radically, the Special theory of relativity is taken to support the idea of a block universe, but then an idealist view of time is a necessary consequence of this construction. In this vein, Gödel pointed out that the Special theory of relativity provided 'unequivocal proof' for the idealist view of time.[74] The idealist view of time holds, roughly, that temporal awareness is built into the human mind either as a pure form of intuition (Kant) or that the passage of time appears to the mind as a change in perceptions (Saint Augustine). As we shall see in the next section, Saint Augustine developed a *subjective* idealist view of time, Kant an *objective* idealist view. How can the idealist view of time at least find support, if not proof, in the Special theory of relativity?

To see this connection, consider the views of Weyl and Eddington. We shall witness a surprising *connection* between physics and philosophy.[75] Minkowski was the inspiration for the block universe. *Kant was the inspiration for an idealist conception of time amongst physicists.* Kant's idealist philosophy of time had been a mere philosophical speculation. The theory of relativity seemed to invest it with

[74] Gödel, 'Remark' (1949), Volume II, 557

[75] This will not be the only connection between physics and philosophy. In the next chapter, we shall see that Laplace's philosophical identification of determinism and causation provided the inspiration for physicists' views on both causality and *a*causality in quantum mechanics.

scientific credibility. The theory of relativity, under Minkowski's representation, seemed to demonstrate that time did not belong to the physical universe. If the events of the universe are stretched out to infinity, like the frames of a film[76], then time and change are a mere human illusion. The theory of relativity seems to have a tremendous philosophical consequence: the physical world is mere *being*, the human world is illusory *becoming*.

Of the physicists of Einstein's generation, Weyl, Eddington and Jeans were the most outspoken defenders of the view that the block universe was a philosophical consequence of the discovery of relative simultaneity. Later generations continued to claim that the Special theory of relativity demonstrated that physical reality was static.

According to Eddington and Weyl, the theory of relativity has destroyed the classical edifice of the natural world, which made time and space real features of the physical world. The passage of time – *becoming* – was seen as the 'real progression of the world in time'. But the Special theory of relativity extirpates time and space from the physical world and relegates them to the status of secondary qualities. Just as colour does not belong to a physical object, but is the transformation of light waves from an object into human perceptual awareness, so the perception of time and space depends on the human conceptual apparatus. Kant was the first to show that time and space were only forms of intuition. Now Kant had been vindicated by Einstein's revolution.

> In the realm of physics it is perhaps only the theory of relativity which has made it quite clear that the two essences, space and time, entering into our intuition have no place in the world constructed by mathematical physics.[77]

[76] The *film analogy* of the block universe has often been observed: the individual frames of the film are stretched out, the past is as fixed as the future. Only the human observer must view the frames in succession, thus creating the illusion of the passage of time. See Popper, *Open Universe* (1988), 32–33; Nahin, *Times Machines* (1993), 103; Čapek, 'Time in Relativity Theory' (1966), 434ff; Frank, *Philosophy of Science* (1957), 158; Wendorff, *Zeit und Kultur* (1980), 464. The idea of a block universe is also reminiscent of the representation of particle tracks in cloud chambers, the first of which appear around 1912 – 'the pictures are simple, silent and still; there is no evidence of motion.' Yet these pictures are the result of atomic motion. See Holton, *Einstein, History* (2000), 80

[77] Weyl, *Space Time Matter* (1918/1952), 3, 227; Weyl remained faithful in his commitment to the block universe and its consequence: an idealist view of time and space for a long time. In a later paper, 'Geometrie und Physik', (1931), 49, Weyl agrees with Kant that time and space are forms of intuition. See also Weyl's contribution 'Electricity and Gravitation' (1921) to *Nature* **106**, 802–4 devoted to relativity. Only in his later life did Weyl tentatively move away from the idea of a deterministic block universe. See Weyl, 'Open World' (1978) and Čapek, 'Myth of frozen Passage' (1965), 447f. It is significant in this context that Eddington opened an article on relativity with the Kantian theme: 'According to the principle of relativity in its most extended sense, the space and time of physics are merely a *mental scaffolding* in which, for our own convenience we locate the observable phenomena of Nature.' See Eddington, 'Gravitation and the Principle of Relativity' (1916), 328; emphasis added. For similar statements see Eddington, 'Gravitation and the Principle of Relativity' (1918), 34. Eddington was not the first British physicist to introduce this 'continental' theme. In

The world of physical reality is a four-dimensional space-time structure, a union of time and space. Reality

> is a four-dimensional continuum, which is neither "time" nor "space". Only the consciousness that passes on in one portion of this world experiences the detached piece which comes to meet it and passes behind it, as *history*, that is, as a process that is going forward in time and takes place in space.[78]

Von Laue showed the possibility of endorsing a Kantian view of time without commitment to a block universe. Cunningham indicated that the succession of points on a world line constituted a history. This meant that a world deprived of human time was not a changeless world. Following the relational view the history of particles – their world line – would constitute time. With his considerations of the arrow of time Einstein began to entertain doubt about the validity of the block universe. But Weyl was much more radical. He relegated even history from the realm of the physical world. Eddington completed this step. There is no change in the physical world. All is static.

> In a perfectly determinate scheme the past and future may be regarded as lying mapped out – as much available to present exploration as the distant parts of space. Events do not happen; there are just there, and we come across them. (...) We can be aware of an eclipse in the year 1999, very much as we are aware of an unseen companion to Algol. Our knowledge of things *where* we are not, and of things *when* we are not, is essentially the same.[79]

---

an earlier article, E. Cunningham had already suggested that phenomena were ordered 'under the categories of space and time', so that 'our measures of space and time are (...) modes of thought (...). See Cunningham, 'The Principle of Relativity II' (1914), 408–9.

[78] Weyl, *Space Time Matter* (1921/1952), 217; bold characters in original

[79] Eddington, *Space, Time & Gravitation* (1920), 51; italics in original. With respect to the eclipse in 1999, remember that this was written in 1920. In his review of Eddington's *Romanes Lectures* (1922), Cunningham (1922), 568–9 captures the spirit of Eddington's thinking very well: 'The world is laid out before us as a changeless whole. Time and space are no more. All is static. Dynamics has been resolved away.' Again, as with Einstein and Weyl, we find some ambiguity in Eddington about the passage of time. In a somewhat obscure passage in *The Nature of the Physical World* (1929), Chap. V Eddington seems to claim both that we have a 'justifiable conception of 'becoming' in the external world', and that the passage of time is 'a condition of consciousness.' See also Sklar, *Physics and Chance* (1993), 409–11 for a discussion of this passage. In another context Cunningham writes: a four-dimensional static view of the universe 'is inseparable from a mechanical determinism in which the future is unalterably determined by the past and in which the past can be uniquely inferred from the present state of the universe.' Cunningham, *The Theory of Relativity* (1923), 213, quoted in Čapek, *Philosophical Impact* (1962), 159. With respect to the view that the future is already determined (mapped out) it is important to note that physicists at that time had a strong tendency to equate causation and determinism. The following Eddington statement is indicative of this identification: 'Ten years ago every physicist of repute was ... a determinist. He believed he had come across a scheme of strict causation regulating the sequence of phenomena.' Eddington, *New Pathways in Science* (1935), 72–3

In the jigsaw puzzle of these fundamental notions, relative simultaneity dovetails nicely with an objective idealist view of time, without any commitment to the block universe. But add the latter piece and you purchase more than a static view of physical reality. An endorsement of the block universe does not only entail an idealist notion of time, it also entails a commitment to *determinism*.

Eddington explicitly introduces the assumption of a *deterministic world*. Determinism (Box 4.3) is an essential ingredient of the picture of the block universe. For Eddington the true analogy of the four-dimensional world, with its union of space and time, was a solid block of paper. If the four-dimensional continuum can no longer be objectively divided into past, present and future, since time depends on the motion of the observer, and *Now* has no objective meaning, then, as J. Jeans states, 'the whole history of the universe, future as well as past, is already irrevocably fixed (...) and inescapable determinism reigns.'[80] We see here clearly, as we observed before, how some of the fundamental notions evolved differentially. The notion of determinism becomes really problematic with the discovery of quantum events. We will discuss the notions of determinism and causation in the next chapter. For the present discussion it suffices to note that Jean's notion is not the only alternative. Eddington and Jeans embrace a *static* view of determinism, according to which there is no change in the physical universe. This is the block universe captured in the analogy of the filmstrip or the particle tracks in a cloud chamber. But Cunningham hints at an alternative, *dynamic* form of determinism. The world lines propagate at a finite velocity: they have a history. The world lines undergo a dynamic evolution, according to deterministic laws. But this involves interactions between world lines, which may lead to unpredictable contingencies.

As the Minkowski space-time view was difficult to comprehend, physical understanding produced new models to present the four-dimensional world. The *map analogy* became an important feature of the Special theory of relativity literature. Terms like *light cone*, *lamination* and *slicing* were also introduced to capture the thinking behind Minkowski's four-dimensional world. Let us consider these attempts at physical understanding in turn.

- *Maps*. The history of the universe is a map of the world lines of all particles through four-dimensional space-time.[81] There is no indication of the *flow* of time. World lines of particles at rest are presented as *straight* lines. Particles moving with a constant velocity are indicated by straight lines but inclined at an angle of less than 45° with respect to the stationary particles. Accelerated

[80] Jeans, *Physics and Philosophy* (1943), 119 but see also Davies, *About Time* (1995), 278; Bunge, *Causality* (^{3}1979), 65 and Bondi, 'Relativity' (1952), 660

[81] Eddington, 'Gravitation and the Principle of Relativity' (1918), 17; Cunningham, 'Einstein's Relativity Theory of Gravitation II' (1919), 375; Weyl, 'Geometrie und Physik' (1931), 50; Ames, 'Einstein's Law of Gravitation' (1920), 206–216. A modern proponent of the *map* analogy and its association with the block universe is Gold, 'The Arrow of Time' (1962), 403–410 and 'The World Map' (1974), 63–71.

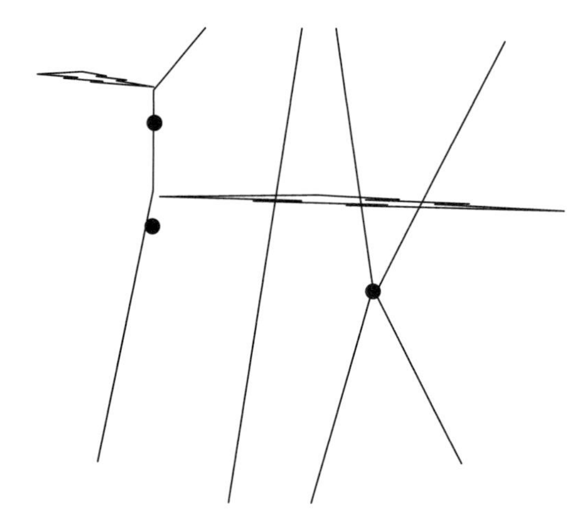

Fig. 4.9. Adapted from Misner/Thorne/Wheeler, *Gravitation* (1974), 6. 'The crossing of straws in a barn full of hay is a symbol for the world lines that fill up space-time.' The events in space-time are indicated by *black dots*. Only the intersections of world lines can claim physical reality and are observable. The first event, from *left* to *right*, is the absorption of a photon. The second event is the reemission of the photon. The third event represents the collision of two material particles

particles are indicated by curves. Note that the knowledge of nature consists in the knowledge of the *intersections* of world lines.[82] (Fig. 4.9)
If we concentrate on a particular point of a world line, at a particular event in space-time, we observe that *light cones* stretch out from each particular event in space-time. One light cone opens up into the future, another into the past.

- Eddington's solid block of paper[83] captures the idea of a four-dimensional continuum and the observer-dependence of the separation of space and time. Each

[82] Einstein, *Relativity* (1920), 95; 'Zur Grundlage' (1916), §3; Eddington, *Space, Time & Gravitation* (1920), 87; 'Gravitation and the Principle of Relativity' (1916), 328; 'Gravitation and the Principle of Relativity' (1918), 17; Ames, 'Einstein's Law of Gravitation' (1920), 211; Reichenbach, *Philosophy of Space and Time* (1956), §45; Schlick, 'Raum und Zeit in der gegenwärtigen Physik' (1917), 181; Friedman, *Foundations* (1983), 24

[83] Eddington, *Space, Time & Gravitation* (1920), 36, 51; 'The Relativity of Time' (1921), 803. Eddington's analogy of a solid block of paper has struck a chord with physicists. The analogy was taken up by Born: 'In itself the four-dimensional space-time continuum is structureless. It is the mutual relations of the world points disclosed by experiment that impresses a geometry with a definite metric on it.' Born, *Einstein's Theory of Relativity* (1962), 335. Costa de Beauregard invites us to imagine 'spacetime as a sort of book, the pages of which are the layered spacelike surfaces ε (...), which our conscious 'now', or 'attention to life', is 'reading', each in turn, in the 'right' order.' Costa de Beauregard, *Time* (1987), 155. In Minkowski space-time, 'all light cones have the same "width" and are "tilted" at the same "angle", Friedman, *Foundations* (1983), 186. See also Norton, 'Philosophy of Space and Time' (1992). In modern cosmology the term 'foliation' is used, see Barrow/Tiper, *The Anthropic Cosmological Principle* (1986), 627

observer, depending on the state of motion, will slice the block differently. The lamination planes at different angles indicate the respective observer's perspective on simultaneity. The Special theory of relativity only allows *relative* simultaneity. For any observer, events are simultaneous if they lie on a plane, which lies perpendicular to the observer's path through space-time. (Fig. 4.10)

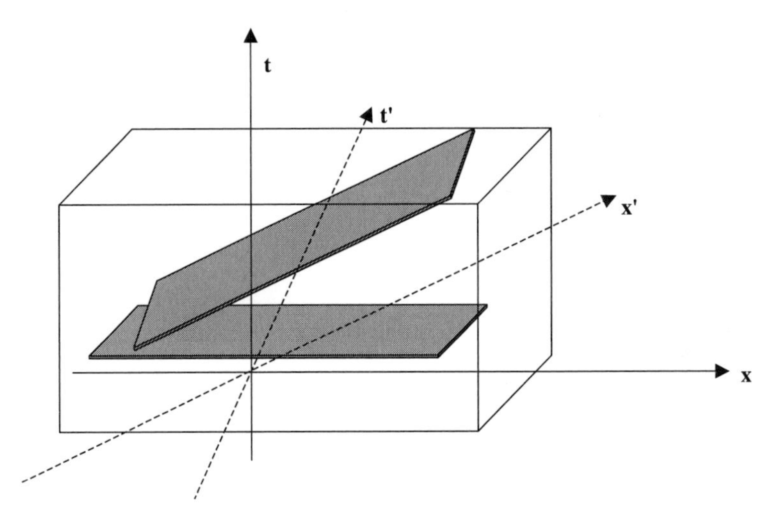

Fig. 4.10. Lamination of space-time block by different observers

- *Light Cones.* The most popular representation of the geometry of the four-dimensional world is that of *light cones*. Future-directed and past-directed light cones issue from every event in space-time. The boundaries of the light cones are formed by the light signals, which diverge from a particular event in space-time or converge onto such an event. This makes sense, since it is a postulate of the Special theory of relativity that no signal can travel faster than light. So the world lines of the photons constitute the boundaries, which no other world line of a material particle can cross. Light cones, which converge onto a given event, constitute the *past* of this event. Light cones, which diverge from a given event, constitute the *future* of this event. (Fig. 4.11)

Let us now concentrate on a particular event, *E* at the space-time point near the *Here-Now*. Two observers, moving into their forward light cones, pass the event, *E*. One observer is at rest, and the other observer is moving with constant velocity relative to the first observer. Each observer must specify four data to describe the event. The stationary observer will use the coordinates (x, y, z, t), the moving observer will use the coordinates (x', y', z', t'). What events these observers judge as happening simultaneously is relative to their state of motion. (Fig. 4.12a,b)

As the simultaneity planes of the two observers do not coincide, the two observers do not judge the same events as being co-present with *E*. There is thus no absolute sense of *Now*. Hence the separation of the four-dimensional order into

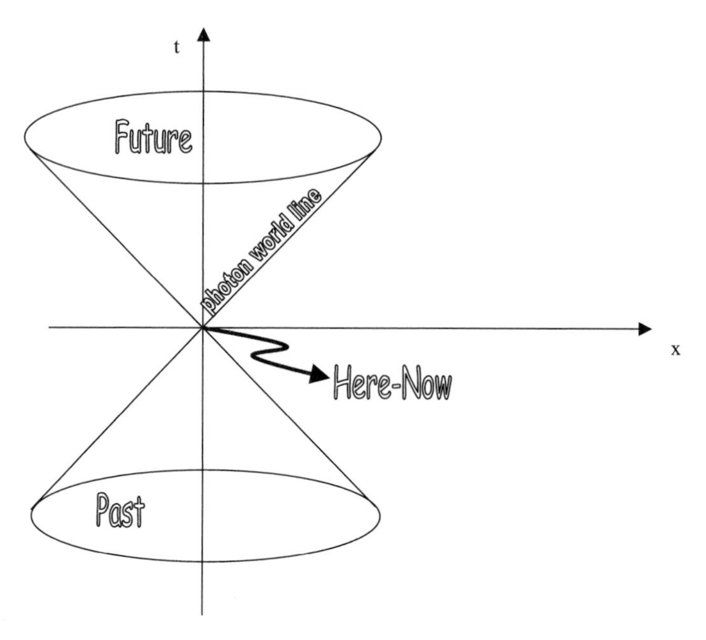

Fig. 4.11. The Light Cone Structure of Space-Time

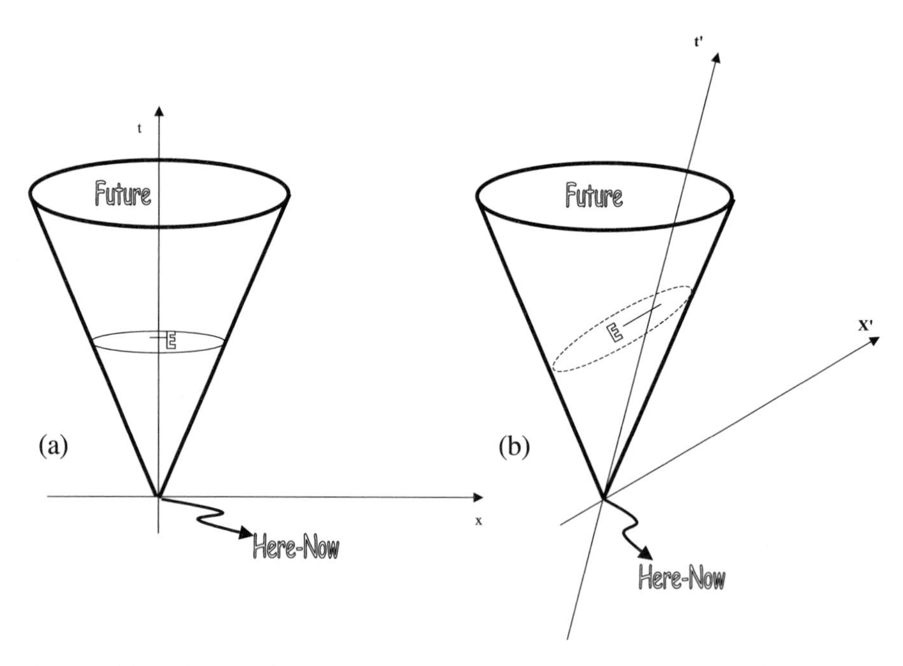

Fig. 4.12. (**a**) Stationary Observer and simultaneity plane to E. (**b**) Moving Observer and simultaneity plane to E

space and time coordinates depends on the state of motion of each observer. The lamination of space-time yields different *Nows*. Space and time seem to be relegated to the observer and the Special theory of relativity seems to vindicate the Kantian view that temporal and spatial judgements have their source in the minds of the observer.[84]

An idealist philosophy of time lay ready for the physicists to employ in the service of their purpose. The four-dimensional space-time continuum could be sliced at different angles, depending only on the state of motion of the observer. This was the consequence of the new notion of relative simultaneity. The separation of the natural world into spatial and temporal components lost its objective meaning. It was natural for the physicists to borrow philosophical elements from the larger cultural background to complete their relativistic worldview. Kant became the inspiration *for an idealist conception of time amongst physicists*. In fact, it is a logical consequence of the deterministic block universe. In this strange Kant-Einstein-Minkowski jigsaw, let us review the main tenets of an idealist philosophy of time.

4.3 Idealist Views of Time

4.3.1 Saint Augustine

Saint Augustin (354–430 AD)

Saint Augustine's reflections on the notion of time can be interpreted as an early conception of an idealist view on time. Saint Augustine was led to a discussion on the notion of time by reflection on such questions as "What was God doing before he made heaven and earth?" Saint Augustine rejects this question as nonsense because he argues that when there was no creation, there was no time. But this answer then leads to the question, "What is time?" Saint Augustine seems to have an interesting suggestion in mind: for there to be *time*, there must be some creation or, in other words, there must be some material events happening. So, where there is a total void, there is no time. God then is the creator of time for Saint Augustine. But this only answers part of the question: time is co-existent with creation but not co-eternal with God. Given that some form of creation (= material universe) is necessary for the existence of time, *what is time itself?* Saint Augustine gives a further clue: 'no time

[84] In his famous Cologne lecture on 'Raum und Zeit' (1909), §3 Minkowski had already emphasised that the *Here-Now* separates events into a 'before-after'-relation: events belonging to the past light cone necessarily occur *earlier* than events at *Here-Now*. Equally, events belonging to the future light cone necessarily occur *later* than events at *Here-Now*. So it is far from clear whether Minkowski would have accepted the block universe as a consequence of the Special theory of relativity, even if he accepted an idealist view of time.

is co-eternal with God' – he writes – 'because God never changes; whereas if time never changed, it would not be time.' (*Confessions*, Book XI, §14) This seems to imply that there is only *time* where there are *physical changes*. From this Saint Augustine concludes:

- If nothing had passed, there would be no past time
- If nothing were going to happen, there would be no future time
- If nothing were, there would be no present time.

So time seems to reside in changing events or, in other words, in the passage from future to past events. If the crucial aspect of time is the passage from future to past, when does time actually exist? The *past* no longer exists and the *future* does not yet exist. So maybe the only time, which really does exist, is the *present*.

But when is the *present*? As words cascade from a speaker's mouth, his words recede irretrievably into the past. Does the present than reside just in the words of the speaker as he utters them at the moment of speaking? No, even the utterance of one word takes time and involves a passage from the future to the past. Does the present reside perhaps in the utterance of just one syllable? But even this takes time. As long as some event has duration we can divide it into future and past. The present, then, concludes Saint Augustine, has no duration.[85]

Still, Saint Augustine continues, we seem to be able to *measure* time. We are aware of units of time or periods of time. But we cannot measure the past because it has gone, nor can we measure the future because it has not yet arrived. The conclusion is that we can be aware of time and measure it only while it is passing. (*Confessions*, Book XI, §16) But how do we measure the present time, if it has no duration (§21)?

Saint Augustine dismisses Plato's suggestion that time is constituted by the movement of heavenly bodies, because time continues even when there is no such movement. He appeals to a passage in the Bible (Joshua 10:13) according to which 'the sun once stood still in answer to a man's prayer so that he could fight on until victory was his, the sun indeed stood still but time continued to pass.' (§23) In addition, we always measure the movement of bodies *in* time, never time by reference to the movement of bodies, including the sun. We seem to have a problem: we cannot measure the duration of moments in the past or future because they have either gone or not yet arrived. Nor can we measure the present because it has no duration. Yet we do measure time.

Saint Augustine was right. During his lifetime the passage of time was measured by the use of shadow clocks and water clocks (clepsydras). The mechanical clock was not to be invented for almost another thousand years. Its uncertain origin is dated between 1271 and 1300. Yet Saint Augustine's solution to his puzzle makes no appeal to shadow or water clocks. He arrives at the conclusion that time is measured in each person's mind. The measurement of time is not objective. Things and events pass from the future into the past. What is measured are the impressions, which

[85] This is of course very much a mathematical idea of duration: between any two numbers we can insert other numbers, as long as we allow a sufficient rich set of numbers.

things and events that pass leave on the perceiver's mind. So we do not say, 'past time is long', because the past no longer exists. We say that a long past is a long remembrance of the past. And a long future is a long expectation of the future. The duration of the present moment is captured by the attention, which the mind devotes to it (§28). Saint Augustine uses the analogy of a passing sound – a noise – to illustrate the *passage* of events from the future to the past. Imagine you stand on a hill overlooking a tennis court some distance away. Two tennis players are hitting the ball across the net. You will *see* them hitting the ball before you will *hear* the hitting of the ball. Sound travels at a much slower velocity than light. When you see the ball being hit, the sound still lies in your future. You cannot measure it. When the sound waves reach you, you can register the sound. Then the sound passes into the past. You will no longer be able to measure it.

Saint Augustine's *idealist* view of time has severe limitations.

1. Time has no objectivity. There is no clock time. Yet Saint Augustine starts his reflections with the observation that there can only be time where there is physical change. Events must be ordered according to a 'before-after'-relation. Saint Augustine's idea that time resides in changing physical events anticipates Leibniz's *relational* theory of time. According to the relational view, as we have seen, time is the order of succession of events. The human mind records the succession of these physical events and thus forms a notion of time. Saint Augustine's view can be described as a relational-idealist view of time. He shares with Leibniz the view that events are ordered according to an objective 'before-after'-relation. This relation does not depend on human awareness. But unlike Leibniz he lapses into idealism because he surreptitiously replaces the 'before-after'-relation by the 'past-present-future' relation. The latter depends on human consciousness because it depends on conventional units of time. But in this respect Saint Augustine also differs from Kant. Saint Augustine makes the awareness of the 'past-present-future' relation dependent on individual minds, while Kant builds it into the structure of the human mind.
2. Saint Augustine's insistence that the passage of events from future into past leaves impressions on the minds of individual perceivers makes him the originator of the concept of *psychological* time.[86] Modern psychology has studied the psychological experience of time as a function of both the influence of a person's age and the influence of chemical substances. This is an interesting discovery but it does not answer Saint Augustine's original question, 'What is time?'
3. Saint Augustine's assumption that time is mathematically divisible into infinity has been questioned by modern physics. There may exist a shortest moment of time – the Planck time. We do not have to accept Saint Augustine's conclusion that the present is durationless, and hence that time has no objective existence. Saint Augustine's finding that the present is durationless is based on the mathematical idealization of the infinite divisibility of time. But this is not an actual

[86] Psychologists have found that the subjective impression of temporal duration can be influenced by drugs and by the complexity of the signal, see Ornstein, *On the Experience of Time* (1969); Fraser *ed.* (1966), Part III; Treisman, 'The Perception of Time' (1999)

characteristic of physical time. It is thought that there exists a minimum physical amount of time, i.e. a *chronon*. This sets the stage for temporal atomicity. Just as there can be no continuous amount of energy but quanta of energy, so there can be no infinite divisibility of time. There is a quantum of time whose magnitude has been calculated to be 10^{-23} or 10^{-24} seconds. This would be the shortest possible time.[87] Latest theoretical calculations in a field called *quantum gravity* put an even shorter limit on the quantum of time and add that there is also a quantum of space.

Saint Augustine is led by logical reasoning to the conclusion that time cannot exist as a feature of the physical world. Yet there are objective elements in our understanding of time, which cannot be regarded as purely conventional.[88] Thus it is purely conventional that the day has 24 hours or that some months have 30 days. But apart from these *conventional* units of time there are *natural* units of time, which are independent of human conventions. The longest and shortest days of the year, the summer and winter solstice, the orbit of the earth around the sun provide natural units of time, on which many of the conventional units of time are based. Saint Augustine admits that there is a succession of events stretching from the future into the past (§27) and this leaves an impression on the mind. This succession of events is objective. However this does not give us a metric to measure the duration of time between two events. Saint Augustine suggests that the mind provides such a time metric. This choice is conventional, since even in Saint Augustine's time water clocks and shadow clocks existed which could have given a rough measure of the duration between two events. These were very primitive clocks but their accuracy would have been better than the attention, memory and expectation, which serve as metrics in Saint Augustine's philosophy of time. The choice of the metric is purely conventional but metrics differ in their accuracy. What is not conventional is the occurrence of the events and that they are separated by intervals between the earlier and the later event. Saint Augustine makes an implicit distinction between two relations. There is the 'before-after'-relation: there is change in the physical world. This material change affects the human mind in that we become aware of time. The *past-present-future-relation* requires human awareness. It belongs to the domain of human conceptualizations. Saint Augustine locates this relation in

[87] We arrive at the shortest possible time span by dividing the shortest natural length, 10^{-15}, which corresponds to the diameter of the proton or electron, by the speed of light in vacuum, i.e. 3×10^8 m/s. For more on the quantum of time, see Salecker/Wigner, 'Quantum Limitations of the Measurement of Space-Time Distances' (1958); Campbell, 'Time and Change' (1921), 1106 and 'Atomic Structure' (1921). Some physicists express doubts about the existence of the *chronon*, see Davies, *About Time* (1995), 187; Čapek, *Philosophical Impact* (1962), 230–8, 386 discusses some of the conceptual difficulties of a 'chronon theory'. According to recent calculations the quantum of time may be as short as 10^{-43} seconds. The essential aspect in this approach to quantum gravity is the *discreteness* of time and space at the Planck scale, see Smolin, *Three Roads* (2001), 62–3 and Chap. III.

[88] See van Fraassen, *Introduction* (1970), 76–7; Zwart, *About Time* (1976), 4; Čapek, 'Time in Relativity Theory' (1966), 434–54 and 'Doctrine of Necessity' (1951)

the human mind. As Saint Augustine treats time as an extension of the individual human mind, his notion of time becomes subjective. Saint Augustine's notion of time is a *subjective-idealist view of time*, since time is measured in individual minds as a result of impressions left on them through passing events. He does not conceive of the notion of *physical time*, which became prominent with Galileo and found its ultimate expression in Einstein's Special theory of relativity. According to the notion of physical time, time is what our clocks measure. This is not a fully-fledged notion of time, since it leaves out both psychological and social aspects of time. But it is a thoroughly objective notion of time.

4.3.2 Immanuel Kant

Immanuel Kant (1724–1804)

Kant defends an *idealist* view of time but not in the subjective sense, in which Saint Augustine upheld it. According to Saint Augustine, time can only be measured if the mind (of the individual) can record an impression left on it by the passing of external events. But he failed to explain how the mind could be an accurate chronometer for the external order of physical events. He was a pioneer of the study of *internal* time.

Kant drew a distinction between the temporal order, in which objects of the external world appear to the human mind, and the order of the noumenal world. Of the latter he simply denies that we can have any knowledge – the noumenal world is not knowable to us.

> If we abstract from our *mode* of inwardly intuiting ourselves (...) and so take objects as they may be in themselves, then time is nothing. It has objective validity only in respect of appearances, these being things, which we take *as objects of our senses*. It is no longer objective, if we abstract from the sensibility of our intuition, that is, from that mode of representation, which is peculiar to us, and speak of *things in general*. Time is therefore a purely subjective condition of our (human) intuition (which is always sensible, that is, so far as we are affected by objects), and in itself, apart from the subject, is nothing. Nevertheless, in respect of all appearances, and therefore of all the things which can enter into our experience, it is necessarily objective.[89]

Kant denied, against Leibniz, that the concept of time was derived from experience, and he denied, against Newton, that time had any claim to absolute reality. In his view, the concept of time does not inhere in objects but merely in the subject, which intuits them. Time and space are pure forms of intuition.[90] This position leads to an *objective* idealist view of time.

---

[89] Kant, *Critique of Pure Reason* (1781/1787), B51/A35

[90] See Whitrow, *Natural Philosophy of Time* (1980), 49–51. Karl Pearson, *The Grammar of Science* (1892) adopts a Kantian viewpoint when he declares that both time and space are modes of perception. 'Of time as of space we cannot assert a real existence; it is not in things, but in our mode of perceiving them.' Pearson, *Grammar* (^{3}1911), 211

Kant proposed his view as a solution to the puzzle of time (and space) because he found both the Leibnizian and Newtonian views unsatisfactory. The Newtonian view gave rise to the antinomy of time (*Critique* B454). The Leibnizan view presupposes what it claims to demonstrate.

Leibniz gave an empirical explanation of how the notion of space and time originate. Men consider that 'several things exist at the same time and they find in them a certain order of co-existence.'[91] By implication, when they observe 'this order in relation to the successive position of bodies' [92], they form the notion of time.

Kant made some fundamental objections against empirical notions of time and space, derived from experience (*Critique* B37–53). He says that the experience of things or objects in a spatial and temporal order already presupposes the availability of the notions of space and time. We can imagine a *space* without objects but no objects without spatial arrangements. Equally we can imagine *time* without events but no events without temporal arrangements. We have no empirical experience of time and space per se; 'they are not objects of experience but "transcendental conditions" which make experience possible'[93] Time and space are necessary conditions *a priori* of the possibility of experience. Furthermore, time and space are not general concepts. Kant says that we represent to ourselves only one space (*Critique* A25) and one time (A32, A189).[94] Time and space are pure forms of intuition.

Kant's position amounts to two claims regarding time:

- The affirmation of the *empirical* reality of time. By this he means the objective validity of time in respect of all objects, which may be given to our senses. All objects we experience, we experience as temporally ordered.
- The denial of all claims as to the absolute reality of time. Kant denies that time belongs to things absolutely or is a property of things. His main reason is that time could not be observed as a property of things (A36), nor in itself (B56).

Kant observes that if humans are deprived of their sensibility, then the representation of time vanishes. This is not only an idealist view of time. It is an *objective* idealist view of time. According to Kant time does not inhere in the objects but merely in the subject, which intuits them (*Critique* A35). But *all* human subjects possess *this* form of intuition. It is built into the structure of the human mind.

---

[91] Leibniz, *Philosophical Writings* (1973), 230

[92] Leibniz, *Philosophical Writings* (1973), 220, cf. Mach, *Mechanik* (^{9}1933), 217

[93] Čapek, *Philosophical Impact* (1962), 48

[94] It is good to remember the Newtonian background to Kant's philosophy. Kant denies that time and space have absolute reality, in the Newtonian sense. Newton held that spatial events took place within some *spatial container*, called *absolute space*; and temporal events can be measured with respect to a *river of time*, which Newton calls *absolute time*. According to Newton both absolute time and space exist irrespective of the occurrence of spatial and temporal events. Kant shares with Newton the pre-relativistic idea that there is only one space and one time. What Kant denies is that time and space exist irrespective of the existence of human beings.

Kant speaks of time or the representation of time. But how is time to be *represented*? It is clear that he excludes the subjective experience of time, as Saint Augustine defended it. But does the representation of time mean *clock* time, *human* time or merely some ordering of events according to the 'before-after'-relation? Clock time and human time have conventional aspects and are based on natural units of time. If time and space are pure forms of intuition, does this simply mean that a metric of time could not exist without human awareness? Or are we also to deny that natural events are subject to a 'before-after'-relation irrespective of human awareness?

We can only have knowledge of the external world as it appears to the human mind. And the human mind processes the appearances through the categories, which ensure that subjective sense impressions become objective knowledge. Instead of claiming that time is nothing with respect to things in themselves Kant's position would have required of him to remain agnostic. Natural events in themselves *may* be subject to a 'before-after'-relation, but within the Kantian viewpoint, humans cannot know. But even the representation of time involves a genuine 'before-after'-relation. A temporal order is produced through the application of the category of relation [including *substance*, *causation* and *reciprocity*] to the appearances. In pre-relativistic fashion, this temporal order is independent of the reference frames of different observers.[95] This application of the category of relation is to be understood as follows.

- All appearances of succession in time are only changes in the determinations of substance, which itself abides (A189/B233).
- All changes occur according to the law of causation, that is of the connection of cause and effect (A189/B233). For Kant the law of causation is a rule, by which we determine something according to the succession of time. The rule states: the condition under which an event invariably and necessarily follows is to be found in what precedes the event (B246).
- Finally, things can appear to humans as existing simultaneously, only insofar as they interact with each other. Again, Kant would conceive of this simultaneity in the pre-relativistic sense of absolute simultaneity.

Kant's view of time is not *clock* time, nor is it *human* time in the sense of calendar dates. It is an objective, mind-dependent metric of time, as the temporal order of appearances is produced, objectively, through the application of the category of relation.

Kant thought that the impossibility of perceiving both spatial and temporal events without some pre-existing spatial or temporal form dealt a blow to Leibniz's empiricist explanation of the origin of these notions. He also considered that his idealist view of time was confirmed by a consideration of the application of the concept of time to the whole universe. Or by the treatment of time as a cosmological concept, which far exceeds our experience. (Although Kant makes no reference to

---

[95] Mittelstaedt, *Philosophische Probleme* (1972), 37; Kant, *Critique* B218ff (analogies of experience)

Newton in this connection, this could be a criticism of Newton's absolute time and space, understood as metaphysical concepts. Kant states repeatedly that 'time cannot be perceived in itself.' [*Critique* B233]) Kant tried to show that a metaphysical notion of time involved us in *antinomies*. (Antinomies are contradictions between two theses, which in themselves are coherent. Only when conjoined do the theses produce a contradiction.) To give a flavour of Kant's reasoning consider both the *thesis*

- 'The world has a beginning in time, and is also limited as regards space' (i.e. finite in time and space),

and the *antithesis*

- 'The world is infinite in both time and space' (*Critique* B454ff).

Both can be consistently established in separation, but in conjuction. Kant concluded that the concept of time was inapplicable to the whole universe, insofar as this transcends our experience.[96]

Kant's idealist view of time was motivated by his rejection of an empirical explanation of the genesis of the notion of time (Leibniz) and by a consideration of the inconsistencies, which arise when we deal with purely metaphysical or transcendent notions. Leibniz and Mach saw in Newton's notions of absolute space and time only untestable speculations. Kant seems to have followed Leibniz in this interpretation. This then led to his thesis that time and space were pure forms of intuition. In his *Principia Mathematica* (1687) Newton is at pains to stress that his notions of absolute space and time are subject to empirical testing. They do not lead to direct empirical tests but they can be indirectly inferred from Newton's thought experiments. So Newton claims. We have already seen that instead of dismissing Newton's notion of time as an idle speculation, there are good reasons for treating it as a philosophical presupposition for the formulation of his laws of motion. Kant's theory of objectivity is formulated in the framework of Newtonian mechanics, with its assumption of *absolute time* and *space*. Although Kant rejected the absoluteness of time, he shared with Newton and all pre-relativistic thinkers the idea of the universality of time. We represent to ourselves only *one* time and *one* space. Einstein has taught us that this is the result of our myopia. We experience a world of macro-objects. We deal with everyday experiences. The world of macro-objects does not reveal the relativity of spatial and temporal intervals. Einstein's discovery that observers measure different temporal and spatial intervals, depending on their reference frames, was taken by many *relativists* (as the proponents of the relativity theory were once called) as further evidence of the correctness of the idealist view of time.

In a certain sense Kant is right that material things or phenomena do indeed not possess the temporal properties, which we ascribe to them. Phenomena do not carry dates and do not possess the properties of past, present and future. But irrespective

[96] See Whitrow, *Natural Philosophy of Time* (1980), 27ff; Falkenburg, *Kants Kosmologie* (2000)

of these temporal properties, appearances stand in temporal relations of succession with respect to each other: some appearances are *before*, some appearances are *after* others. If we replace Kant's empiricist terminology of appearances and phenomena by a realist terminology of material objects, or space-time events, then the relation of succession is independent of our perception or awareness of this temporal relation. *World lines have histories.* In this sense, time does not vanish when human consciousness vanishes. According to our interpretation Kant's idealist view is equivalent to the refusal to accept the succession of events ('before-after'-relation) in the material world as constituting time.[97]

4.3.3 Transience: On the Passage of Time

> *The passage of time is relative; it depends on the observer's velocity.* (*Scientific American*, January 2000)

Saint Augustine has defined the *philosophical* problem of time. Logical reasoning about time seems to lead to the conclusion that time cannot exist. Saint Augustine recognised that where there were events, there was time. John E. McTaggart produced a famous argument, and distinction, according to which he denied the reality of time. According to McTaggart's idealist view of time it makes no sense to attribute the properties 'past', 'present', 'future' to physical events. Hence time is unreal.[98] Here again we encounter a static view of time: events are ordered according to the 'earlier-later'-relation. Events do not change with respect to being past, present, future.[99]

McTaggart argued that there were two ways of talking about events in time. A sequence of events ordered according to the 'earlier-later'-relation (or 'before-after'-relation) will be called *B-series*. A sequence of events ordered by the degree to which they are future, present or past will be called *A-series*. It is obvious that the *A*-series depends on human conceptualisation. An *A*-determination of an event ascribes a position in the *A*-series to a particular event – it ascribes a date to this event. *A*-statements, unlike *B*-statements, have different truth-values at different times. For instance, it will be true *today* to say that Scott will come tomorrow. But tomorrow it will be false to say that Scott will come (the day after) tomorrow. These changing truth-values do not affect *B*-statements. If *A* comes before *B*, then it is logically impossible to reverse this order.[100] So we have a contrast between two different views of time.[101]

---

[97] It is a matter of dispute whether Kant's view of time contradicts the Special theory of relativity, as Reichenbach affirms, or whether it is compatible with it, as Cassirer claims.

[98] McTaggart, 'The Unreality of Time' (1908)

[99] See Seddon, *Time* (1978), §§1, 2, 6; for further discussion, see Zwart, *About Time* (1976), 44–47; Whitrow, *Natural Philosophy of Time* (1980), 345–8, 370–1; Schlesinger, *Aspects of Time* (1980), Chap. 2,3; Horwich, *Asymmetries in Time* (1987), 18–25; Nahin, *Time Machines* (1993), 72–4; Shimony, 'Implications of Transience' (1998); Weingard, 'Space-Time' (1977)

[100] According to the Special theory of relativity this claim is restricted to *time-like* separated events.

[101] Seddon, *Time* (1978), §§7, 1

Tensed View of Time	Static View of Time
Events change with respect to being past, present, and future. As a consequence of changing their temporal properties, they change their *A*-determinations.	Events just are. They do not become past, present, and future. They always retain their particular *B*-determinations with respect to all other events.

But according to McTaggart[102], the belief in temporal properties – that, for instance, 'past' is a predicate of events – is incoherent. Temporal properties are mutually exclusive and yet every event possesses all three at different times. Every event at one point is future, then present, then past. This leads to an infinite regress: we can always ask, 'At what time does an event possess these properties?'

There is no change in the *B*-series: events will forever be ordered according to the 'before-after'-relation. McTaggart accepts that time requires change. If time is real, events must change. But there is no change in the *B*-series. So events must change with respect to *A*-determinations. But it is incoherent to believe that events change with respect to *A*-determinations. So time is unreal.

These attempts are unsatisfactory, since apart from arguments of logical consistency they do not pay heed to the constraints from science. Saint Augustine reduces time to *psychological* time. McTaggart's argument builds on the ambivalence between the 'before-after'-relation and the 'past-present-future'-relation. We may hope to say something about *physical* time and *philosophical* time. Psychological time is just the way the passage of time appears to the individual and this may depend on psychological and biochemical factors. Physical time is clock time according to Galileo and Einstein. Both notions are grounded in physical events. Galileo measured time intervals for balls rolling on inclined planes against the amount of water that would flow from a container during the ball's trajectory. Einstein used the transmission of light signals to synchronise clocks between distant observers.

Can the past, present and future exist as well as the transitions between them, without the presence of human consciousness? Or is this situation similar to the one in which a physical event occurs – say the emission of waves of a certain range of wavelengths – which only becomes a certain type of event – 'music' – because of the presence of creatures equipped with higher forms of consciousness? In other words: Is the *passage of time* a primary quality in the physical world or a secondary quality in the mind of the observer?

The *passage of time* is a beguiling image. The image suggests that there is an entity called *time*, a stream, which undergoes a continuous movement into the future, leaving a riverbed of records in the past. In this image time is like an eternal river, which marks the passage of time. This is essentially the Newtonian metaphor: absolute, true and mathematical time, in Newton's enduring phrase, 'flows equably without relation to anything external.' In the Newtonianimage, time is a primary

[102] Seddon, *Time* (1978), §9, see also Nahin, *Time Machines* (1993), 73; Schlesinger, *Aspects of Time* (1980); Horwich, *Asymmetries in Time* (1987)

quality. The universe has a clock. For Leibniz and Mach, too, time is a primary quality, though not in the Newtonian sense. For Saint Augustine and Kant time is a secondary quality. In the Special theory of relativity the passage of time became inextricably entwined with the state of motion of an observer, more precisely of a clock, attached to a reference frame. There are many rivers and they do not flow 'equably'. In fact, the block universe interpretation seems to suggest that there is no flow at all. The physical world is static, and the observers 'create' time by slicing the world block according to their coordinates. In the Minkowskian image, time is a secondary quality. But the Special theory of relativity is fundamentally committed to the view that there is change in the physical universe. For physical events propagate at a finite velocity, which is always smaller than the velocity of light, *c*. The propagation of physical events at a limited velocity means that there is a succession of events in the physical universe, which is governed by the 'before-after'-relation. *The very assumption of the finite velocity of light in the Special theory of relativity seems to contradict the idea of the block universe.* Saint Augustine's fundamental philosophical question: 'What is time?' and his inability to give a satisfactory answer revolves around the distinction between the 'before-after'-relation and the 'past-present-future'-relation. The Leibnizian identification of time with the succession of events seems to commit us to the *minimalist* notion of time based on the 'before-after'-relation. Time is the order of successive events. Then the perception of becoming and *transition*, the passage from past to future, seem to require human awareness. When time is associated with human awareness of time, as in the idealist view, the familiar temporal aspects of time – *past*, *present* and *future*, *transition* and *becoming* – seem to constitute what is meant by time. Conceptual awareness divides events into *past*, *present* and *future* and offers the image of a *passage* of time from past to future. In this image time is a secondary quality. But the passage of time can be construed as a primary quality. If time is a succession of events in the physical world, a change from an earlier to a later state, then this change of physical events constitutes the passage of time. And human conceptual awareness only helps to differentiate these events first into past, present and future and further into calendar dates. The events themselves do not change with respect to being future, present and past. The date at which an event occurred is not a further physical property, which must be specified. Events carry no dates.[103]

In the absence of any conceptualised awareness, events just seem to happen. They seem to be related to each other along an 'earlier-later'-line. Event *A* is earlier than event *B*, which is later than event *C*. On such a scale, an event being much later than some earlier event means that between these two events many other events have happened – or t_2 is not immediately preceded by t_1. The existence

[103] See Smart, 'Time and Becoming' (1980), 11 for further discussion of the indexical character of the terms 'future', 'present', 'past'; the concepts are anthropocentric in the sense that they have significance only with respect to human thought. By contrast, terms like 'earlier' and 'later' and 'simultaneous' are devoid of all traces of anthropocentricity and apply to the physical universe. See Smart, 'The Space-Time World' (1963), 131–148; also Smart, 'The River of Time' (1949), 483–94. For a modern discussion of these questions, see Le Poidevin, 'Time, Tense and Topology' (1996), 467–81

of natural clocks would give an indication of how much earlier. For instance, the construction of the Egyptian pyramids is separated from the construction of the Louvre pyramid by so many revolutions of the earth around the sun. There is no need for a conscious observer. There are already natural clocks in the universe, which order objects according to the 'before-after'-relation. McTaggart pointed at a fundamental flaw in the identification of time with the 'before-after'-relation. It seems to result in a *static* view of the universe.

Does the 'before-after'-relation necessarily imply a world of *static* being? The 'before-after'-relation gives us a tenseless way of speaking about the world. If event E_1 is *before* event E_2, then it will remain so to eternity. But this does not exclude physical change and becoming in the universe. Recall that world lines constitute the history of events, and that events experience *transitional states*. For a proper appreciation of the block universe, it is important to realise that 'timelessness is not the same as tenselessness.'[104]

The 'before-after'-relation between events has often been associated with a *static* view of the universe, as it is encapsulated in the idea of the block universe. This may be an impoverishment of a physical view of the universe. Are present things not also influenced by their past such that their past may at least partly affect their present state? Think of evolutionary accounts of nature. Are there no *transitional* phenomena in the physical world?[105]

All events are related by a 'before-after'-relation but this does not preclude that some events are related by much stronger links than 'before-after': causal links, energy links, functional links – all these tell us more about how two or several events are related than the 'before-after'-relation. These are structural links existing in the natural world. Leibniz, Kant and Mach may be right that there is no entity called *time*, which, like a river, flows, independently of actual events. Nor are dates physical properties, which events possess. McTaggart has shown that this assumption leads to inconsistencies. This does not exclude the possibility of an *intermediate* position. It denies that the 'before-after'-relation must lead to a *static* picture.

If this is the case then it is important to realize that *transitional changes* between system states are an important feature of physical and biological systems. The past state of a system may have a significant influence on its present state. Kant's evolutionary theory of cosmic history argues this point. It assumes a deterministic evolution of cosmic history from chaos to order through the workings of mechanical laws. In today's cosmological theories, the observable pattern of galaxies and their dynamic evolution (Hubble's law) is projected backwards in time to a Big Bang event, followed by an inflationary expansion. This raises the fundamental question of the arrow of time. Darwinian evolution is a statistical model of how random genetic changes in an organism and natural changes in the environment will favour some

---

[104] Smart, 'The Space-Time World' (1963), 139; see also Shimony, 'Implications of Transience' (1998)

[105] Benjamin, 'Ideas of Time in the History of Philosophy' (1966), 21 and Whitrow, *Natural Philosophy of Time* (1980), §§7.5, 7.10 argue that the very essence of time lies in its transience or that time has a transitional nature.

species and disfavour others. The present state of the organism will have a significant impact on its future success in an ecological niche. So the 'before-after'-relation can be enhanced by an incorporation of *transitional changes* without yet invoking human awareness. Would there be an indication in such systems that a present state is a *transitional state*?[106] There are many transitional states in Nature (in astronomy, physics, and biology) whose nature is governed by laws, which operate independently of our knowledge of them. The history of physical systems leaves *traces*. A biological system grows from infancy to maturity to death. Its present state is shaped by its past state and its interaction with the environment, and it evolves towards a future state. Its present state is marked externally by a date, imposed by human convention, but also internally by its physical and biological condition. Its transience is a function of its internal conditions. Its past conditions projected it into its present condition; its present conditions will project it into some, not completely determined, future condition. So there need be no conscious anticipation of the future for its present state to be transitional. The recognition, noted earlier, that world lines evolve along a history, includes such a transitional view. At any point in time, the state of a physical system is in a transitional state from an earlier to a later state. Only when natural systems suffer a maximum of entropy or a biological death, does the transitionality cease to exist. The question is whether the transition between states is completely determined by its earlier state. If it is its trajectory into the future will then be completely deterministic. It would be tempting to conclude that the future is already there: in Eddington's words that physical reality is a block universe. However, we have already encountered one reason to assign a *history* to the world lines of particles: the very assumption of the finite propagation of material events in the Special theory of relativity seems to contradict the idea of a static, timeless universe. The succession of events or the history of world lines already establishes time, according to the relational theory. If the succession of events is deterministic, at least we get a *dynamic*, not a static form of determinism. The proponent of the block universe may derive comfort from the Laplacean demon: a sufficiently capable mind will *see* the past and the future stretched out to eternity, as if it already existed. Eddington and Jeans, as we observed, explicitly introduced a deterministic assumption into the view of the block universe. Quantum theory, however, seems to suggest that the trajectory of events is not completely determined by their past. If objective indeterminism lies at the root of the physical universe, then there is room for objective becoming, novelty and transience.[107] We have to gain a better grasp of the idea of the block

[106] Collingwood, *Autobiography* (1939), 98 characterizes transitional states in history in the following sense: 'If P_1 has left traces of itself in P_2 so that an historian living in P_2 can discover by the interpretation of evidence that what is now P_2 was once P_1, it follows that the 'traces' of P_1 in the present are not, so the speak, the corpse of a dead P_1 but rather the real P_1 itself, living and active though encapsulated within the other form of itself P_2.' H. Bergson stressed the dual feature of presentness having novelty and a dynamic cohesion with the anterior phases; see Čapek, *Philosophical Impact* (1962), 338–41; see also Popper, *Quantum Theory* (1982), 185f on transitions between quantum states without observers.

[107] Čapek, *Philosophical Impact* (1962), 333–8; see also Box 4.3 on determinism

universe by looking at the conceptual tool, which enabled its renewed flourishing: the Minkowski *space-time* interpretation of the Special theory of relativity.

4.4 Minkowski Space-Time

> Spacetime contains a flowing "river" of 4-momentum.
>
> Ch. Misner et al., *Gravitation* (1971), 130

Einstein destroyed the notion of absolute time: there is no preferred reference frame in which the *correct* time for the duration of events can be read and in which two events are *really* simultaneous. Einstein made the existence of time dependent on the existence of measurable events – *physical time* is *clock time*. But it was the German mathematician Hermann Minkowski who, in a lecture in Cologne in 1908, drew the most radical conclusion for the understanding of space and time. As he stressed, his conclusion was based on experimental evidence and led to a completely new view of space and time. In a famous phrase Minkowski announced that the notions of space and time, which hitherto had been separated (even in Einstein's famous 1905 paper on relativity), were to be united into a notion of *space-time*, a notion that alone was to enjoy independence.[108] Many prominent physicists then used this notion of space-time, as we have already seen, to defend a *static view of the universe* – or what has become known as the *block universe*. The notion of the block universe commits its proponent, as we have argued, to an idealistic view of time. Kant's views on space and time, which existed in the philosophical heritage of the early 20th century, gained renewed prominence when the physicists drew what they took to be philosophical consequences of the discoveries associated with the Special theory of relativity. The objective, idealist view of time, worked out by Kant, seemed to fit like a philosophical glove to the notion of a block universe, inspired by Minkowski's union of space and time.

We have already given some graphical representation of this new notion of *space-time*. Equipped with some further Minkowski insights, we can complete this picture.

According to the classical physicist, observers in two different reference frames, moving uniformly relative to each other, will assign different space co-ordinates but the same time co-ordinate to the same event. Mr. Tompkins, enjoying dinner on the train, will assign as space co-ordinates of this event the co-ordinates of the carriage in which dinner is being served. Mr. Tompkins, say, sits at the table opposite the bar. He will remain at this table during his dinner. The table remains at the same place in the train carriage with respect to the bar. For an observer on the embankment, however, Mr. Tompkins will begin and finish his dinner at different

[108] Minkowski, 'Raum und Zeit' (1909/1974), 54. The idea was born before Minkowski – an article in *Nature* **106** (1921), 693 attributes the idea to the Irish mathematician Hamilton. Galison, 'Minkowski's Space-Time' (1979) describes the emergence of Minkowski's space-time idea.

spatial locations – the locations, which the train covers during its journey, while Mr. Tompkins is having his dinner. Even for Mr. Tompkins the spatial co-ordinates could change if he chose to measure the place of his dinner in different reference systems. He would eat his soup and his dessert at the same table in the dining carriage but at widely separated points of the railway track.[109] How long does the dinner last? Let us say Mr. Tompkins spends one hour at his dining table. Observers positioned on the embankment who time Mr. Tompkins's dinner will measure the same time interval between soup and dessert as Mr. Tompkins.

However, according to the Special theory of relativity, temporal and spatial intervals change when we pass from one reference system to another. In classical physics it was sufficient to have one time co-ordinate, t, and a set of changing space co-ordinates $(x, y, z; x', y', z')$, depending on the direction in which the object was moving. But in relativistic physics, both temporal and spatial co-ordinates change, so that the classical asymmetry between time and space co-ordinates disappears.[110] Any description of any event requires, not only spatial co-ordinate set $(x, y, z; x', y', z')$ but also temporal co-ordinate sets (t, t').

Minkowski calls a spatial point existing at a temporal point a *world point* (x, y, z, t form its value system). These coordinates are now called 'space-time coordinates'. The collection of all imaginable value systems or the set of space-time coordinates Minkowski called *the world*.[111] This is now called the *manifold*. The manifold is four-dimensional and each of its space-time points represents an *event*. Minkowski considered that at each world point, there existed a 'substantial point'. But this is misleading. Two light beams might interact and this would be an event in space-time. Idealising even further, we could regard events as locations of possible happenings.[112] We want to be able to recognise an event at every moment of its existence. Such an event will change in its spatial and temporal coordinates (these changes are marked by dx, dy, dz, dt). As a picture of the 'eternal existence' of such an event, as Minkowski says, we obtain a *world line*. If we do this for every event,

[109] Gamow, *Mr. Tompkins* (1965), 15–6; Langevin, *La Physique Depuis Vingt Ans* (1923), 218, 310, 334; Einstein/Infeld, *Evolution of Physics* (1938), 206–8

[110] Langevin, *La Physique* (1923), 279

[111] Minkowski, 'Raum und Zeit', (1909/1974) 55. Minkowski's vocabulary has changed since his time. For excellent modern discussions of the space-time world, see Holton, 'Metaphor' (1965); van Fraassen, *Introduction* (1970), 167–9; Sklar, *Space, Time and Spacetime* (1974), 56–61, 251–60; Galison, 'Minkowski's Space-Time' (1979); Sklar, *Philosophy of Physics* (1992), Chap. 2; Norton, 'Philosophy of Space and Time' (1992/1996), 179–232; Sexl/Schmidt, *Raum-Zeit-Relativität* (1978), Chaps. 7–9. An excellent early introduction to this topic can be found in Langevin, *La Physique Depuis Vingt Ans* (1923).

[112] Langevin, *La Physique* (1923), 274 defines an *event* as a space-time location at which 'something happens or something exists'. Modern physicists agree. In their classic text-book *Gravitation* (1971), 6, Ch. Misner, K. Thorne and J. Wheeler write: 'Characterize the point by what happens there! Give a point in spacetime the name "event".' See also Reichenbach, *The Philosophy of Space and Time* (1928/1958), §45. Friedman, *Foundations* (1983), 32 regards space-time events as the locations of actual and possible events. Idealizing further, Sklar, *Space, Time and Spacetime* (1974), 56 calls events 'the locations of possible ideal events.'

the whole world can then be resolved into such world lines. Minkowski adds that physical laws can be understood as interactions between world lines.

To every *event* is attached a reference system, i.e. a system of temporal and spatial axes.[113] It is customary to describe these features of space-time, *as if* to each reference system a group of observers was attached. (We shall follow this convention where it does not obscure the points.) But this is not necessary, and can even be misleading. The human observer can be replaced by ideal clocks and rods, without altering the result of the theory in any way. Then every event is determined, with respect to its position in space and time, by four co-ordinates given by a particular reference system (x, y, z and t). If we have two events attached to a certain reference system, then they will differ in space and time by differential spatial points, dx, and differential temporal instants, dt. To a couple of events correspond a spatial distance and a temporal distance.

Thus *time* can be defined by all the events, which succeed each other at a certain point in space-time. And *space* can be defined by all the events, which happen simultaneously on a plane parallel to the x-axis of the particular coordinate system. This definition reminds us of the relational theory of time and space, due to Leibniz and Mach, but with the important difference that the notions of absolute time and space have been abandoned.

Minkowski speaks of the *individualisation* of time and space.[114] How are the world lines of particles, as seen by various observers who introduce their *individualised* coordinate systems, to be represented graphically? The creation of individual coordinate systems is what Eddington called the lamination or slicing of the four-dimensional space-time world (Fig. 4.10). For an observer, in whose coordinate system a particle is at rest (a stationary event), the world line of the particle must run parallel to the time axis, t, always keeping the same x-value (the y- and z-axes are neglected in each case). For such an observer, a uniformly moving event is represented as a world line inclined at an angle from the t-axis. And an accelerated event is represented as a curved world line. (Fig. 4.5)[115]

However, we have left out an important detail: according to the relativistic understanding of the universe, no material particle can travel faster than the speed of light. This aspect must be incorporated into these *space-time diagrams*. As we have already seen, these diagrams are drawn with the speed of light as unity ($c = 1$). Then a distance of 300 000 km per second on the x-axis equals the extension of one second on the t-axis. This means that the world line of a photon is tilted away from the vertical time axis by 45°. Since c is taken to be a limiting speed, the collection of possible world lines must never tilt more than 45° away from the t-axis (Figs. 4.6, 4.7). If we add the other space dimensions and take the origin, $x = 0, t = 0$ of the diagram to be the *Here-Now* (from which world lines converge from the past

[113] Langevin, *La Physique* (1923), 274ff; Nahin, *Time Machine* (1993), Technical Note 307; Minkowski, 'Raum und Zeit' (1908), 56, 57f; Born, *Einstein* (1965), 251; Whitrow, *Natural Philosophy* (1980), §§6.1, 6.2

[114] Minkowski, 'Raum und Zeit' (1909/1974), 57

[115] Minkowski, 'Raum und Zeit' (1909/1974); 57f; Nahin, *Time Machines* (1993), Technical Note 4; Sklar, *Space, Time, Spacetime* (1974), 261–72

and diverge into the future), then we obtain three regions. The upward region is called the *future light cone* of *Here-Now* and the downward region is called the *past light cone* of *Here-Now*. But there is also a region, which cannot be reached from inside a particular light cone. It is called *Elsewhere*. This region cannot be reached by material particles or signals confined to their light cones, for material particles cannot cross the world line of the photons. This is a consequence of the Einsteinian postulate of the constancy of light. It has an important consequence for the *causal* connection of events. As every event can be reached by causal signals and can emit causal signals, every event in space-time has its own light cone (see Fig. 4.13).[116]

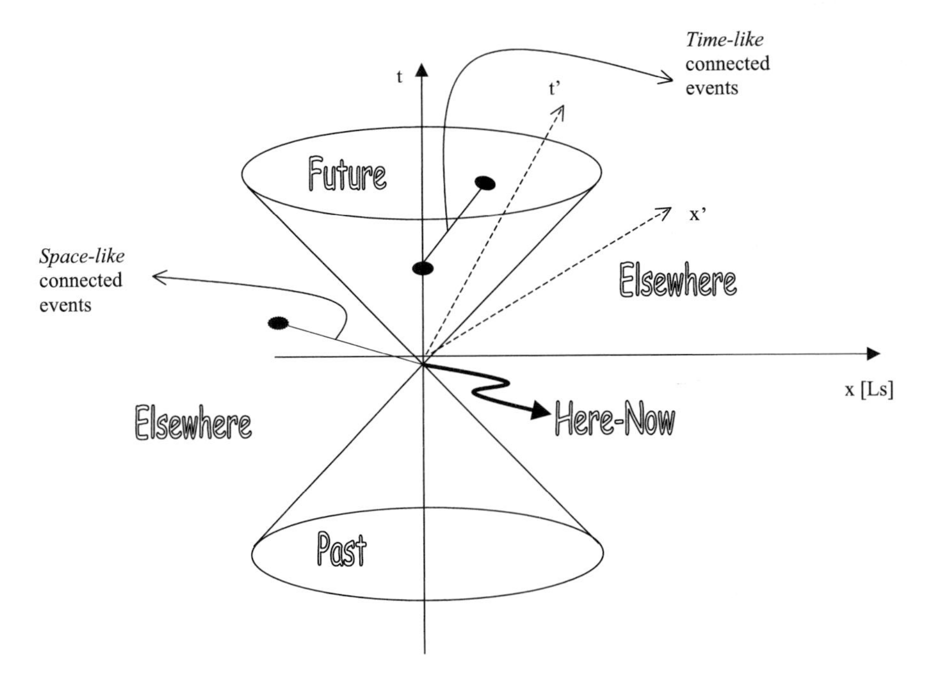

Fig. 4.13. Representation of future and past in Space-time diagrams, where *time-like connections* are possible from the *Here-Now*. The *Elsewhere* regions lie outside the reach of causal links and have a *space-like separation* from *Here-Now*

Events in space-time can be separated in two different ways:

1. When they have a *time-like separation*, the events are close enough together in space and far enough apart in time so that signals or particles propagating more slowly than the speed of light can get from one to the other and there is at least a potential causal link between them.
2. When they have a *space-like separation*, they are too far apart in space and too close together in time for any signal travelling at the speed of light or less to

116 Friedman, *Foundations* (1983), 126; Nahin, *Time Machines* (1993) 307

connect them. World lines connecting them would tilt at more than 45° from the vertical. They would represent world lines at speeds in excess of the speed of light, *c*. There could be no causal connection between them at least in any conventional sense.[117]

It is important to realise that where there is a *causal connection* between events (*time-like* connection), the order of events is the same for all observers; only the amount of time measured between these events will vary from observer to observer. The reason for this invariant order lies in the existence of the Lorentz transformations (Box 4.1). As we shall see below, this will become one of the arguments against

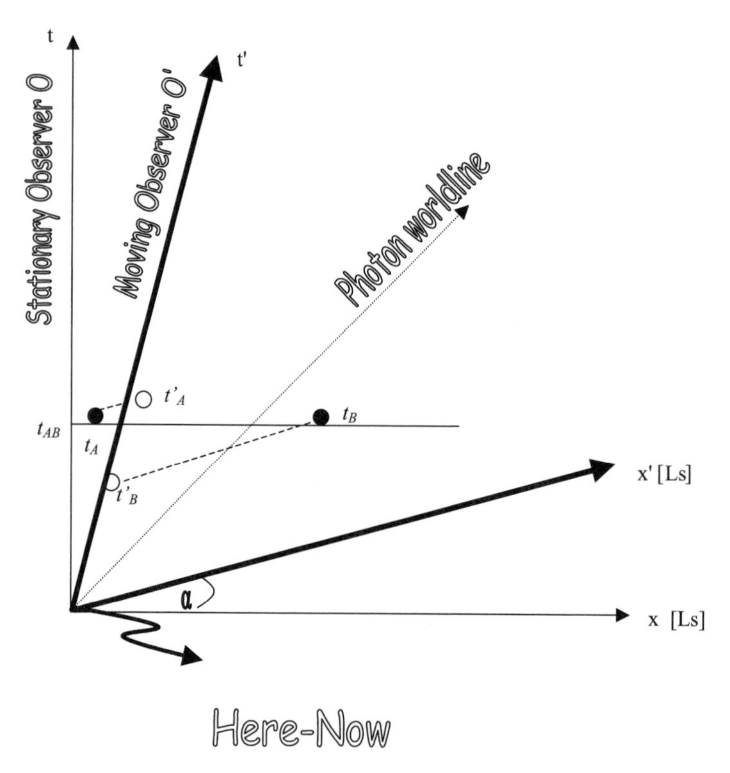

Fig. 4.14. The Relativity of Time Order for *space-like separated* events

[117] Nahin, *Time Machines* (1993), 306–16; Sklar, *Philosophy of Physics* (1992), 32f; de Beauregard 'Time in Relativity Theory' (1966), 426f; Čapek, 'Time in Relativity Theory' (1966), 440–3; Friedman, *Foundations* (1983), 159–65; Mittelstaedt, *Philosophische Probleme* (⁴1972), Chap. I. A typical example of a *space-like connection* comes from quantum mechanics. In quantum mechanical experiments certain spin correlations between pairs of particles are observed. For instance, in the simplest case, if particle *A* has spin up, then it will be observed that particle *B* has spin down. The spatial separation of the particle pair is however too great for causal signals to be exchanged between them, which could explain the 'spooky' distant correlations between their spin states.

the notion of the block universe. Only when there is a *space-like* separation between events, and hence no causal connection, can the time order be reversed for different observers (Fig. 4.14, Box 4.2).[118]

Box 4.2:

What is meant by *causation* here? As we shall see in the next Chapter, prior to the advent of quantum mechanics, physicists took causation to mean the predictability of events in the future. In other words, they made no distinction between causation and predictive determinism. Given the present condition of a physical system in which the events take place, and the knowledge of the general mechanical laws, which govern the system, it seems that the trajectory of the events can be predicted with reasonable certainty. The laws take the form of *differential* equations. We will call this as the *functional view of causation*. As we shall see, Laplace, who identified determinism and causation, popularized this view. This identification influenced generations of physicists. We had already occasion to note that determinism and causation are not the same (Chap. 2, Interlude A). Assiduous observations of the coming of the tides on a beach would enable many observers to predict the tides at least for a few days. But this does not mean that the observer can causally explain why the tides arise. Equally, the causal explanation of atomic events does not mean that the behaviour of these events can be predicted with deterministic certainty. The Laplacean identification held such a firm grip on the physics community that the philosophical convictions departed in opposite directions, once quantum mechanics had cast doubts on the deterministic predictability of quantum systems. Einstein and Planck came to hold the view that the unpredictability of individual atomic events meant that quantum physics did not properly understand the atomic realm. Heisenberg and Bohr came to embrace the view that the unpredictability of atomic events meant that the notion of causation did not apply in the atomic realm. It was only when some physicists realised that the notions of causation and determinism needed to be separated that causal explanation of atomic events became envisageable. In the atomic realm causation comes without determinism. We will discuss this as the *conditional view of causation*.

---

[118] Sklar, *Philosophy of Physics* (1992), 37; Čapek, 'Time in Relativity Theory' (1966), 441; Sexl/Schmidt, *Raum, Zeit, Raum-Zeit* (1978), 86–91; Bohm, *Theory of Relativity* (1965), 155–60; Joos, *Theoretical Physics* (1951), Chap. X, §7; Arntzenius, 'Causal Paradoxes' (1990); Nehrlich, 'Special Relativity' (1982). We observed already in a previous footnote that Minkowski may not have been committed to the idea of a block universe. In his lecture 'Raum und Zeit' (1909/1974), 61 he points out that every world point in the past light cone of *Here-Now* is 'necessarily' earlier than Here-Now and every world point in the future light cone of Here-Now is 'necessarily' later than *Here-Now*. As far as I can see, von Laue, *Das Relativitätsprinzip* (1913), 57 was the first to read Minkowski's 'before-after'-relation in a causal sense. *Here-Now* can be causally influenced by events in the past light cone and can causally influence events in the future light cone. Similar considerations can be found

In Minkowski space-time diagrams, the simultaneity of events is represented as the class of events, which lie parallel to the *spatial* axes of the respective reference frames. In three-dimensional light cones, these become simultaneity (hyper-)planes. The simultaneous events for any observer are those events, which lie on a simultaneity plane, which must be drawn parallel to the space axes and at an angle to the respective time axes of the observers. For stationary observer O, event t_A happens at the same time as event t_B (Fig. 4.14). These two events happen simultaneously at $t = t_{AB}$. For O, the simultaneous events are those, which lie parallel to the x-axis, that is, events $t_A - t_B$. For moving observer O', moving at uniform velocity with respect to observer O, the simultaneous events are those, which lie parallel to the x'-axis, that is, events $t'_B - t_B$. These events lie at an angle α from the x-axis. These two observers will judge the order of events differently. At t_{AB} events t_A and t_B happen simultaneously in reference system $(x - t)$ but in $(x' - t')$ event t'_B happens *before* event t'_A. How is this curious reversal of time order possible? The answer is that events t_A and t_B are *space-like* connected. Between simultaneous events there can be no causal link because of the finite velocity of the propagation of causal signals. Let A be a football match and B live commentary of the match. In $(x - t)$ there can be no causal influence between these two simultaneous events. The commentator must announce a goal after it has happened: t_B and t_A lie outside each other's light cone. In $(x' - t')$ the commentator cannot yet report the goal; for him it has not yet happened. The event t'_A lies in the future light cone of event t'_B.[119]

If there could be a causal connection between *space-like* separated events, signals would have to move at speeds greater than the speed of light. This would allow us to construct the case of a *relativistic lottery swindle*[120] (Fig. 4.15). Consider Alice, who follows the lottery draw on Saturday night. Alice occupies the co-ordinate system (x, t). Her colleague Zoë moves with great velocity relative to Alice's inertial frame.

---

in Weyl, *Raum Zeit Materie* (1921), 174–6; Eddington, *Space, Time & Gravitation* (1920), 50–1; Pauli, 'Relativitätstheorie' (1921), §6; Langevin, *La Physique Depuis Vingt Ans* (1923), Chap. V; Reichenbach, *The Philosophy of Space and Time* (1958), §29. Thus Nahin, *Time Machines* (1993), 117 is mistaken in crediting M. Čapek with the discovery 'that the temporal ordering of potentially causal events in both the future and past of one observer is invariant for any other observer...'. But, as we shall see, Čapek's work was important in the criticism of the block universe as a consequence of Minkowski space-time.

[119] Sexl/Schmidt, *Raum, Zeit, Raum-Zeit*, (1978), 89–91. An analogy may help to understand this reversal of time order. The British Prime Minister Tony Blair was at an EU summit in Seville, Spain, when the England-Brazil match for a place in the quarter finals of the Football World Cup took place in Japan (2002). Mr. Blair wanted to watch the match live but with English rather than Spanish commentary. His aides set up a system, which allowed him to watch the Spanish TV pictures with BBC commentary. There was only one hitch. The BBC words arrived five seconds before the Spanish TV pictures. So when Brazil scored the winning goal against England Mr. Blair learnt about the goal before he saw the pictures (reported in *The Sunday Times*, June 23, 2002).

[120] Sexl/Schmidt, *Raum, Zeit, Raum-Zeit*, (1978), 86f; cf. Sklar, *Space, Time, Space-time* (1974), 285; Friedman, *Foundations* (1983), 161–2

Zoë's reference frame is marked by her own coordinates (x', t'). Alice and Zoë have concocted a lottery swindle, involving strange radio waves, which travel at speeds faster than light. Alice's plan is to send the winning numbers to Zoë via superluminal radio waves (wriggly lines). Because they are superluminal signals, outside her light cone, Zoë receives the message at a point on her simultaneity plane, x''. After receiving the winning numbers, Zoë immediately returns the message to Alice, again using faster-than-light (superluminal) radio signals. Alice receives this message just before the closing time of the lottery draw. She ticks the right number and wins the jackpot. Allowing signals faster than light would lead to *causal* anomalies like this one: sending messages into your own past. As Alice in this fictitious case is allowed to send messages at speeds greater than the limit speed of light, and hence at speeds forbidden by the Special theory of relativity, these radio waves are inclined at angles below the co-ordinate system of her colleague's co-ordinate system and travel into *Elsewhere*. In other words, the lottery swindle involves *space-like* separated events, yet causal (superluminal) signals are allowed to be exchanged. This would make the lottery swindle possible.

These facts about Minkowski space-time will complicate the position of the proponents of the block universe who are tempted by a *philosophy of being*. Such a philosophy takes its inspiration from the slicing of the space-time world into what Minkowski called the individualisation of time and space. A Kantian idealist view of time seems to be the only philosophical consequence. However, as we argued before, all the evidence must be taken into account before we can infer a philosophy of being from Minkowski space-time. One complication for the philosophy of being is the important distinction between *time-like* and *space-like* separated events. The order of *time-like* connected events is the same for all observers.

There is another complication. In the chapter on Nature, we developed the view that the *invariant* is a candidate for the real, while the real must be what remains invariant across reference frames. This spells another complication for the philosophy of being: the existence of an invariant in Minkowski space-time.

What remains invariant[121] in relativistic space-time can however no longer be the spatial and temporal measurements separately. In *classical mechanics*, the spatial distance and the temporal interval between two events, E_1 and E_2, are separately invariant for each inertial frame. If S is the *spatial* distance between E_1 and E_2 in system x, y, z, then, following Pythagoras: $S^2 = (x_2 - x_1)^2 + (y_2 - y_1)^2 + (z_2 - z_1)^2$. (In differential terms: $ds^2 = dx^2 + dy^2 + dz^2$). The *temporal* interval $t_2 - t_1$ = constant, since in classical mechanics all observers measure the same temporal intervals between events and the universe is governed by one universal clock. If we change the reference system to x', y', z', for instance by rotation, then the *spatial* coordinates for the two events are different, but the distance between them remains the same:

[121] Langevin, *La Physique* (1923), 306, 329ff; Costa de Beauregard, 'Time in Relativity Theory' (1966), 426; Čapek, 'Time in Relativity Theory' (1966), 440f ; Eddington, *Time, Space & Gravitation* (1920), 70–6; Misner *et al. Gravitation* (1973), §13.4; Sklar, *Space, Time, Space-Time* (1974), 261–4; for a numerical computation of the earth-sun space-time distance, see Davies, *About Time* (1995), Chap. 8, §3

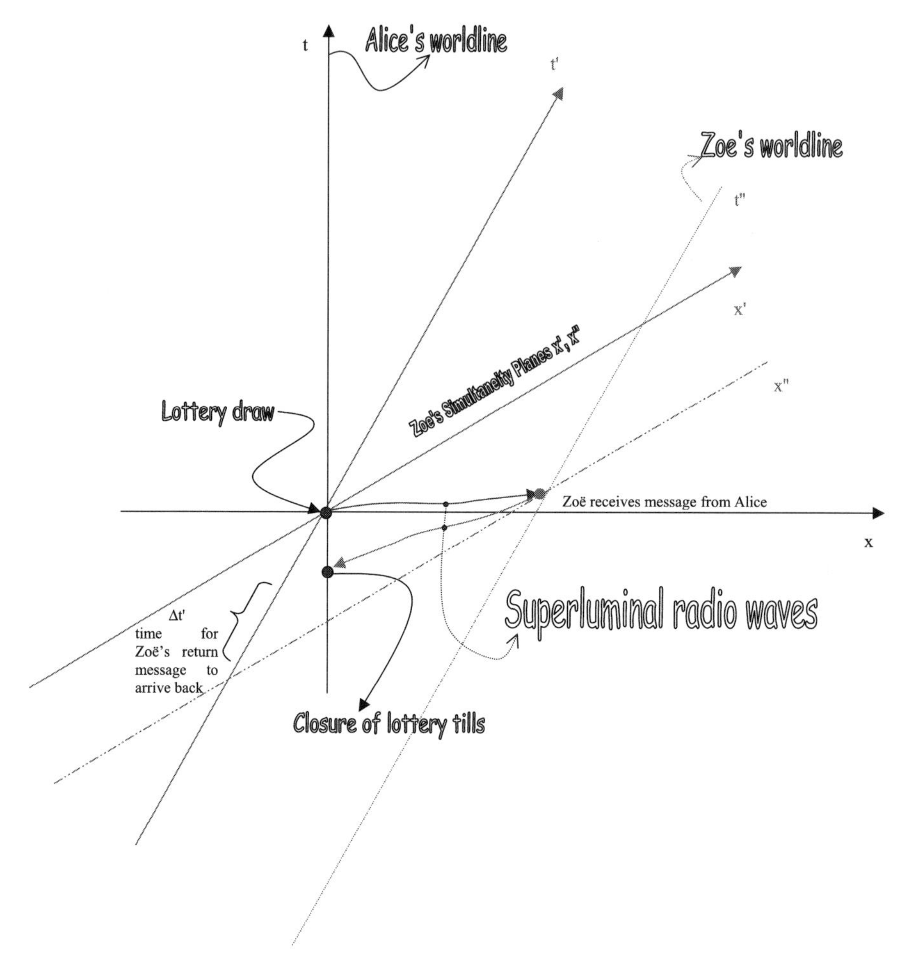

Fig. 4.15. Alice's superluminal lottery swindle in a fictitious world, in which causal signals can travel faster than light

$S'^2 = (x_2' - x_1')^2 + (y_2' - y_1')^2 + (z_2' - z_1')^2$. (In differential terms: $ds'^2 = dx'^2 + dy'^2 + dz'^2$). Again, $t_2' - t_1' =$ constant.

In *Minkowski space-time*, the invariance does not belong to the spatial distance and the temporal interval separately, because now events happen at space-time points, for which each observer assigns different spatial and temporal coordinates. However, there is another quantity called *world interval I*, or the *space-time interval* ds, which remains invariant. This space-time interval ds does not dependent on the inertial reference system, in whose coordinates it is expressed. The expression is a measure of space-time distances between events. It is defined in the following alternative ways:

$$ds^2 = dx^2 + dy^2 + dz^2 - c^2dt^2$$
$$ds^2 = +dt^2 - dx^2 - dy^2 - dz^2(c = 1)$$
$$I^2 = s^2 - c^2(t_2 - t_1)^2$$

I can be written in other forms, but this does not change the invariance of I.[122] It is a consequence of the Special theory of relativity that different observers, if in relative motion with respect to each other in space-time, will assign different space and time coordinates to the same events. This view of the individualisation of spatial and temporal coordinates led to the disappearance of the objective *Now* and the adoption by a majority of physicists of a physical block universe and an idealist view of time.

The observers attached to different systems, respectively, assign different individual spatial and temporal separations for two events, but they measure the same space-time interval, i.e. $(ds)^2 = (ds')^2$.[123] Minkowski expressed this idea in terms of his *postulate* of the absolute world (or *world postulate*). It states that 'only the four-dimensional world in space and time is given by phenomena, but that the projection in space and in time may still be undertaken with a certain amount of freedom.'[124] The four-dimensional space-time world exists absolutely and is not dependent on observers. Only the lamination of the space-time block grants the observer some freedom to introduce different reference frames.

If the invariant is a candidate for the real, as Minkowski with his world postulate seems to assume, then the invariance of space-time intervals between events, on which all observers agree, must be real, i.e. frame-independent, according to this criterion. The space-time intervals, ds, give us geometric information about the structure of space-time. The quantity, s, is the proper time of a clock moving along with the world lines.

The *time-like* and *space-like* separation between events, mentioned earlier, can be expressed in terms of the space-time interval, I.[125] Using the metric ds^2, *time-like* world lines, in the interior of the light cones, have positive intervals, $[ds^2 > 0]$.[126]

[122] See Minkowski, 'Raum und Zeit' (1909/1974), 58; Čapek, 'Time in Relativity Theory' (1966), 439f–40; Nahin, *Time Machines* (1993), 312 writes: $(ds)^2 = (dt)^2 - (dx)^2 - (dy)^2 - (dz)^2$; Whitrow, *Natural Philosophy* (1980), 272 writes: $ds^2 = dt^2 - \frac{dx^2+dy^2+dz^2}{c^2}$. For a good analogy of the invariance of the space-time interval, ds, see Spielberg/Anderson, *Seven Ideas* (1995), 235

[123] The physical significance of ds for *time-like* world lines is that it represents *proper time*; see Sexl/Schmidt, *Raum-Zeit* (1978), 114; Whitrow, *Natural Philosophy* (1980), 277

[124] Minkowski, 'Raum und Zeit' (1909/1974), 60; see Galison, 'Minkowski's Space-Time' (1979) for an interpretation of the world postulate.

[125] Nahin, *Time Machine* (1993), 313; Čapek, 'Time in Relativity Theory' (1966), 440–3; Cook, *Observational Foundations of Physics* (1994), 33, 35; Sklar, *Philosophy of Physics* (1992), 37; Reichenbach, *Philosophy of Space and Time* (1958), 143–7; Costa de Beauregard, 'Time in Relativity Theory' (1966), 425–30

[126] *Time-like* world lines have positive intervals I, if we write $ds^2 = dt^2 - dx^2$. However, there is a degree of convention at work. If we choose to write $ds^2 = dx^2 - dt^2$, then $I^2 < 0$. The same applies to *space-like* intervals.

Space-like world lines, in the exterior of light cones, i.e. in *Elsewhere*, have negative intervals, [$ds^2 < 0$].

- If event E_1 is the emission of a photon at one point in space-time, event E_2 is the absorption of that photon at some other point in space-time, then the interval between them is always zero. It is always zero for any events connected by light. The world line of any photon is said to have a *null interval*, which will always be on the surface of the light cone: $ds^2 = 0$. For a photon, travelling at the speed of light, time stands still.
- When $ds^2 < 0$, we find that the spatial separation between two events is greater than their temporal separation. The two events stand in a *space-like* separation. It is not possible to send a causal signal between these events so that they are not causally connectible. For instance, events beyond the visible limits of the expanding universe are so far away that any signals have not yet reached us. There could be no exchange of causal signals between events on earth and such distant events. Such a spatial separation does not need to occur on a cosmic scale. In quantum mechanics, the physics of the atom, certain types of experiments are set up, which involve *space-like* connected events. In these so-called *EPR* experiments, which are also important in the consideration of causation in quantum mechanics, subatomic particles or photons are sent along two different arms of an experimental apparatus. In one such experiment, the particles were approximately 12 metres apart.[127] When measurements of certain physical properties (polarization) are carried out on a particle in one arm, it is found that *instantaneously* the particle in the other arm, 12 meters away, displays correlated properties. This is puzzling because the two arms are too far apart for any causal signals to be exchanged between the two particles. No causal signal can travel faster than the speed of light: the spatial distance between the two arms is too far for signals to be exchanged between the two particles whose properties are correlated instantaneously upon measurement.
- When $ds^2 > 0$, we find that the spatial separation between two events is smaller than their temporal separation. The two events stand in a *time-like* separation. It is possible to send causal signals between these events because their spatial separation is small enough for a causal influence to propagate from event E_1 to event E_2. These events are confined to the same light cone.

The Minkowski space-time diagrams have played a significant part in recent attempts to justify or reject the *block universe* as a *philosophical consequence* of the space-time conception of the universe. Many physicists have embraced the block universe. A few have made an attempt to show that it really is a philosophical consequence of the postulate of relative simultaneity and Minkowski space-time. For, as we shall see, it is by no means certain that the block universe *is* a philosophical consequence of the Special theory of relativity. What is certain, however, is that the opinion of both physicists and philosophers is divided. For one camp the Special

[127] Aspects, 'Experimental Tests' (1982); today *EPR* correlations are measured across distances of 12 kilometres.

theory of relativity leads us to a *philosophy of being*. For another camp the Special theory of relativity leads us to a *philosophy of becoming*.

4.4.1 Philosophical Consequences I: The Philosophy of Being

> To accommodate everybody's nows (...) events and moments have to exist "all at once" across a span of time.
>
> P. Davies, *About Time*, 1995, 71

Gödel pointed out that the Special theory of relativity with its notion of the relativity of simultaneity offered 'new and surprising insights into the nature of time.' In particular, he held that through this theory one obtained 'unequivocal proof' for the *idealist* view of time. How can this be? Gödel argued that the relativity of simultaneity and, by implication, assertions of temporal successions lose their objective meaning, 'in so far as another observer with the same claim to correctness can assert that A and B are not simultaneous.'[128] If simultaneity is relative, the lapse of time will be affected: it cannot 'consist of an infinity of layers of "now" which come into existence successively, in an objectively determined way.'[129] Thus Gödel denies, in agreement with modern physics, the Newtonian character of time.

> Each observer has his own set of nows and none of the various systems of layers can claim the prerogative of representing the objective lapse of time.

Although Gödel puts emphasis on the *idealist* view of time, as a consequence of relative simultaneity, he also embraces, unlike von Laue, the *block universe*.[130] For Gödel associates the idealist view of time with the aforementioned denial of the objectivity of change. The *passage of time* is a subjective human experience. Thinking back on our interpretation of Kant and Saint Augustine, this rendering of the idealist view needs the following modification. The passage of time is subjective in Saint Augustine's sense if each observer records the passage of time in his or her own mind; it is subjective in Kant's sense if *the* human observer, as opposed to some other differently equipped observer, measures the passage of time. Gödel takes the Special theory of relativity to have provided 'proof' of the ideality of time. The onus of proof falls on the notion of relative simultaneity. We lose the objective division of space-time into 'events, which have already occurred' and 'events which have not yet occurred'. This division is now effected by the observers who travel with relative speed to each other and slice the static space-time world according to their state of motion. Each space-time point has its own light cones. When viewed by an observer, it has its own separation into past, future and elsewhere. Space-time is the world of static events. There seems to be no process of temporal *becoming*. Everything is

[128] Gödel, 'Relativity and Idealistic Philosophy' (1949), Volume II, 557; Čapek, 'Time in Relativity Theory' (1966), 438f; Costa de Beauregard, 'Time in Relativity Theory' (1966), 429f; Sklar, *Philosophy of Physics* (1992), 37; Stein, 'Paradoxical Time Structures' (1970)

[129] Gödel, 'Relativity' (1949), 558. Recall that for Kant there is only one time (*Critique* A189, A110, B308) and it flows (B291)

[130] Nahin, *Time Machines* (1993), 103ff

already there. As we perceive the passing of time, we become conscious of ever more of Minkowski's world points (events) that lie on our individual world lines.[131] For Newton, time was a river that flowed equably and eternally in its riverbed, without reference to anything external. Humans lead their existence on the banks of the river, on which all objects are in relative motion. But experiments, this was Newton's hope, could at least indirectly determine the existence of absolute time. Minkowski's four-dimensional space-time world inspires the opposite analogy: The river of time is stagnant. Humans now swim upwards the river, thus having the illusion of change and becoming.[132]

Those who adopt this *philosophy of being* (Minkowski, Weyl, Eddington, Einstein, Costa de Beauregard, Davies and Barbour) take space-time and its material contents to be spread out in four dimensions. 'Change is only relative to the perceptual mode of living beings.' Living beings are 'compelled to explore little by little the content of the fourth dimension, as each one travels a *time-like* trajectory in space-time.' But *everything is already written*, not only in the past cone but also in the future cone. The future is already there. 'Nature "will take" one of the alternatives open to her, and it is this that we must imagine inscribed, even though we do not know what "it will be".'[133] The *philosophy of being* comes with a commitment to static determinism (Box 4.3).

This *philosophy of being* is not just a curiosity, beguiling a minority of physicists just after the relativity revolution. Whilst a neutral observer may suspect that philosophers would lend enthusiastic support to the *philosophy of being*, it is rather the other way round. A long line of physicists endorsed the block universe, which was only supported by a minority of philosophers. For the physicists the idea of the block universe seems to hold a strange attraction. A younger generation of physicists continues to see in the theory of relativity 'proof' of the *static* view of the universe: O. Costa de Beauregard, T. Gold, C.W. Riejdijk, P.W. Atkins, P. Davies and most recently J. Barbour have all endorsed either an idealist view of time or a fully-fledged version of the block universe.[134]

---

[131] Nahin, *Time Machines* (1993), 107, 208, 137f

[132] This analogy was suggested by Costa de Beauregard, *Time* (1987), 23; see also Frank, *Philosophy of Science* (1957), 159. Costa de Beauregard, *Time* (1987), 150 is a firm believer in the block universe: "Our awareness of time is such that 'we', at our conscious 'now', are exploring 'forward' our personal time-extended history, as it exists all at once – 'at once', of course, *not* being synonymous with 'at the same time!' "

[133] Costa de Beauregard, 'Time in Relativity Theory' (1966), 429–30; cf. Costa de Beauregard, *Time* (1987), 159, 'Burning Question' (1980), 94–7

[134] Gold, 'The Arrow of Time' (1962), 403–410, 'Cosmic Processes' (1966) and 'The World Map and the Apparent Flow of Time' (1974), 62–72; Rietdijk, 'A Rigorous Proof of Determinism' (1966), 341–44 and 'Special Relativity and Determinism' (1976), 598–609; Atkins, 'Time and Dispersal' (1986), 80–8; Davies, 'Time Asymmetry and Quantum Mechanics' (1988), 99–124 and *About Time* (1995); Barbour, *The End of Time* (1999). It should be noted that, strictly speaking, Barbour does not derive the block universe from the Special theory. Nor does he equate the denial of time with the denial of temporal becoming. Barbour derives the idea of a block universe from the attempt to unify quantum theory and the General

Those who adopt a *philosophy of being* must consider that physics has a direct impact on our philosophical conception of the world. They hold that the inference from the space-time representation of the world to a static space-time world is legitimate. We have already seen that the Minkowski space-time conception of the universe persuaded influential physicists like Eddington, Einstein, Gödel, Jeans and Weyl of the superiority of a *static* view of the universe. The loss of the objective division of space-time into past, present and future, Costa de Beauregard writes, encapsulates 'a small philosophical revolution.'[135] The mighty prestige of physics, buttressed by the empirical success of the theory of relativity, is brought to bear on the philosophical notion of time. The Special theory of relativity seems to provide us with the tool to 'prove' that the *passage of time* – temporal change and becoming – is a human illusion.

The American physicist Thomas Gold does not even appeal to the notion of relative simultaneity to argue that the flow of time is a subjective illusion. The four-dimensional space-time *representation* of the world, introduced by Minkowski, is enough to show the redundancy of the notion of the passage of time. Following Einstein and Eddington, he suggests that the 'physical world can be regarded as a *map* of all the world lines of all the particles.' The world lines have no arrows in a forward direction. The laws of physics are time-symmetric. The asymmetry, observed in complex systems, is a statistical effect, due to the large-scale motion of the universe. Thus Gold takes the four-dimensional *representation* of the world to be an objective statement about the physical world. As the passage of time does not fit into this representation, the flow of time must be a subjective impression.

Such an inference from representation to objectivity would at least have to be justified, especially as the tool of four-dimensional space-time is available to argue for the establishment of a *philosophy of being*. A static world of being was already inherent in the Laplacean deterministic universe. But this was more a metaphor than a 'proof'. The difference between the classical '3 + 1' view (3 spatial dimensions clearly separated from the one temporal dimension) and the modern 4-dimensional view is that now the static world seems no longer to be a metaphor but the philosophical consequence of a well-confirmed scientific theory. The basic instrument of proof is that the choice of simultaneity planes is dependent on the inclination of the coordinate axes, which is a function of the motion of the reference system. From the statement that simultaneity planes are relative to coordinate systems, several authors[136] have attempted to establish – rigorously – that the *future is already*

---

theory of relativity (see §4.3 below). According to this approach at a fundamental level the universe is timeless. Time becomes an emergent property as we approach our familiar macro-world; see Butterfield, 'Critical Notice' (2002).

[135] Costa de Beauregard, 'Time in Relativity Theory' (1966), 429

[136] Shortly after Rietdijk's 'A Rigorous Proof of Determinism' (1966), Hilary Putnam published his paper 'Time and Geometry' (1967), in which the openness of the future is also denied. These publications received a great amount of attention – see for instance Nahin, *Time Machines* (1993), 119–20 [including further references]; Sklar, *Space, Time and Spacetime* (1974), 274–5 and 'Time, Reality' (1981); Čapek, 'Relativity and the Status of Becoming' (1975); McCall, *A Model of the Universe* (1994), 33–4 and 'Time flow' (1995), 155–72;

determined and that *all past and future events are real.* But how can it be *proved* that relative simultaneity implies determinism and the eternal reality of all events and therefore the block universe?

Box 4.3: *Static* and *Dynamic* Determinism

From a consideration of Einstein's notion of the block universe we have learnt that determinism may either be *static* – as envisaged in the block universe – or *dynamic*, as in Cunningham's view that world lines have a history, a view to which the later Einstein inclined. Both are forms of *ontological determinism* because they make claims about how the four-dimensional world is constituted, irrespective of the knowledge states of human observers. According to William James (ontological) determinism professes, "that those parts of the universe already laid down absolutely appoint and decree what the other parts shall be. The future has no ambiguous possibilities hidden in its womb; the part we call the present is compatible with only one totality" (Quoted in McCall, *Model*, 1994, 12). In view of the arguments presented by the *philosophy of being*, we may attempt the formulation of a dynamic ontological form of determinism.
A *pure* ontological definition of determinism can also be derived from the four-dimensional representation of the Special theory of relativity: Minkowski space-time. It characterizes determinism as a restriction on the availability of trajectories, between two time slices t_1 and t_2, along which a physical system will evolve, given the laws and boundary conditions, under which the system operates. Ontological determinism must require that, given time slice t_1, there is only *one unique trajectory* for the system to get to t_2 and beyond. The *uniqueness* of the geodesics from past into future – the inertial world lines, whose shape depends on the state of motion of the particle – can be determined from the point of view of every relative simultaneity plane. The geodesics are governed by the laws of motion, expressible in terms of differential equations. The time-slices feed the boundary conditions into the equations. In this scenario, *ontological determinism is the unidirectional, linear, non-splitting tracing of geodesics onto the canvass of space-time.* At any spatio-temporal stage of the word line the geodesics have a definite past and future. This unique history is the same in *all* reference frames. An invariant exists in all reference frames: the space-time interval ds, which expresses the 'distance' between two (time-like or space-like connected) events. Interactions between world lines are permitted and do occur.

Jammer, 'A Consideration of the Philosophical Implications of the New Physics' (1979), 49; Stein, 'On Einstein-Minkowski Space-Time (1968) and 'On Relativity Theory and the Openness of the Future' (1991); Maxwell, 'Are Probabilism and Special Relativity Incompatible?' (1985), 23–43 and 'Are Probabilism and Special Relativity Compatible?' (1988); Balashov, 'Relativisitic Objects' (1999). Predating these publications is Frank's rejection of a derivation of predetermined futures from Minkowski space-time, see his *Philosophy of Science* (1957), 158–62; see also Jeans, *Physics and Philosophy* (1943), Chap. V; Bondi, 'Relativity and Indeterminacy' (1952); Freundlich, 'Becoming' (1973); Lucas, *Time* (1973)

Box 4.3: (continued)

But as the laws of interaction are mechanistic, there is no probabilistic branching of the geodesics. No observer, no predictive abilities have been mentioned. The spatio-temporal stages of space-time, which show the history of geodesics, can be measured by clocks, which are either attached to the trajectory (recording the *proper* time of the system at successive spatio-temporal stages) or from external clocks (recording a dilated time for the system undergoing linear translation).

The general idea is this[137]: Events, which exist *now*, must be regarded as real. Saint Augustine already expressed our ordinary beliefs that past and future events do not possess reality. The Special theory of relativity has, however, introduced a complication into this view. The *Now*, even if we deny that it is purely subjective, has become frame-dependent. So we must introduce the coordinate system of observer O, who will regard all events as real, which lie on O's *space-like* simultaneity planes. These are all the events simultaneous with O's *Now*. But what is now for one observer will not be now for another. There will be a second observer O', in relative constant motion with respect to O (Fig. 4.16). If the observers are *space-like* separated from the event, then the time order will not be unique for them. The second observer's reference frame can be made coincident with that of O, by a convenient choice.[138] That means that both coordinate systems can be made to coincide at the origin. Both observers will be in each other's present. There will be many events, which are simultaneous and hence real for one observer, which lie in the past or future for a second observer. Take an event E_1, which is already *past* for one observer but still in the future of the other observer. This event cannot be determinate for one and indeterminate for the other observer. It must be determinate for both observers. For what is real for one observer must be real for another observer in his presence. Take an event E_2, which is *now* for one observer but lies in the future of the other observer. Then it is real and determinate for one observer; so it must be real and determinate for the other observer. Hence *all* events must already be determinate and equally real for all observers.

Consider this argument in more technical detail. The space-time representation of the universe is given in Fig. 4.16. It will always allow us to find a distant observer, O_1 for whom an event, which is *now* or *future* for some other observer O_2, will already exist in his past. O_1 is in relative constant motion with respect to O_2. They synchronize their clocks at time t such that $t_1 = t_2 = 0$. Then O_1 will consider certain events to be simultaneous: those events, which lie on a *space-like* hyperplane, x_1, which lies at an angle α to O_1's time axis. Because O_2 is stationary with respect to O_1, observer O_2 will regard different events as simultaneous, since for this observer

[137] See Weingard, 'Relativity and the Reality of Past and Future Events' (1972); Putnam, 'Time and Physical Geometry' (1967); Sklar, 'Time, Reality' (1981), 129–30

[138] Joos, *Theoretical Physics* (1951), 247; Minkowski, 'Raum und Zeit' (1909/1974), 61; Frank, 'Das Relativitätsprinzip' (1910), 488; Langevin, *La Physique* (1923), Chap. VI; Prigogine, *End of Certainty* (1997), 168

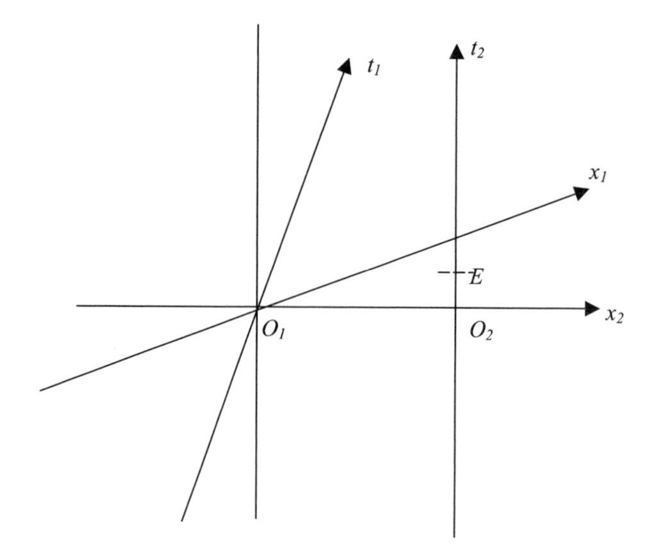

Fig. 4.16. *Determinism and Relativity*: For O_1 event E is already in the past. For O_2 E is still in the future. As event E is determinate from the point of view of O_1 it cannot be indeterminate from the point of view of O_2. For both it is equally determinate and hence equally real

the *space-like* simultaneity plane, x_2, forms an angle of 90° with respect to O_2's time axis. Then there will be an event, E, which will belong to the past for observer O_1, lying below O_1's simultaneity plane. The same event, E, will however reside in the future of observer O_2, lying above O_2's simultaneity plane. This has a puzzling consequence for the ontological status of E. If E is determinate for observer O_1, since it lies in O_1's past, how can E be indeterminate for observer O_2, even though it lies in O_2's future? It seems that the proponents of the block universe provide the only sensible answer. *All* events must be regarded as determinate and real at *all* times. O_2 cannot influence event E, although it lies in this observer's future. According to the Special theory of relativity, there is always an observer like O_1 for whom each *future* event in O_2's frame of reference is already a *past* event. Our ordinary conception must be mistaken. If an event is already determinate for one observer, it must be determinate for all observers. Equally, if an event is already real for one observer, it must be real for all observers. The physical world is a block universe. The passage of time is a human illusion.

This result is counterintuitive, as so many implications of the Special theory of relativity. It contradicts commonsense intuition. *Intuitively*, we regard the past as frozen and unreal. But the future is open and equally unreal. It is subject to our modifications, determinations and interventions. We regard the future as an area of *objective* possibilities and becoming. Commonsense intuition can, however, not be trusted.

The 'proofs' of the unreality of time from the Special theory of relativity are in the best tradition of the block universe. They presuppose that scientific theories can decide philosophical questions, just as Bohr proclaimed.

In this instance, a four-dimensional *representation* of reality is used as a statement about the *nature* of reality. On a realist understanding of scientific theories, there is prima facie nothing wrong with this step: our models and theories describe and explain the structure of reality; the structure of the model represents the structure of reality.[139] But there is a snag. The structure of the ancient universe, for instance, could be presented according to a *geocentric* or a *heliocentric* view, but these representational alternatives alone did not yet tell us anything about the structure of reality. In the face of representational alternatives the realist requires further tests to ascertain, which of two representations is the better one. Minkowski and Einstein claimed that the four-dimensional representation of the world was superior to the '3+1' representation. Yet, in the case of the Minkowski interpretation of the Special theory of relativity both a *philosophy of being* and a *philosophy of becoming* have been claimed to follow from the theory. Clearly, no empirical test is available to decide between these two philosophies.

But we can examine the assumptions, which underlie the apparent proofs. In the case of *space-like* connected events simultaneity between such events can only be of a conventional, not of a fundamental kind.[140] That is, by an appropriate choice of coordinate systems, the temporal order between such events can be reversed. The temporal order is not constrained by the Lorentz transformations. Such temporal relations are merely conventional and cannot have physical significance.[141] It is true, of course, that what is real for one observer must be real for another observer. (This is Putnam's principle of the transitivity of relativity.) But the situation under discussion is different. When Putnam's principle is based on the conventionality of temporal order between *space-like* connected events, no conclusion as to the reality of these events follows. From the static representation no claim about ontological determinism follows.

In an earlier chapter of this book, we found that a good criterion for the real was the invariant. If we go by this criterion, the *philosophy of being* has some more justifications to provide. For the temporal relations between *space-like* connected events

---

[139] As Bohm, *Special Theory of Relativity* (1965), 180 observed, the Minkowski diagram "is a kind of map of *events* in the world, which can correctly give us the order, pattern, and structure of real events, but which is not in itself the world as it actually is.... Nevertheless, a good map has a *structure* that is in certain ways similar to the structure of the world." (Italics in original)

[140] Joos, *Theoretical Physics* (1951), 241; Čapek, 'Relativity and the Status of Becoming' (1975); Čapek, *Philosophical Impact* (1962), 188 declares that the succession of causally unrelated events has no operational meaning; cf. Frank, *Philosophy of Science* (1957), 158–64; Godfrey-Smith, 'Special Relativity' (1979); Sklar, 'Time, Relativity' (1981); Prigogine, *End of Certainty* (1997), 168–9

[141] Weingard, 'Relativity' (1972), 119. Harris, 'Simultaneity' (1968) argues that the Special theory of relativity underlines the indeterminacy of future events.

are not invariant and can therefore, by this criterion of reality, not be candidates for what is regarded as real.

By contrast, consider a *dynamic* representation. According to it, world lines form histories due to the finite velocity of the propagation of all signals. A history here is a succession of events. Such successions can be considered under causal and entropic aspects. Then the only way we can have information about the existence of things is either if they exist 'now', on our simultaneity axis, or if we find traces of their past existence in our past light cone. There can be no causal connection with events lying outside our light cone. We can therefore envisage the compatibility of the Special theory of relativity with a *philosophy of becoming*.

Let us consider again the distinction between the 'before-after'-relation and the 'past-present-future'-relation. It may take us to a *philosophy of becoming*. Minkowski's dissolution of the world into world lines, spatio-temporal events, is a far cry from an observer's *impression* of objective changes in the world. Most observers have an objective sense of *becoming*. But according to the *philosophy of being* what the observer sees as a *coming into being* is only a *coming* into human conceptual awareness. Therefore the passage of time is a human illusion. As Adolf Grünbaum[142] stresses, what qualifies a physical event at a time t as belonging to the present, to the *Now*, is that a mind possessing organism, M, experiences the event at time t.

> What qualifies a physical event at a time t as belonging to the present or as now is *not* some physical attribute of the event or some relation it sustains to other *purely physical* events. Instead what is *necessary* so to qualify the event is that at the time t at least one human or other *mind-possessing* organism M is conceptually aware of experiencing at that time either the event itself or another event simultaneous with it in M's reference frame. And that awareness does not, in general, comprise information concerning the date and numerical clock time of the occurrence of the event.[143]

Grünbaum's position has often been associated with an endorsement of the block universe. It is easy to see why: qualifying an event as belonging to the present requires conceptual human awareness. But Grünbaum does not equate the existence of a *timeless* universe with the existence of a *changeless* universe. We can envisage a universe devoid of the 'past-present-future'-relation, but this can still be a *dynamic* universe. Different notions of time are involved in this argument. If by the phrase 'passage of time' we mean the 'transient now', the transition of events from the *future* through the *present* into the *past*, then human conceptual awareness is required.

---

[142] Grünbaum, 'The Meaning of Time' (1971); Čapek, 'Time in Relativity Theory' (1966), 437, 452; Whitrow, *Natural Philosophy* (1980), §6.1

[143] Grünbaum, 'The Meaning of Time' (1971), 206–7; for a similar formulation see Grünbaum, 'Are Physical Events Themselves Transiently Past, Present and Future?' (1969), 146 and 'Anisotropy of Time' (1967). Grünbaum's thesis produced a lively debate in the pages of the *British Journal for the Philosophy of Science*; see Dobbs, 'The "Present" in Physics' (1969), 317–324 and 'Reply to Professor Grünbaum' (1970), 275–78; Ferré, 'Grünbaum *vs.* Dobbs: The Need for Physical Transiency' (1970), 278–80. Baker, 'Temporal Becoming' (1975) provides a reflective discussion of this debate.

But physical events may occur at certain *clock times* – clock times constituted by natural processes like planetary motion. These are independent of conceptual human awareness. Then they are ordered according to the 'before-after'-relation.

> It is of the essence of the relativistic account of the inanimate world as embodied in the Minkowski representation that there is change in the sense that different kinds of events can (do) occur at different times: the attributes and relations of an object associated with any given world-line may be different at different times (e.g., its world-line may intersect with different world-lines at different times). Consequently the total states of the world (when referred to the simultaneity criterion of a particular Galilean frame) are correspondingly different at different times, i.e., they change with time.[144]

Grünbaum's argument presupposes a basic distinction between an *anthropocentric* notion of time – characterized by 'becoming', 'transience' and quite generally the availability of calendar dates – and a *physical* notion of time. The latter is *clock* time, characterized by the 'before-after'-relation and the exchange of light signals. This distinction already lies at the heart of the relational view of time: *physical* time is the order of the succession of events; *human* time is the abstraction of a model of time from the observation of temporal succession. But Grünbaum employs this distinction as an argument against the block universe. The language of the *passage of time* requires the existence of human conceptual awareness. It is *tensed* language: 'past-present-future'. The language of physical occurrence does not require human observers. It is *tenseless* language: 'before-after'. *Tenselessness* therefore does not equate *changelessness*.

As Leibniz pointed out, the minimalist notion of time turns into a maximalist notion: the 'before-after'-relation, which is based on natural units of time, is enriched with conventional units of time (seconds, hours, days, weeks, months and years). Human conceptual awareness allows a slicing of the series of events into *past, present* and *future*. Not only did the dinosaurs exist *before* the pharaohs, they did so *many years before* them. Not only do we have, in Saint Augustine's words,

---

[144] Grünbaum, *Philosophical Problems of Space and Time* (^{2}1973), 325. We have already seen that Cunningham, Reichenbach and Schlick asserted well before Grünbaum that world lines acquire histories. In his chapter on the 'Space-Time World' J. Smart (1963), 138–40 also rejects the equation of *tenselessness* with *changelessness*. Such considerations have a venerable history. Hugo Bergmann, *Kampf um das Kausalgesetz* (1929), §14 pointed out that the notion 'now' had no legitimate place in physics, while past and future were distinguished by the thermodynamic arrow of time. Grünbaum not only defends a 'temporal' reading of Minkowski space-time diagrams, he also rejects the view that Weyl was a proponent of the block universe. See Grünbaum, 'The Meaning of Time' (1971), 214 and *Philosophical Problems of Space and Time* (^{2}1973), 327–8. The fact is that many physicists were simply unable to solve Einstein's problem. Their views on what the admissible consequences of the physical discoveries are, remain inconsistent. A close reading of Einstein, Eddington and Weyl reveals, as we have seen, their commitment to the block universe and their endorsement of an idealist notion of time. Only in his later writings did Weyl consider the impact of quantum mechanics on the question of determinism; see Weyl, 'Open World' (1976).

a long remembrance of the past, we have numerical measures of *how long ago* a past event happened.

Grünbaum's insistence on the presence of human conceptualisation as a necessary condition for the awareness of an event 'coming into being' or as 'belonging to the present' trades on the systematic ambiguity in our notion of time. When he attributes 'timelessness' to the physical world, he has the full-blown notion of time in mind. Physical events do not carry dates as physical properties. But when he holds that timelessness does not mean changelessness, he has the minimalist Leibnizian conception of time in mind. Although events do not carry dates, they succeed each other in a 'before-after'-relation. And where there is *change*, according the relational view of time, there is time, for there is an order of the succession of events.[145] In his quote Grünbaum alludes to a feature, which we have already stressed. *Transitional* phenomena in nature exist, independently of the existence of conceptual awareness. If the history of world lines and the transitional nature of material existence are taken into account, the Minkowski representation of the Special theory of relativity becomes compatible with a *philosophy of becoming*.

We have entertained some doubts whether the *philosophy of being* is a genuine consequence of the Minkowski representation of the Special theory of relativity.[146] To show that it is not a genuine consequence, it is not enough to present a *philosophy of becoming* as a mere alternative. Rather, we have seen that a fundamental postulate of the Special theory of relativity – the finite velocity of light – confers a history on the world lines. The relativity of simultaneity, on which proponents of the block universe base their arguments for a *philosophy of being*, is itself derivative of this fundamental postulate. Between *time-like* connected events there is a unique, irreversible, succession of events – the spatio-temporal distance between them can never be zero or smaller than zero. Between *space-like* connected events, relative simultaneity is merely a representational convenience and does not provide 'proof' of a block universe.

Given the deficiency of a *philosophy of being*, its alternative must muster some positive arguments to show that the Special theory of relativity is compatible with a *philosophy of becoming*. These positive arguments cluster around the irreversibility of causal propagation, the transitional nature of time and the emergence of novelty, and finally the idea of cosmic time.

The *philosophy of being* may not be *the* consequence of four-dimensional space-time, yet it holds a powerful sway over the physicists' minds. An answer to the

---

[145] What Leibniz meant by the expression 'order of the succession of events' is made clear by the Special theory of relativity: *time-like* connected events are irreversible but are restricted to the interior of the observer's light cones; *space-like* connected events are not causally connected and can either see their order reversed or can be make simultaneous with the origin of another observer's coordinate system by an appropriate choice of the coordinate system. See Joos, *Theoretical Physics* (1934), 247; Minkowski, 'Zeit und Raum' (1909/1974), 61

[146] For some related and unrelated points about the inadequacy of the block universe, see Čapek, *Philosophical Impact* (1962), 385–6 and 'Myth of Frozen Passage' (1965); see also Horwitz *et al.*, 'Two Aspects of Time' (1988)

question, posed in a newspaper column, whether the essence of life was time, solicited the following response:

> According to Einstein's theory of general relativity, the universe that all life occupies is a four-dimensional continuum in which the one dimension of time cannot be separated from the three of space. More useful to the purpose is the distinction between "time" and our consciousness of it.
>
> Baffling as it appears to almost everyone not versed in contemporary theoretical physics, it is far from unorthodox for scientists categorically to deny any meaning or reality whatever to time (...). Our subjective sense of "nows", "thens" and "futures" is revealed as precisely that – solipsism and illusion with no place in the scheme of things.[147]

4.4.2 Philosophical Consequences II: The Philosophy of Becoming

The Minkowski space-time concept may have far-reaching implications for the comprehension of the world around us. In the minds of many physicists it demonstrated that the passage of time was a mere human illusion. The physical universe is static. All events, which will ever exist, are already real for some distant observer. The combined authority of an older generation of physicists – Eddington, Einstein, Gödel, Jeans, Minkowski, Weyl – and a newer generation – Costa de Beauregard, Barbour, Davies, Gold and Rietdijk – has given the conception of the *block universe* scientific respectability. A highly successful scientific theory seems to prove a philosophical point of view. Those who wish to embrace a *philosophy of becoming* must deny that the block universe is the only rational consequence, which follows from the principle of relative simultaneity. They must affirm that 'to accommodate everybody's nows' does not imply that 'the division of time into past, present and future' is physically meaningless.[148] In particular, they must throw doubt on the idea of a *proof* of the conception of the block universe from the result of relative simultaneity. That means that all the conceptual and empirical facts about the Special theory of relativity, and newer cosmological theories, must be taken into account before we can accept a 'proof' of the block universe. Some of these facts, however, seem to be more compatible with a *philosophy of becoming*. Although the authority of the master physicists discouraged a plea for a *philosophy of being*, at least some physicists, like Paul Langevin in the 1920s adopted a *dynamic* interpretation of the Special theory of relativity. Taking up Cunningham's suggestion that world lines have a history, Langevin added that this history was irreversible. The language of irreversibility belongs to thermodynamics. As we have seen Einstein, too, reverted to irreversibility to express doubts, towards the end of his life, about the idea of a static universe. In the main, it seems to have fallen to philosophers to explore the compatibility of the theory of relativity with the *philosophy of becoming*. As the

---

[147] This comment appeared in Notes & Queries of *The Guardian* (25 January 2001). The article is signed by Dr. Paul Underhill of the Open University. Paul Davies reiterates this view in a recent article in the *Scientific American* (September 2002).

[148] Davies, *About Time* (1995), 71 but Davies is tempted to adopt the block universe.

crucial argument against the objective status of the *transience* of time is that the relativization of simultaneity implies a relativization of the succession of events, the task is to show that it does not follow from relative simultaneity that the succession of events loses its objective meaning.[149] In other words, the relativity of simultaneity does not mean that the succession of events is purely relative to an observer's viewpoint. A *dynamic* interpretation of the Special theory of relativity leads to a *philosophy of becoming.*

Let us review some of the facts from the Special theory of relativity, which would support a *philosophy of becoming.*

Consider first *irreversible processes.* In his objection to Gödel, Einstein had pointed out that a *time-like* world line from *B* to *A* satisfied the fundamental 'before-after'-relation (Fig. 4.8). It is in fact a fundamental result of the Special theory of relativity that the entropy of a system is frame-independent.[150] The entropy of closed systems is their universal tendency, loosely speaking, to develop over time from a state of relative order to a state of maximal disorder. Frame-independence of entropy means that from every reference frame the same tendency of closed systems to increase their disorder will be observed. Entropic order is not reversible by a convenient choice of reference frame. It is a difficult and separate question whether entropy is the basis of the direction of time.[151] Even if the arrow of time cannot be identified with entropic processes, such processes are irreversible in every reference frame. Entropic processes bear witness to change and the succession of events from a lower to a higher state of entropy. This is an *invariant* of the theory and can lay claim to being part of objective reality. By this criterion, then, the transience of entropic states is evidence of real physical becoming, on which all observers in every reference frame will agree.

Consider, secondly, the argument from the *irreversibility of causal propagation.* The propagation of a signal from an earlier to a later space-time event is an irreversible process. The irreversibility of world lines not only means that they acquire a history. It also means that there is an asymmetry between an earlier and a later space-time event, such that causal signals can be sent from the earlier to the later event.

Causal propagation only concerns *time-like* connected events. At least as far as *time-like* connected events are concerned, their order of succession is the same for all observers. Even proponents of the *philosophy of being* agree that between *time-like* connected events, the 'before-after'-relation is irreversible. Observers in relative motion to each other cannot agree on the set of events, which they call 'simultaneous'. But this does not mean, contrary to what Gödel affirms, that the notion of simultaneity loses its objective meaning. As long as observers do not reside

[149] Čapek, 'Time in Relativity Theory' (1966), 439ff, 'Relativity and the Status of Becoming' (1975); Whitrow, *Natural Philosophy* (1980), §§7.5, 7.6; Newton-Smith, *Structure of Time* (1980), Chap. III (ii) and 'Space, Time' (1988); McCall, *Model* (1994), Chaps. I, II, 'Time Flow' (1995)

[150] Einstein, 'Relativitätsprinzip' (1917), §15

[151] See Sklar, *Philosophy of Physics* (1992), Chap. 3 and *Physics and Chance* (1993); Savitt *ed.*, *Time's Arrow* (1995); Price, *Time's Arrow* (1996)

outside of each other's light cones, they can make use of the Lorentz transformations (Box 4.1) to communicate their respective temporal and spatial measurements to each other. Equipped with the Lorentz transformations, each observer can *predict* two things: a) the other observer's temporal and spatial measurements for the duration of events and b) the other observer's determination of the simultaneity plane of events. There is no *subjective* ingredient in these calculations.[152] All observers moving relative to each other at constant speed will agree on the same temporal order – but disagree on the duration and simultaneity of events!

The succession of causally related events is a topological invariant, independent of our choice of reference frame. The invariance belongs to the space-time interval ds and concerns the relations $ds^2 = 0$ (the null interval) and $ds^2 > 0$ (the *time-like* interval). In other words, as Langevin pointed out[153], world lines of this kind are irreversible. They may coincide in space, but cannot be made to coincide in time. The order of events cannot be inverted by a change of reference system.

By contrast, in the case where $ds^2 < 0$, the case of a *space-like* separation of events without possibility of causal links, there is no definite order of succession of events. These space-time events cannot belong to the same world line. They are truly independent. By an appropriate choice of the reference system, they can be made to coincide in time.[154] It was the existence of *space-like* separated events, which particularly motivated Rietdiejk and Putnam to attempt a 'proof' of the universal determinism and eternal reality of space-time events. We have already observed that proponents of the *philosophy of being* place the burden of their proof on the arbitrary reversibility of the temporal relations between *space-like* connected events. From such conventionality no serious conclusions about the reality of events in the physical world can be derived.

> When two events are in each other's absolute elsewhere, so that they can have no physical contact, it makes no difference whether we say they are before or after each other. Their relative time order has a purely conventional character, in the sense that one can ascribe any such order that is convenient, as long as one applies his conventions in a consistent manner. As we have seen, observers, moving at different speeds, and correcting for the time Δt, taken by light to reach them from a point at a distance r by the formula $\Delta t = r/c$, will arrive at different conventions for assigning such event as before, after, and simultaneous with some event taking place in the immediate neighbourhood of the observer. But as long as there is no physical contact, which is the basis of the relationship of causal connection of

[152] Langevin, *La Physique* (1923), Ch VI; Bohm, *Special Theory of Relativity* (165), 151. For similar considerations, see Reichenbach, *Philosophy of Space and Time* (1957), 146; Stein, 'On Einstein-Minkowski Space-Time' (1968), 12 and 'On Relativity Theory and the Openness of the Future' (1991)

[153] Langevin, *La Physique* (1923), 287; Čapek 'Time in Relativity Theory' (1966), 440–7; Mittelstaedt, *Philosophische Probleme* (⁴1972), Chap. I; Bohm, *The Special Theory of Relativity* (1965), §§XXVIII, XXXI; Bunge, *Causality* (1979), §§4.2, 5.5

[154] Langevin, *Physique* (1923), 285f; Reichenbach, *Philosophy of Space and Time* (1958), Chaps. II, III; Čapek, 'Time in Relativity Theory' (1966), 443f, 446–7, Stein, 'On Einstein-Minkowski Space-Time' (1968), 5–12

> events, it does not matter what we say about which is before and which is after. On the other hand, as we have seen, where such causal contact is possible, the order of events is unambiguous, so that the Lorentz transformation will never lead to confusion as to what is a cause and what is an effect.[155]

Both the proponents of the *philosophy of being* and the *philosophy of being* agree on the fundamental postulate of the Special theory of relativity: the invariance of the velocity of light in all inertial reference frames. This means that light cones, emanating from events, have an invariant structure. Light cones do not tilt. And it is agreed that physical trajectories or causal signals travel inside the light cones between events.[156] On the other hand, if we allow '*space-like* curves to represent possible physical trajectories', then well-known paradoxes arise: the lottery swindle, involving superluminal signals between Alice and Zoë and the indeterminate temporal order, which Rietdijk, Putnam and Weingard regarded as 'proof' of the block universe.

More can be said in favour of a *dynamic* interpretation of the Special theory of relativity, especially if we attend to the *transitional nature of events*. This is a *third* argument in favour of a *philosophy of becoming*.

The relativisation of simultaneity has led many physicists to the conclusion that *Now*, the present, has no objective status. The *transitional* nature of time, the concept of *becoming* and its relation to past, present and future has been located on the level of subjective consciousness.[157] But we begin to see that this view may draw hasty philosophical conclusions from the theory of relativity. Entropic processes are frame-independent and invariant. The asymmetric, causal propagation within light cones is subject to a very basic 'before-after'-relation, which does not require conceptual awareness. According to the relational theory of time, such entropic and causal successions of events constitute time.

The relational view of time sees the 'before-after'-relation between events as the fundamental sequence, which lies at the root of the notion of time. But a 'before-after'-relation may lead to a rather undynamic view of the universe. Events are juxtaposed like beads on an abacus. In the four-dimensional view, the universe is criss-crossed by world lines of particles, the intersections of which are the only observable phenomena. But physical systems have a past, which affects their present and has implications for their future. As we argued earlier, physical systems go through *transitional* states.[158] The interaction of the components of a physical system in its past state can lead to qualitatively new properties and states of the

[155] Bohm, *Special Theory of Relativity* (1965), 159–60; Frank, *Philosophy of Science* (1957), 158–64

[156] Friedman, *Foundations* (1983), 161; Whitrow, *Natural Philosophy of Time* (1980), §7.6; Čapek, 'Time in Relativity Theory' (1966), §4

[157] Whitrow, *Natural Philosophy of Time* (1980), §7.6; Čapek, 'Time in Relativity Theory' (1966), §9

[158] We may think here of quantum jumps or the transition of quantum systems from preparation to measurement; cf. Costa de Beauregard, *Time* (87), Part IV and the contributions in Savitt *ed.*, *Time's Arrows* (1995); Čapek, *Philosophical Impact* (1962), 338–41; Popper, *Quantum Theory* (1982), 185–6

system.[159] We are familiar with such *transitional* phenomena from our daily experience with the world around us. A seed grows into a young plant. The young plant grows into a tree. At each stage, this growth process is in a transitional state from a past state through a present state to a future state. Fossils records equally tell a story of change and the succession of events. If such transitional phenomena are taken into account, should we say that the ultimate significance of time resides in its *transitional* nature?[160]

The *transitional* nature of time – in the sense of the relational view as the order of the succession of events or spatio-temporal events and not as just a subjective awareness of the passage of time – gives rise to the appearance of genuine *novelty* in the universe.[161] For the *philosophy of becoming* this is an important aspect of a dynamic interpretation of four-dimensional space-time. In the static interpretation the emergence of genuine novelty plays no part, as the universe simply *is* and *becoming* is a human illusion. The notion of physical becoming – transitional states between physical systems, the emergence of novelty – is the litmus test of a dynamic interpretation of space-time.

How can this transitional aspect of physical time – the emergence of genuine novelty – be included in a dynamic interpretation of space-time? Those who believe in the reality of temporal relations in the physical universe will turn to *irreversible* processes as the root of the flow of time.[162] The irreversibility is not exhausted by the entropic processes, mentioned earlier. Entropic processes lead to a maximum of disorder. The irreversible processes we have in mind here lead to genuinely new systems of order. Traditionally, physicists have distinguished between the *time-reversible* processes on the micro level of atomic particles and *time-irreversible* processes on the macro level of everyday objects. They take the laws of nature to be time-symmetric: the laws display no arrow of time. The temporal symmetry of microscopic processes led them to believe that temporal becoming and the arrow of time were macroscopic illusions. On a fundamental physical level, there was no distinction between future, present and past because the laws, which govern such systems, lack an arrow of time.

But if temporal asymmetry is shifted to the level of microscopic processes then both temporal asymmetry and the emergence of novelty will exist at a much more fundamental level. Such a shift is suggested by non-equilibrium thermodynam-

[159] This is strongly argued by Bohm, *Causality and Chance in Modern Physics* (1957); Popper, *Quantum Theory* (1982), §§23–25; from the point of view of thermodynamics, see Prigogine, *End of Certainty* (1997)

[160] Whitrow, *Natural Philosopy of Time* (1980), §7.5; Zwart, *About Time* (1976), 47–8; Bunge, *Causality* (^{3}1979), §§4.2, 5.5; Dieks, 'Special Relativity and the Flow of Time' (1988), 456–60; Prigogine, *End of Certainty* (1997); Popper, *The Open Universe* (1982), §§1, 11, 19, 26; Smolin, *Three Roads* (2001), Chap. 4

[161] Barbour, *The End of Time* (1999), 45, 69–70, 333; Čapek, *Philosophical Impact* (1962), Chap. XVII; Popper, *Open Universe* (1982), Chap. IV

[162] Prigogine, *End of Certainty* (1997), Prigogine/Stengers, *La Nouvelle Alliance* (1979); see also Haken, 'Laws and Chaos' (1995), 227–47; Savitt *ed.*, *Time's Arrow* (1995); Rosenberg, 'Statistical Causality' (1972/1974)

ics. Instability and probability are not the result of our coarse-grained view of the world, but are inherent features of the fundamental microscopic processes such that "macroscopic irreversibility (is) the manifestation of the randomness of probabilistic processes on a microscopic scale."[163] Classical physics deals with 'simple' mechanical systems, in which time could be a mere geometrical parameter. But with the discovery of complex systems, time becomes an 'emerging' property.[164] Irreversible processes produce self-organizing order: new *dissipative* structures. Irreversibility leads to the flow of time.[165]

The notion of the *flow* of time is not meant in a metaphorical sense but in a thoroughgoing physical sense: 'the flow of time depends on a history of events.'[166] This history of events is expressed in world lines but also in their interactions. This intersection of world lines harbours the possibility of genuine novelty (dissipative structures).

Recent developments in cosmology have provided a fourth argument in favour of a *philosophy of becoming*: the possibility of *cosmic time*. The idea of the block universe emerged as a result of the need to adopt relative simultaneity. When reference frames carry clocks, which indicate different times depending on motion, it may seem that the physical universe does not partake in time. Time and the passage of time are a human illusion. But from the point of view of the General theory of relativity, the Special theory is only an approximation. It applies to very limited local parts of the universe. When the General theory is applied to the universe as a whole, a cosmic time scale is permitted. Such a cosmic time scale measures the history of the universe. However, all cosmological models have to assume that the universe looks the same from every perspective (cosmological principle).[167] The cosmic background radiation, a cold afterglow of the Big Bang, discovered in 1965, provides the physical basis for the conception of such a cosmic time. The cosmic background radiation allows cosmologists to formulate a special and unique reference frame, from which a global time ordering, with the universal division of time into past, present and future would be possible. The requirement for

---

[163] Prigogine, *End of Certainty* (1997), 60; there is strong experimental evidence for microscopic chaos; see Gaspard *et al.*, 'Experimental Evidence' (1998); Dürr/Spohn, 'Brownian Motion' (1998); Sklar, *Physics and Chance* (1993), Chap. 7

[164] Prigogine, *End of Certainty* (1997), 60

[165] Prigogine, *End of Certainty* (1997), 160; see also Schlegel, 'Time and Thermodynamics' (1966); Watanable, 'Time and the Probabilistic View of the World', (1966). *Dissipative structures* can be described as 'systems capable of maintaining their identity only by remaining continually open to the flux and flow of their environment; see Briggs/Peat, *Turbulent Mirror* (1989), 139

[166] Prigogine, *End of Certainty* (1997), 170

[167] The idea of a *cosmic* or *universal* time is frequently discussed in relation to the discovery of the cosmic background radiation and its implications for cosmology. See Barrow/Tiper, *The Anthropic Cosmological Principle* (1986), 126, 194, 216 (fn 304), 601, 627; Costa de Beauregard, *Time* (1987), 23–4, 127; Davies, *About Time* (1995), 128; Ferris, *The Whole Shebang* (1998), 159; Shallis, 'Time and Cosmology' (1987), 71; Sklar, *Space, Time and Spacetime* (1974), Chap. V; Liebscher, *Einsteins Relativitätstheorie* (1999), 105–6, 151

such a unique reference frame, permitted by the General Theory of Relativity, is that from its point of view the cosmic microwave radiation looks rather uniform in all directions across the universe. Very slight variations or *ripples* have been measured in the microwave afterglow[168] revealing information about non-uniformities in the very early universe, which produced the formation of stars and galaxies. Cosmic time would give rise to two corrections in our views of time. On the one hand it would correct Newton's view of absolute time as being independent of material processes and the same for *all* observers. Rather there is one unique reference frame to which all individual observers would refer. On the other hand, it would correct the Einstein-Minkowski view of relative time as being defined exclusively by the reference frames, to which individual clocks are attached. It would still be true, *from the point of view of the Special theory of relativity*, that there 'are as many times and spaces as there are Galilean reference systems.'[169] But an inference from the validity of the Special theory of relativity to the existence of a block universe would not be legitimate. For it would be possible to refer all observers, who remain attached to their individual reference frames, to a *cosmic* time and history.

The existence of such a cosmic time is also supported by recent discoveries about the flatness of the universe and its eternal expansion. The 19th century image of the Heat Death may come true after all. New cosmological data[170] indicate that the universe will expand forever and dissipate all the available energy into a cold cosmic sea of radiation. There may have been a Big Bang but there will be no Big Crunch. The universe expands in one direction. Once the expansion rate is known, it will form the basis of a cosmic clock. Leibniz will have been vindicated. The universe is a clock and it is made of cosmic matter.

It is interesting to note that Arthur Eddington, whom we encountered as a staunch defender of the block universe, conceded the possibility of cosmic time. Relativity theory, he wrote, 'is not concerned to deny the possibility of absolute time, but to deny that it is concerned in any experimental knowledge yet found.'[171] This experimental knowledge seems to have been found now.

Eddington's concession is in perfect agreement with the view of the scientist-philosopher that philosophical notions should remain open to modifications due to new experimental discoveries.

Why then did a long line of physicists embrace the notion of the block universe? The answer is that the Special theory of relativity was developed and adopted in a cultural climate, in which determinism and the idealist view of time lay unquestioned before the physicists' eyes. They were presuppositions, which guided physicists in their attempt to make physical sense of their discoveries. Moreover, the Kantian model of time seemed to find empirical support in a central discovery of Special theory of relativity: the relativity of simultaneity.

---

[168] Rees, *Just Six Numbers* (2000), 117–21; The Once and Future Cosmos, *Scientific American* (2002)

[169] Pauli, *Theory of Relativity* (1921/1981), 14–5

[170] *Nature* **404** (2000), 939–40

[171] Eddington, *Space, Time & Gravitation* (1920), 163

However, things are not as straightforward, as the proponents of the block universe would make us believe. When more of the empirical and conceptual facts of the Special theory of relativity are taken into account, the evidence seems to point more in the direction of a *philosophy of being*.

4.4.3 The Emergence of Time

The Special theory of relativity has not solved the enigma of time. On the contrary, the *problem of time* has come into sharp focus in quantum cosmology and quantum gravity. The need to develop a quantum theory of gravity arises from the incompatibility of two fundamental physical theories: the Quantum theory and the General theory of relativity. The desire for unification calls for the removal of this inconsistency. The Special theory of relativity had left a fixed Minkowski space-time structure, in the absence of all gravitational fields. As Einstein did not consider accelerated motion or the effect of matter on space-time, an aftertaste of privileged reference frames lingered on, against the intention of the generalized relativity principle. In his *General* relativity, Einstein endeavoured to make the equations of motion covariant. They had to be the same in *all* reference frames. At the same time, instantaneous action at a distance, built into the Newtonian conception of gravity, was anathema to Einstein's field conception of physics. Einstein therefore identified space-time with the existence of matter and energy fields in the universe. Space-time becomes a *dynamic* entity, in which clocks and rods experience the effects of space-time locally. In dynamic space-time light cones are allowed to tilt, for instance in the vicinity of Black Holes. The Lorentz transformations no longer hold. Clocks run differently from neighbourhood to neighbourhood in dynamic space-time. Photon geodesics are no longer invariant. The assumption of the constancy of light in vacuum no longer holds universally. The space-time interval now reads $ds = \sum_{\sigma\tau} g_{\sigma\tau} dx_\sigma dx_\tau$. The theory of general relativity makes space-time dynamic and concentrates on the large-scale features of the universe. However, the theory breaks down in extreme quantum conditions, near Black Holes and the Big Bang. Standard quantum theory deals with small-scale features of the universe but it retains an external time parameter. Its fundamental equation is the Schrödinger equation. As we shall discuss in Chap. 5 it describes the deterministic evolution of quantum systems. But it relies on a traditional Newtonian or Minkowskian understanding of time. To overcome the inconsistency between these two theories, space-time must be quantized. The matter-energy fields are then subject to quantum fluctuations, which are neither smooth nor continuous. Hence the external time parameter of the Schrödinger equation can no longer be used. The smooth space-time structure of the General theory of relativity gets distorted by the quantum fields. A quantum theory of gravity – a theory of quantized space-time – has the advantage of unifying gravity with the quantum phenomena. Things can get more ambitious. Quantum cosmology treats the whole universe as a quantum system. Quantum theories of gravity and quantum cosmology all face the puzzle of the existence of space, time and space-time. Several contenders are at play.

String theory has all the trappings of a Theory of Everything. It starts from a fundamental quantum theory of strings. The *canonical approach* starts from a quantization of the General theory of relativity. And some of its versions are set up as quantum cosmology. Both approaches deal with such minute dimensions that they leave a large hiatus between the familiar features of the classical world and the unfamiliar features of the quantum world. Their task is to explain the *emergence* of classical features from a deeper-lying reality.

Our concern is with time. From this perspective the work of string theory is disappointing. The vibrational states of the fundamental strings in a multidimensional space are said to produce the familiar properties of classical particles. But string theory presupposes a fixed space-time background. The strings vibrate against this background. String theory leaves space, time and space-time unexplained. In this respect the canonical approach is a more audacious attempt. It comes in various versions, two of which are philosophically interesting with respect to time.

Loop Quantum Theory.[172] Its first step is to quantize general relativity. Then it descends to the level of the unobservably small Planck scale. On this level we reach the smallest possible units: the Planck length (10^{-33} cm), the Planck time (10^{-43} s), as well as the Planck energy, mass and volume. These are *discrete* moments, which cannot shrink further. They are therefore quantized but they are invariant across all reference frames. As no fixed background of space and time is assumed, the world becomes a dynamic network of processes. Motion and change are primary. Events, not things, lie at the root of this relational universe. The Leibnizian ring of loop quantum theory is not accidental. Only the ontology has changed. Leibniz spoke of space as the coexistence of events, and of time as the order of this coexistence. Leibnizian events were of course classical, macroscopic events. But if these macro-events can be shown to originate in microscopic events, relationism reappears in terms of fundamental quantum events. In string theory, the fundamental entities are one-dimensional strings and higher-dimensional branes. In loop quantum theory, quantized loops constitute the fundamental entities. These quantized loop states are dynamical and evolve in spin networks. The spin networks of quantized loop states create space and time at the Planck scale. They are analogous to the coexistence of events, which constitute space in Leibnizian relationism. The spin networks then evolve, much like the Leibnizian events. As they evolve from one configuration to the next, they constitute time. Time is therefore still the order of the succession of coexisting quantum loops. The network of evolving relationships between quantum events constitutes microscopic space and time.

In attempts to formulate a theory of quantum gravity loop quantum theory is a minority approach. If it succeeded, it would be a fully relational theory of space-time. But some obstacles stand in its way. First it is not a truly fundamental new unifying theory. It may eventually be absorbed into a more powerful theory. Its value lies in its attempt to provide a truly *relational* theory of space and time

[172] Smolin, *Three Roads* (2001); Rovelli, 'Quantum spacetime' (1999); Butterfield/Isham, 'Emergence' (1999) and 'Spacetime' (1999) discuss the numerous difficulties involved in these programmes.

on the quantum level. Its commitment to relationism makes it philosophically attractive. The human world, however, knows nothing of the atomic structure of space and time. Space and time may be relational but they appear to be smooth and continuous. Loop quantum theory needs to explain how classical space and time emerge from the underlying quantum space-time.

The *emergence* of space, time and space-time is the central issue, which all approaches to a quantum theory of gravity must tackle. The recently developed programme of *decoherence* may offer some useful insights into the process of emergence.[173] Decoherence has become important both in reflections on the emergence of space-time and the interpretation of quantum mechanics. To see how *decoherence* works, consider another version of the canonical approach, the quantum theory of cosmology.

Quantum Cosmology. Its centrepiece is the Wheeler-de Witt equation, $H\Psi = 0$, which no longer contains any time parameter, t. This universal wave function describes a *timeless* superposition of matter-energy quantum states for the whole universe. So the time of our familiar world and the space-time of the physicist must be *emergent* properties. Such properties emerge at a new level, far beyond the Planck scale. At this higher level they are no longer reducible to the quantum fields, from which they originated. Familiar examples of emergent properties are dissipative structures, chemical properties of molecules, human consciousness and the evolution of life from inanimate matter. More mundanely, the solidity and shape of a homemade cake are emergent properties. The emergence of time cannot be a process *in* time, because such a way of speaking presupposes the existence of an external temporal parameter, against which emergence could be timed. But it is the emergence of space, time and space-time, which require explanation. Herein lies the importance of the programme of decoherence. First there is a mathematical procedure, in which the Schrödinger equation arises from the Wheeler-de Witt equation by a mathematical process of approximation. The second element is decoherence. It explains the emergence of classical properties, like time, the localization of macroscopic objects and the spatial structure of molecules, as a *physical* process.

The basic idea is this: for reason, which we will discuss in the next chapter, quantum theory is *the* universally valid theory. Classical theories are only approximations. Adopting the language of quantum mechanics, decoherence assumes that there is a universal *entanglement* of all objects in the universe. For present purposes we can regard this entanglement as another example of the interrelatedness of all physical systems in the universe. In the classical world, this quantum nature of objects does not appear to human observers. Classical properties, which human observers register, are the result of a dissipation of the quantum correlations into the surrounding environment. This makes objects and their properties *appear* as localized. The universal entanglement between quantum objects leaks out when quantum systems are coupled to the environment. This physical process is called *decoherence*. When quantum systems decohere, they acquire classical properties.

[173] See Joos, 'Classical Spacetime' (1986); Zeh, 'Emergence of Classical Time' (1996), *Direction of Time* (1992), Chap. 6.2; Kiefer, 'Decoherence' (1996), 146–56, 'Decoherence' (2000)

In the next chapter we will see how this works for quantum systems. Classical space-time emerges from a superposition of quantum states in a timeless quantum world. This superposition of quantum states gets decohered through interaction with environmental factors. This occurs because the environment performs a 'continuous' measurement on the superposition of quantum states. Relevant variables ('the system') get separated from irrelevant ones ('the environment'). Although the universe possesses no environment, it is assumed that some variables (like matter states) can act as environment to the relevant variables, which constitute the 'system'. It is only in these relevant subsystems that classical behaviour emerges. According to the programme of decoherence, the irrelevant variables never really disappear. They are not truly destroyed so that even classical systems remain correlated. But these quantum correlations become observationally unobservable. When classical particles and their properties emerge, space, time and space-time emerge in a relational sense.

If space-time is an emergent property from a timeless quantum universe, does this mean that space, time and space-time lose their true significance? Decoherence ensnares us in a philosophically felicitous loop. On the fundamental quantum scale, time does not exist. At least this is what the Wheeler-de Witt equation tells us. (Loop quantum theory conceives of space and time as forming through evolving spin networks at the Planck scale.) When space-time emerges through a process of decoherence, it becomes available for treatment in Newtonian mechanics, the Special theory of relativity and philosophy. Even though space-time is an emergent property with no independent existence at the fundamental Planck level, any inference to an idealist conception of time is still mistaken. We do not spurn the cake just because its shape and solidity are emergent properties. Physical and philosophical time are not less worthy of consideration just because they emerge from a static quantum world. For emergence is a physical process, explained through a process of decoherence. 'The practical role of time remains unchanged as long as there are dynamic laws for relative motions.'[174]

If classical space-time emerges in the physical process of decoherence, Minkowski space-time must be a geometric representation (a map) of the homogeneous distribution of matter-energy states throughout the universe. It cannot exist over and above the material happenings in the universe. It is an approximation of a more fundamental timeless quantum world. Space-time exists due to the emergence of matter states, not due to the emergence of observers. We noted earlier that spatial and temporal dimensions have only perspectival reality. Space-time itself, in its Minkowskian instantiation, possesses invariance. But if space-time is only an approximation, then this invariance is relative to the Special theory of relativity. But this is acceptable: that the invariant is the real is no more than a *claim* about reality.

[174] Zeh, 'Was heißt: es gibt keine Zeit?' (2000); author's own translation

5

Causation and Determinism

"If Gessler had ordered William Tell to shoot a hydrogen atom off his son's head by means of an α particle and had given him the best laboratory instruments in the world instead of a cross-bow, Tell's skill would have availed him nothing. Hit or miss would have been a matter of chance."

A. Eddington, quoting Max Born in *New Pathways in Science* (1935), 82

5.1 Laplace and the Classical World

5.1.1 Rising Shadows

In 1874, as a young man, Max Planck went to Munich to consult the physicist Philipp von Jolly about his prospects of a career in physics. Von Jolly saw no reason to encourage the young man. Physics, he explained, had reached a high degree of maturity. The recent discovery of the principle of the conservation of energy had crowned its achievements. Mechanism had developed in the 17th century and thrived to perfection in the 18th and 19th centuries. The development of electromagnetism and thermodynamics had completed the magnificent edifice of physical knowledge. With the principle of the conservation of energy physics had acquired its final, stable form. According to von Jolly, no major new discoveries were to be expected in physics. Admittedly, there would still some 'specks of dust and bubbles' to be fitted into the edifice of physics but 'theoretical physics was approaching a degree of completion which geometry had possessed for hundreds of years.'[1] Von Jolly was not a man of great philosophical foresight. He did not see that field physics and thermodynamics were undermining mechanics. Within a few years of his advice to Planck, the science of physics was set to undergo even more profound modifications. These revolutionary changes would have profound effects on the way

[1] Planck, 'Vom Relativen zum Absoluten' (1924), 169; author's own translations. In his 1871 inaugural lecture at Cambridge University, J.C. Maxwell expressed a similar view, so did A.A. Michelson, *Light Waves and their Uses* (1903), 23–4.

science and its results were interpreted. Planck himself was to become a scientific revolutionary and a scientist philosopher. The understanding of new empirical findings increasingly resisted their interpretation in terms of classical physics. The edifice of physical knowledge was not complete. Planck, fortunately, ignored von Jolly's advice and continued his study of physics. Von Jolly's confidence rested on the extraordinary success of physics. Success makes blind. What von Jolly described as 'specks of dust and bubbles' were in fact new far-reaching discoveries. They would affect the most fundamental notions of physics and have profound philosophical consequences. The serious questioning of the notions of causation and determinism at the beginning of the 20th century had its roots in the 19th century. In 1859, Gustav Kirchhoff and Robert Bunsen introduced spectral analysis. In 1897, Joseph John Thomson discovered the existence of electrons inside the atom. In 1900 Max Planck postulated a new fundamental constant in the analysis of atomic phenomena. Only a few years later, in 1902, Ernest Rutherford and Frederick Soddy formulated the transmutation theory of radioactivity and formulated a statistical decay law for radioactive elements. In 1913, Niels Bohr proposed a structural model of the atom, which was based on Rutherford's discovery of the atomic nucleus. And in 1917 Albert Einstein formulated his theory of spontaneous and induced emissions and absorptions, in which *chance* entered in a fundamental way into the picture of the physical world.

The atom, like light, changed the nature of physics. It also challenged, like Einstein's postulate of the finite propagation velocity of light, the philosophical assumptions, which had provided the framework for the construction of the physical world picture.

Many physicists contributed to the understanding and explanation of atomic structure. For several decades their attempts to understand the atom was accompanied by serious worries about the nature of causation and determinism. Contributions written in the great scientific journals of the day testify to the severity of the impact of the new scientific discoveries on established fundamental notions. To understand the philosophical consequences, which the cracking of the atom had, we must first understand the nature of these discoveries. And we must understand the conceptual presuppositions before we can consider the philosophical consequences, which the scientists themselves drew from their discoveries. Causation is one of the most prolific topics in philosophy. During centuries philosophers have developed various models of causation, the best of them in conformity with at least a limited selection of the scientific facts. This practice has led to a small number of alternative models of causation. But causation is a difficult notion. Often, deeply entrenched presuppositions about causation, presumably borne out by scientific fact, have encouraged scientists and philosophers alike to hold unquestioned views about the natural world. The notion of causation shows again that philosophical presuppositions can both constrain and misguide us about Nature. Let us not take the fundamental notions for granted. A careful study of the experimental evidence, especially experimental facts about the atomic realm, may lead us to the most adequate model of causation, at least for the purpose at hand. Such a model will have to be compatible with the quantum-mechanical evidence. The evidence provides the *constraints* against which the various conceptual responses to the new discoveries can be tested. We are dealing with Bohr's challenge to philosophy. This sets the agenda for this chapter.

5.1.2 Laplace's Grip on the Classical World

> If the whole prior state of the universe could occur again, it would again be followed by the present state.
>
> *J.S. Mill, A System of Logic (1843), Bk. III, Chap. VII, §1*

From the discovery of the laws of thermodynamics, the field concept, atomic structure and evolutionary theory to the formulation of the Special Theory of Relativity, the modern world has witnessed the demise of the mechanistic worldview. As we have seen, the cluster of elements making up the notion of Nature undergoes conceptual change differentially under the strain of empirical evidence. One of the unquestioned pillars on which the mechanical worldview rested is the identification of causation with determinism. It is this identification, which survived, largely intact, into the 20th century. We have discussed the conjunction of determinism and the block universe. With the quantum theory this last pillar of the mechanistic and even cosmic worldview begins to crumble.

Pierre-Simon, Marquis de Laplace (1749–1827)

One of the chief conceptual developments in the wake of quantum mechanics – the physics of the atom – is the separation of the notions of causation and determinism. Contrary to popular belief, quantum mechanics does not abandon causation. The common belief that quantum mechanics has led to a world of *a*causality is due to a failure to shed the classical identification of causation with determinism. This belief is widespread. It was nourished by on over-concentration on certain types of experiments. It had the seal of approval of Heisenberg and Bohr, the most prominent proponents of quantum mechanics. From the impossibility of determining the exact present conditions of an atomic system Heisenberg infers that quantum mechanics shows definitely 'the invalidity of the causal law.'[2] In a review of the relationship between physics and philosophy in the 20th century, the British physicist James Jeans echoes this sentiment. 'As discontinuity marched into the world of phenomena through one door, causality walked out through another.'[3] In order to understand what these authors mean by the demise of causation, we have to look at determinism.

---

[2] Heisenberg, 'Über den anschaulichen Inhalt' (1927), 197

[3] Jeans, *Physics & Philosophy* (1943), 127

For Heisenberg and Jeans rely on an identification of causation with determinism, which is largely due to Laplace. The ability to determine from the present state of a physical system, in terms of specific parameters, its state at a later point in time, in conjunction with the knowledge of lawful regularities under which the parameters of the systems evolve, is one way of defining *determinism*. It is tacitly assumed that this understanding also provides the meaning of *causation*. This leads to a *functional view of causation*: one of the most influential models of causation during the reign of the classical worldview. But one of the philosophical lessons of the new discoveries is that the conceptual map of quantum mechanics allows for the coexistence of the notions of indeterminism and causation. This leads to a new model of causation, popular amongst scientists and some philosophers: the *conditional model of causation*. Yet even today the notions of causation and determinism are still frequently used interchangeably.[4]

Before we spell out the functional view of causation, let us look at its origins. The functional model, the identification of causation with determinism, is a central feature of classical physics and its philosophy. It has its origins with the founding fathers of modern science and was eternalised by Laplace's demon. In an often-quoted passage, Laplace derives determinism – the ability of a superhuman intelligence to predict future events – from the *axiom of a universal causal concatenation* of all events. This axiom, which Laplace adopts from Leibniz, is the Principle of Sufficient Reason. It states, in Laplace's words:

> The present events have a profound link with the preceding events, which is based on the obvious principle that a thing cannot begin its existence without a cause, which precedes it.[5]

This statement of the universal law of causation has had a respectable tradition in philosophy. Leibniz maintains

> That everything is caused by a determined destiny is as certain as $3 \times 3 = 9$. For destiny consists of the interdependence of everything as in a chain, and will take place infallibly as much so before it has occurred, as when it has occurred.[6]

[4] See Cushing, *Philosophical Concepts* (1998), 288, 290; Spielberg/Anderson, *Seven Ideas* (1995), Chap. 3; Pagels, *The Cosmic Code* (1994), 75; Penrose, *The Emperor's New Mind* (1990), 273–8; Margenau, 'Probability' (1978), 'Causality' (1978); Mittelstaedt, *Philosophische Probleme* (1972), Chap. V; Nagel, *Structure of Science* (1961), Chap. X; Jeans, *Physics and Philosophy* (1943), Chap. V; Eddington, *New Pathways* (1935), 74, 85; Rosenfeld, 'Idea of Causality' (1942); Frank, *Das Kausalgesetz* (1932), 167f, 239, 249, 287, 293, 296; Frank, *Philosophy of Science* (1957), Chap. 11; Schrödinger, 'Indeterminism in Physics', (1931), 53; Rae, *Quantum Mechanics* (^{2}1986), 210; Popper's attitude is ambivalent. In one part of Popper, *Open Universe* (1982), 149, causation and determinism seem to be identified, although in earlier parts of the book (1982, 4, 19, 23) Popper asserts that the principles of causation and determinism are not the same.

[5] Laplace, *Thèorie analytique* (1820), Introduction; cf. Laplace, 'La probabilitè des causes' (1774) ; see also Čapek, *Philosophical Impact* (1961), Chap. VIII

[6] Quoted in Mittelstaedt, *Philosophical Problems* (1976), 134–5

For John Stuart Mill, the universality of the law of causation consists precisely in the requirement that some cause (a set of antecedents) precede every effect (consequent). Mill also saw a link between cause and effect. This could not be one of invariable succession, since day and night follow each other invariably, yet the one does not cause the other. Rather, Mill held that the link between cause and effect had to be an invariable sequence of an unconditional kind.[7] That is, the set of antecedent conditions (cause) would, in all circumstances, uniquely lead to the appearance of the consequent conditions (effect). Or the existence of the effect was *conditional* on the prior existence of the cause. From the statement of this axiom, the Laplacean principle of determinism does not yet follow. For it may well be that some cause always produces the same effect (under the same circumstances, heat makes water always boil at the same temperature), yet this says little about the future state of the *whole* universe. To derive the predictability and retrodictability of future and past states of the whole universe from knowledge of the present boundary conditions and universal laws, Laplace must make a further assumption:

> We ought to regard the present state of the universe as the effect of its antecedent state and as the cause of the state that is to follow.

It is only under this further assumption of uniqueness of states that Laplace can derive his version of causal determinism:

> An intelligence knowing all the forces acting in nature at a given instant, as well as the momentary positions of all things in the universe, would be able to comprehend in one single formula the motions of the largest bodies as well as of the lightest atoms in the world, provided that its intellect were sufficiently powerful to subject all data to analysis; to it nothing would be uncertain, the future as well as the past would be present to its eyes.[8]

This is an extreme version of the mechanistic worldview, which was developed by Descartes and Newton, by Galileo and Boyle. It gave rise to the *clockwork* image of the cosmos. Laplace was not alone in his assumption. Paul Thiry d'Holbach, one

---

[7] Mill, *Logic* (1843), Bk. III, Chap. V, Sects. 2,6

[8] Laplace, *Théorie analytique* (1820), Introduction, VI–VII; the translation is quoted from Nagel, *Structure* (1961), 281–2. Recall that for the French physiologist Claude Bernard, *Introduction* (1865), 69, 87–9 *determinism* is an absolute principle of science. By it he understands 'the absolute and necessary relation between things' in animate and inanimate matter. As may be expected, this Laplacean determinism has attracted numerous discussions. See, for instance, Du Bois-Reymond, *Über die Grenzen des Naturerkennens* (1872); Russell, 'On the notion of cause' (1912); Schrödinger, 'Indeterminism in Physics' (1931); Frank, *Das Kausalgesetz* (1932), Chap. II; Cassirer, *Determinismus und Indeterminismus* (1936), 134–60; de Broglie, *Continu et Discontinu* (1941); Born, *Natural Philosophy* (1949); Nagel, *The Structure of Science* (1961), Čapek, *Philosophical Impact* (1961), Chaps. VIII, IX; Chap. X; K. Popper, *Open Universe* (1982); Earman, *A Primer on Determinism* (1986), 6–8; Cushing, *Philosphical Concepts* (1998). For useful formulations of determinism and indeterminism see Salmon, *Causality and Explanation* (1998), 115–6. For qualifying statements on the uniqueness condition see Maxwell, *Matter and Motion* (1877), Sect. 19 and Bunge, *Causation* (1979), 50

of the leading figures of the French *encyclopédistes* presented the cosmos precisely as a network of interlocking causes and effects. 'The universe', he wrote, 'reveals to us an immeasurable and uninterrupted chain of causes and effects.'[9] The *chain* metaphor, which serves d'Holbach to emphasize the interrelatedness of Nature, is replaced, in the cosmic view of Nature, by the idea of interlocking systems. The chain metaphor reinforces the uniqueness condition. The identification of causation with determinism is only possible under this extra assumption that 'the same cause always leads to the same effect'. Whilst we may primarily have thought of determinism as predictive ability, it now becomes *causal* determinism. If the universe is indeed a vast system of an 'immeasurable and uninterrupted chain of causes and effects' and if posterior events are causally uniquely dependent on prior events, then no other form of determination is possible. Laplace and d'Holbach took the interpretation of classical physics a step beyond the philosophical understanding of its founding fathers: from predictive to causal determinism (Box 5.1). *Predictive* determinism can be presented as an argument scheme, with two premises and a conclusion.

P_1 The initial conditions of a physical system are given in terms of a number of specific parameters characterising its motion (acceleration, momentum, velocity, spatio-temporal location)

P_2 The system will be subject to lawful regularities, especially as specified in Newton's mechanics and its extension to other domains of classical physics. Differential equations are of particular importance.

$\therefore C$ Under these two premises, the past and future trajectories of the system are uniquely specified, in the sense that their probability of occurrence is one, if there are no other disturbances on the system. If an observer has knowledge of P_1 and P_2, the observer will be able to make fairly precise predictions about the trajectory of the system.

Note that this characterisation makes no reference to any causal influences on the system's dynamic behaviour. The laws lay down the trajectory of the system (for instance a planet orbiting the sun), without specifying any particular cause for this determinate behaviour. *Causal* determinism is the stronger view[10], adopted by Laplace and d'Holbach. It adds a further premise to the argument.

P_3 The unique trajectories of the systems are produced by a network of interlocking causes and effects such that *successive* stages of the universe are the unique effects of anterior states; in turn the posterior states become unique causes of further posterior states, lying in the future.

Laplace's version of determinism vacillates between *predictive* and *causal* determinism (Box 5.1). From the causal concatenation of cosmic events, Laplace shifts the focus to their predictability, once the demon is introduced. It is Laplace's *predictive* rather than his *causal* determinism, which has become the focus of many criticisms in the 20th century.

[9] d'Holbach, *Système de la Nature* (1770), Chap. 1

[10] See Bunge, *Causality* (1979), 4; Earman, *Primer* (1986), 4–6

Box 5.1: Forms of Determinism

When we think of determinism a choice offers itself: should we adopt predictive or ontological determinism? *Predictive* determinism is about the cognitive abilities of competent observers to predict the future states of the world. *Ontological* determinism is about the type of state the world will be in, irrespective of the knowledge of competent observers, given that it is in a particular present state. We can make these characterizations more precise. To characterize predictive determinism, let us introduce the idea of predictors [Popper, 'Indeterminism' (1950), *Open Universe* (1982)]. Predictors are predicting machines with the ability to calculate the future and past states of the world from present conditions. Predictive determinism makes the calculating abilities of predicting machines (predictors) to acquire knowledge about the system's future evolution an essential feature of determinism. Insofar as the cognitive abilities of competent observers can be treated as approximations to the predictors, they also, like the predictors, should have the ability to predict all their futures states. This requires the knowledge of boundary conditions and the differential equations, which govern the system.

Laplace's version of determinism can be construed as predictive *or* causal determinism. Laplace's famous demon is an example of *predictive determinism*, since he appeals to a superhuman intelligence to which neither past or future are uncertain. The demon can predict successive states of the universe – or the world lines of particles – because for the demon it is *as if* they already had left their traces on the canvass of space-time. But since Laplace bases his superhuman intelligence on the assumption that 'we ought to regard the present state of the universe as the effect of its antecedent state and as the cause of the state that is to follow', he also assumes a *causal version of determinism*. This allows him to identify causation and determinism. But just as the assumption of a demon may be rejected as untestable, so the stipulation of causal determinism is too narrow, since determinism is not necessarily causal.

It may be argued that cognitive abilities of humans are too unreliable or variant to result in a satisfactory definition of determinism. An *ontological* definition of determinism dispenses with the human observer. Causal determinism is a form of ontological determinism. It should be cleansed of all assumptions of causation for these notions must be kept apart. We could define ontological determinism such that

> the variables of state for S are just the variables in the small subclass of mutually independent variables in terms of which the remaining ones can be defined. Accordingly, the set of laws L constitute a deterministic set of laws for S relative to K, if given the state of S at any initial time, the laws L logically determine a unique state of S for any other time [Nagel, *Structure* (1961), 281]; K is a definite class of properties; S is a system of bodies in isolation from other bodies. For a definition of ontological determinism in terms of possible worlds, see McCall, *Model* (1994), 12–3; Earman, *Primer* (1986), 13–4

Box 5.1 (continued)

This is not strictly speaking an ontological definition of determinism, since it appeals to 'definitions' and mathematical or logical determinations. This definition could be true of a model world, whilst the real physical world is indeterministic. The Schrödinger equation makes a deterministic statement about the trajectory of a quantum system but the outcome of quantum-mechanical experiments is probabilistic. In the last chapter (Box 4.3) we attempted a characterization of ontological determinism in terms of world lines.

Given such an ontological form of determinism, we can make further distinctions [see also Čapek, 'Doctrine of Necessity' (1951)]:

a) A *static* form of determinism. This is Laplace's causal determinism. For the Laplacean demon the world is a map on which all events and their coordinates are already entered. There is no distinction between past, present and future. This is an extreme form of determinism, since both the emergence of novelty and the passage of time are denied. It is a forerunner of the conception of the block universe.
b) A *dynamic* form of determinism. The state of the universe is 'fixed' forever from past to future, because deterministic laws restrict the trajectories of particles to unique trajectories, given the initial conditions of their state. Nevertheless, the world dynamically evolves from state to state. This is precisely the view Kant and Laplace offered of the dynamic evolution of the universe. It is equivalent to the view that world lines have histories.

Walter Nernst (1922), Richard von Mises (1930, 1931) and Philipp Frank (1932) objected to the extreme idealisation of a superhuman being, needed for the formulation of the Laplacean view. This superhuman spirit or demon needs to be able to determine the precise initial conditions and the exact form of the equations for a system of particles. This is beyond human capacity. The force laws, applying to most phenomena in the material world, are far more complicated than those of celestial mechanics. Laplace admits that a human mind remains far removed from the attainment of such an ideal, although certain approximations could be achieved in astronomy. Karl Popper (1982), however, interprets the Laplacean demon as a super-human scientist, surpassing a human scientist only by degrees of predictive ability. Popper then tries to show that the calculating precision, required to exercise Laplacean predictability, cannot be achieved in principle. Ernst Cassirer (1936) felt that Laplace had meant his 'formula' only as a metaphor and that, at any rate, it mixed discursive elements (knowledge of forces and momentary positions) and intuitive elements (comprehension of the future and the past) in a single formula. Louis de Broglie (1941, 59–61) raised three difficulties for Laplacean determinism: (a) the assumption of precise predictability is unrealistic, due to the universal interaction between all bodies in the universe; (b) all our observations and measurements are subject to error and (c) the atomic realm imposes inherent limits on the precision with which the parameters of atoms could be known simultaneously. David Bohm (1957, 158–60; see also Prigogine/Stengers 1979, Chap. II.4) emphasised the inadequacy of Laplacean determinism on the ground of outside contingencies and chance fluctuations, as well as the emergence of qualitatively new causes, new laws and new contingencies in the infinity of time. Ernest Nagel (1961, 281–3) accused Laplace of having committed a non sequitur in his claim that for the superhuman spirit nothing would be uncertain, since a mechanical system may undergo changes due to non-mechanical properties, which cannot be predicted from a mechanical point of view. Laplace could counter, of course, that all non-mechanical properties are reducible to mechanical properties, at least for the superhuman spirit.

While these objections show in various ways the limits of Laplace's conception of what came to be seen as *predictive* determinism, there is a further objection, which grows out of the heart of the Laplacean assumptions. As we have seen above, Laplace accepted Leibniz's Principle of Sufficient Reason, which anticipated the meaning that the law of causation would take for Mill: 'Every event must have a cause.' The whole anterior state of the universe causes the subsequent state of the universe to move along *one* trajectory, at the exclusion of all others. This makes sense in a clockwork universe: the hands of the clock move steadily forward from one position to the next in one direction. If this premise is denied, as it must be according to quantum mechanics, then Laplace's assumption of the unique causal concatenation of all events breaks down. On the one hand, the discovery of *radioactive decay* raised serious doubts about whether the present state of decay of an atom could be causally related to its antecedent state. On the other hand, the present trajectory of an *atom* may leave it with a number of probabilistic chances concerning its future trajectories. A fundamental quantum indeterminism may rule at the root of things. If the evidence does not warrant the statement that every event must have a cause or at least that every cause leads to a unique effect, then one of the premises of Laplace's argument is denied. The derivation of Laplacean determinism must fail.

Nevertheless the association of determinism with predictability has proved useful to the mechanistic worldview. Predictive determinism, even in classical physics, need not require the predictability and retrodictability of the future and past states of the *whole* universe. Natural systems, like the solar system, can be isolated from the rest of the universe to derive a limited knowledge of their future or past spatio-temporal states. Even then it only describes an idealisation. The boundary conditions cannot be known to an infinite degree of perfection, and the universal laws themselves relate parameters in degrees of idealisation and abstraction. The idealised picture is: From the knowledge of boundary conditions and the governance of universal laws it should be possible to make predictions about the future spatio-temporal location of the system under consideration. Equally it should be possible to retrodict past spatio-temporal states. Heisenberg describes this predictive determinism as the principle of Newtonian physics[11] (Box 5.2):

[11] *Der Teil und das Ganze* (1973), 52

Box 5.2:

Initial Conditions & Universal Laws ⟶
Predicting the State of the System in the near Future

Initial Conditions & Universal Laws ⟶
Retrodicting the State of the System in the near Past

Paradigm cases are provided by astronomy. 1) Voyager 2 was launched on August 20, 1977. On its journey it crossed the orbits of Jupiter, Saturn and Uranus at a distance of 100 000 km. Finally, it flew past its target, Neptune, on August 25, 1989 at a distance of less than 5000 km. The calculations of the orbits of Voyager 2 and Neptune are clear cases of *prediction* and essential for the success of this mission. 2) Neptune's discovery provides an illustration of *retrodiction*. The planet's existence had been calculated by two mathematicians, Urbain Leverrier in France and John Adams in Britain. Using their data, Johann Galle in Berlin discovered it in 1846. But Galle was not the first to perceive it.

> Neptune was first seen by none other than Galileo, 234 years earlier. Calculations of Neptune's orbit show that it should have been very close to Jupiter in the sky in January 1613. Galileo's journals have entries showing that he observed an object in the vicinity of Jupiter near Neptune's predicted position on December 27, 1612, and again on January 28, 1613, when Galileo detected a small motion of Neptune with respect to a nearby star.[12]

Given the extraordinary success of classical physics and the powerful grip of the Laplacean identification of determinism and causation on scientists and philosophers alike, it is not surprising to see that the preferred philosophical model of causation of the classical period is the functional model of causation.

5.1.3 The Functional Model of Causation

Laplace appealed to Leibniz's Principle of Sufficient Reason, and, in one fateful quote, made the influential identification of causal and predictive determinism. Later scientists, like von Helmholtz, Planck and Heisenberg, were influenced by Kant's notion of causation.[13] Consequently, they regarded causation as an a priori category, which was prior to all our experience of the world. But although Kant made causation a category of the mind, he did not break the bond between causation and

[12] Zeilik, *Astronomy* (1988), 219

[13] See von Helmholtz, quoted in Warren/Warren, *Helmholtz on Perception* (1968), 201, 208; Planck, 'Kausalgesetz und Willensfreiheit' (1923), 139–68; Heisenberg, *Der Teil und das Ganze* (1973), Chap. 10, 'Philosophische Probleme in der Theorie der Elementarteilchen' (1967), 414–22. Kries, 'Kants Lehre' (1924) assesses Kant's views about time and space for modern physics. Bergmann, *Kausalgesetz* (1929) assesses Kant's view of causation in the light of quantum mechanics. A more recent discussion of Kant's views of time and causation with respect to modern physics is Mittelstaedt, *Philosophische Probleme* (1972).

determinism. Echoing the Newtonian-Laplacean heritage, Kant stipulates that the notion of causation requires

> that something A should be such that something B follows from it *necessarily and in accordance with an absolutely universal rule.*

To the synthesis of cause and effect there belongs a dignity, which cannot be empirically expressed, namely that the effect not only succeeds upon the cause,but that it is conditioned by it and entailed by it.[14] We have already seen in the Chapter on Nature that Thomas Young, Hermann von Helmholtz, Claude Bernard and James C. Maxwell shared this view of determinism or causation as the ultimate maxim of science.[15]

The fundamental notions responded in a differential manner to the new discoveries. These notions must be adequate. The epoch of quantum mechanics has demonstrated, as von Mises put it, that the 'causal principle is *changeable* and will have to be *subjected to the demands of physics*.'[16] Not everyone agreed. So it is not surprising to find a proponent of the functional view of causation on this side of the quantum revolution. Einstein (1927) equated physical causation with the existence of differential equations. Russell, too, argued for a replacement of the law of causation in science by purely functional relations and differential equations. For Russell causal laws provide the ability to make inferences between events in four-dimensional Minkowski space-time. 'The law of causality (...) is of a bygone age, surviving, like the monarchy, only because it is erroneously supposed to do no harm.'[17]

The central notion of the functional view of causation is the differential equation. This is basically an equation, which involves an unknown function and its derivatives, i.e. the rate of change of this function with respect to some variable (such as time or distance).

The expression

$$v = \frac{x}{t}$$

is not a differential equation because it does not express the rate of change of the unknown function, v, with respect to x. But we can easily transform this expression into a differential equation by indicating the rate of change of v with respect to t, as follows:

[14] Kant, *Critique of Pure Reason* (1787), B124; italics in original. This quote is partly the author's own translation. For further discussion of Kant's notion of causation with respect to modern physics, see Cassirer, *Determinismus* (1936)

[15] The functional view of causation, with its proximity to determinism, was also prevalent amongst the founders of modern science; see Burtt, *Metaphysical Foundations* (1924); Crombie, *Styles* (1994)

[16] von Mises, 'Kausale und Statitistische Gesetzmäßigkeit' (1930), 146 (italics in original); see also Jordan, 'Kausalität und Statistik' (1927), 105. English translation: 'Philosophical Foundations of Quantum Theory' (1927), 566

[17] Russell, 'On the notion of cause' (1912), quoted by Mackie, *Cement* (1980), 143; see Rietzler, 'Krise' (1928)

$$v = \frac{dx}{dt}$$

The reduction of the notion of causation to the existence of differential equations seems to offer several advantages. First of all, differential equations play an essential part in the physical sciences. Second, the rate of change with respect to time captures the essential antecedence condition of causation and gives the causal dependence a mathematically precise form. Why differential equations are superior in this respect from ordinary (linear) equations can easily be gleaned from the following example.[18] If we take Newton's second law

$$\boldsymbol{F} = m\boldsymbol{a}$$

it is difficult to recognise any causal features in this formulation. The problem is that this statement is symmetrical. It can be read from left to right – a force, $\boldsymbol{F}$, exerted on a mass, m, will lead to the acceleration, $\boldsymbol{a}$, of this mass. It can also be read from the right to the left – a mass, m, with acceleration, $\boldsymbol{a}$, produces a force, $\boldsymbol{F}$. But it is easy to transform this expression into a differential equation, and thus into a causal form according to the requirement of the functional notion of causation:

$$\boldsymbol{a} = \frac{d\boldsymbol{v}}{dt}$$

$$\boldsymbol{F} = m\frac{d\boldsymbol{v}}{dt}$$

$$\frac{\boldsymbol{F}}{m} = \frac{d\boldsymbol{v}}{dt}$$

In these expressions the force, $\boldsymbol{F}$, can be regarded as the cause, which brings about, as its effect, a change in the rate of change of velocity, $\boldsymbol{v}$, with respect to time, t. This expresses the causal dependence, according to a certain mathematical statement, as required by functionalism. This antecedence of the cause over the effect is expressed in the differentiation of velocity with respect to time.

This functional treatment of the notion of causation has another important consequence. A mathematical dependence can be established between differential equations such that one can be shown to follow mathematically from the other. For instance, Newton showed that Kepler's third law could be derived from his inverse-square law of gravitation. But it is also possible to demonstrate mathematically that Kepler's third law implies that the gravitational force between two bodies varies inversely with the square of the radius between their centres. That is, Newton's law of gravitation is a mathematical consequence of Kepler's law. Born regarded the possibility of such mathematical derivations as the basis of the conception of causation in physics. 'For it is (...) the first and foremost example of a timeless cause-effect relation derived from observations.'[19] How, on the functional view, do

[18] See Frank, *Kausalgesetz* (1932), 142–46; Zilsel, 'Asymmetrie' (1927), 285; Cassirer, *Determinismus* (1936), 345. Bunge, *Causality* (1979), 74–98 for a general criticism of the functional view of causation.

[19] Born, *Natural Philosophy* (1949), 129–32

we causally explain, say, the orbit of a planet around the sun? According to Kepler's third law, the orbital period, T, of a planet P around the sun (or alternatively of a satellite or the moon around the earth) is related to the average distance, A, of that body from the sun (or the earth) by the relation

$$T^3 = A^2$$

(where A is measured in astronomical units, i.e. $1A$ equals the earth-sun distance). If Kepler's third law for planets (or satellites) mathematically entails Newton's law of gravitation, $\boldsymbol{F}_g$, then Newton's law adds an important component. Kepler's law in itself is a functional law, which lacks the causal components. These are added by Newton's law. The gravitational force, $\boldsymbol{F}_g$, is the cause of the planet's acceleration towards the sun (or of a satellite or the moon towards the earth). The planet 'falls' towards the sun and the moon 'falls' towards the earth. The force, exerted by the mass of the sun, becomes the cause, which brings about the effect – the 'fall' of the planet (Fig. 5.1). On this account, planet P is shown moving according to Newton's first law in a straight (dashed) line, with constant velocity, $\boldsymbol{v}$. But P is also subject to a central force, $\boldsymbol{F}_c$, from S, which makes it fall towards S with acceleration $\boldsymbol{a}$. The unbroken circle indicates the actual, idealized, orbit of the planet. It is the vector product of the joint operation of the two other factors, the central force and the inertial motion. The circular line, the shape and size of the orbit, are the *effect* of the composite *cause*, existing of the set of laws and the cluster of initial conditions. In this classic example of a causal story, some of the traditional features of the notion of causation are satisfied. The presence of the massive gravitational body, S, and smaller body, P, in rectilinear motion with respect to S, constitute the antecedent conditions. For a causal account of the present orbit it is not important to inquire into the temporal origin of the factors making up the causal antecedents. The joint antecedent conditions satisfy a conditional priority over the effect. The presence of the set of laws expresses a form of physical dependence of the planetary orbit on the force exerted by a massive gravitational body.

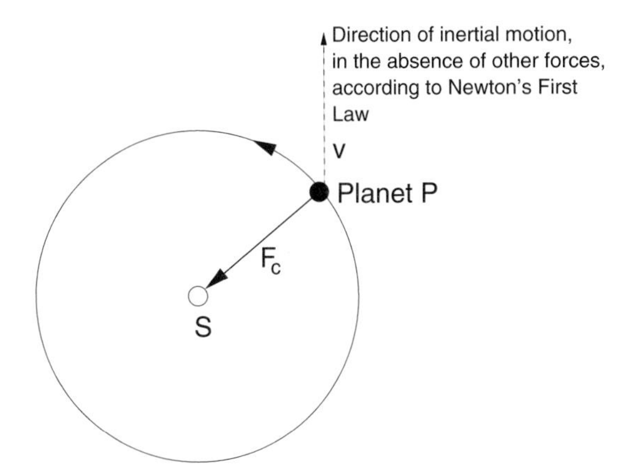

Fig. 5.1. A classic case of causal explanation – idealized, planetary orbits

What about a causal mechanism? In the traditional worldview of mechanism, causation required some fundamental features: *a set of antecedent and subsequent conditions, a mechanical link* or *trace between these sets of conditions, a local interaction between cause and effect* and *a dynamic production of the effect.* Newton made no serious attempt to explain how gravity could actually 'reach out' to the planet, in inertial motion, and pull it into an orbit. It is only a causal account, if we are not too specific about the gravitational pull of body S across the vast distance between S and P. As a causal story is not complete. In this classic example contiguity or the locality condition poses a problem, because gravitation seems to act at a distance.

The step from causation to determinism follows easily from the reduction of causation to differential equations. First there is the idea of lawful regularity, as expressed by Kant and others. This idea can be made precise by associating lawful regularity with the existence of differential equations. As we said, differential equations are concerned with the rate of change of one variable with respect to another. In particular they give the rate of change of one parameter with respect to time. They are therefore 'ideal' vehicles of predictability, which is one form of determinism. Once causation is reduced to temporal dependence – how parameters evolve along the temporal dimension – it becomes identical with predictive determinism.

In the view of functionalism we then arrive at the causal form of physical law.[20] Philip Frank, a scientist philosopher par excellence, provides an excellent account of functionalism. As Frank states: 'The causal laws in physics have a form such that they present the temporal changes of the state variables as certain functions of the present values of the variables.'[21] If we introduce the measurable quantities $u_1, u_2, u_3 \ldots u_n$, and some functions $F_j(u_1, u_2, u_3 \ldots u_n)$ such that these functions mathematically describe a change in these variables by the use of differential equations, the causal law takes the general form:

$$\frac{du_j}{dt} = F_j(u_1, u_2, u_3 \ldots u_n)$$

[where $u_1, u_2, u_3 \ldots u_n$ are the state variables of a physical system at time t_0, e.g. velocity, position etc. and the differential equations, the functions F_j, state the rate of change of the state variables with respect to t_0.] Have we made any progress? This characterisation of the 'causal' law is close to the characterisation of determinism. It is a small step from this characterisation to the denial of causal relationships in quantum mechanics. For what if such a rate of change of the state variables cannot be satisfied? Something is missing from the functional view of causation. Even though the rate of change of a quantity with respect to time has been built into the definition, it detracts us from the original causal situations, which rely on a conditional dependence of the effect on the causal conditions.

The functional view of causation leaves out too many features.

---

[20] Frank, *Kausalgesetz* (1932) 145–6; Rosenfeld, 'Idea of Causality' (1942), 'Causality in Physics' (1971)

[21] Frank, *Kausalgesetz* (1932), 144–5; author's own translation

☞ It tends to regard predictability as a criterion of causation, due to its proximity to the characterisation of determinism.

☞ It reduces the priority of the cause over the effect to a mere temporal relationship. But causal priority is not merely temporal. As Kant pointed out, the 'great majority of efficient natural causes are simultaneous with their effects (...) If I view as a cause a ball which impresses a hollow as it lies on a stuffed cushion, the cause is simultaneous with the effect.' According to Newton's third law if, say, a hand exerts a force on a wall then the wall exerts an equal and opposite force on the hand. The causal impact of the hand on the wall is not temporally prior to the causal impact of the wall on the hand. But there is an independence of the cause with respect to the effect. Or the hand's causal effect on the wall is conditionally prior to the wall's causal effect on the hand. If the hand had not pushed the wall, the wall would not have pushed the hand. As Kant emphasises, what we have to reckon with is 'the order of time, not the lapse of time.' Even when the cause is simultaneous with the effect, we still 'distinguish the two through the temporal relation of their dynamical connection.'[22] Temporally, the causal conditions must be put in place before the effect can arise. But even though the causal conditions are in place, the effect may not be triggered. There could be intervening factors. But if the effect does materialise, then it is conditionally dependent on the cause (even though temporally simultaneous).

☞ There is another, more serious shortcoming of the functional view of causation. It pays no attention to the cluster of causal conditions, which may precipitate or prevent the effect. It ignores the specific form of physical dependence of one set of factors (the effect) on another set of conditions (the cause) – in the ideal case, the causal mechanism. It does not tell causal stories.

This becomes apparent when we revert to the discussion of Kepler's third law. Even though it can be shown, mathematically, that Kepler's law entails Newton's law of gravitation, so that the sun exerts a force, which makes a planet 'fall' towards it, this, essentially, leaves out *why* the planet actually stays in orbit. To explain why planets stay in orbit around the sun, why satellites and the moon orbit the earth, a causal story must be told.

A *causal story* is a causal account, specifying a conditional dependence. It typically refers to a *cluster of initial conditions*, whose presence or absence may have an influence over the occurrence of the effect, and a *set of laws*, which provide the structural regularities, linking the conditions constituting the cause with the factors constituting the effect.

As we shall see, once the quantum revolution had taken hold, at least some physicists abandoned the functional view of causation. They adopted what may be

[22] Kant, *Critique of Pure Reason* (1787), A 203. For a modern discussion, see von Wright, *Causality* (1974), 63–8. Zilsel, 'Asymmetrie' (1927), 280–6 distinguishes between independent and dependent variables and points out that these cannot be distinguished in ordinary functional laws, like Kepler's third law. H. Driesch, *Relativitätstheorie* (1930), 84–5, 90, who argued against the notion of relative time, also offers some interesting critical remarks on the functional view of causation.

termed a *conditional view*, which cuts through the identification of causation with determinism and opens up the possibility of a probabilistic notion of causation.

5.1.4 Keeping Causation and Determinism Apart – Classical Style

We shall see soon that quantum mechanics drives a wedge between causation and determinism. But the separation of these two notions could already have been observed in classical physics. It is only the philosophical presupposition (causation = determinism), which prevented scientists into the 20th century to see this distinction. Consider Kepler's planetary laws. Using the laws of Kepler, it is possible to determine the orbit of a planet, in the future or in the past. But in these determinations the concept of causation plays no role. Kepler's third law, for instance, permits the determination either of the orbital period of a planet (equally a satellite) around the sun or of the average distance of the planet from the sun. In a simplified form the law states that

$$P^2 = A^3$$

in words, the orbital period, P squared, equals the average distance, A cubed. So if a planet has a period of, say, eight years, it will have an average distance from the sun of four times the earth-sun distance. Knowing either of these two parameters, it is possible to determine the other. But we would not say that the temporal period (8 years) causes the average distance. Nor would we say that the average distance causes the orbital period, since average distance is an algebraic, not a physical property. The Keplerian equation states a mathematical equality between orbital period and average distance, from which the one can be determined in terms of the other. But the question of causation does not arise. Clearly the 'precise' determination of future or past states of a system from a present state under the use of universal laws, as required by determinism, does not presuppose any knowledge of a causal dependence between the factors involved. This can be seen from another example.

At the same time, as Kepler developed his 3 planetary laws, the Dutchman Willebrord Snell (1591–1629) discovered the law of refraction, which now bears his name. Snell's law is a further example of a law in classical physics, which could have shown that causation and determinism are not the same. The law states the determination of the refracted beam from the knowledge of the incoming beam and vice versa. The law states the mathematical equality

$$n_1 \sin \phi_1 = n_2 \sin \phi_2$$

between incoming and refracted beam (Fig. 5.2). Again, the notion of causation does not enter the equation. There is no conditional dependence, no temporal asymmetry between cause and effect expressed in this law. The incoming beam does not cause the refraction. It can be asked: 'What *causes* the planets to perform Keplerian elliptical orbits'? 'What *causes* light beams to follow Snell's law'? The appeal to the law of universal causation will not suffice. Surely, amongst all the factors which

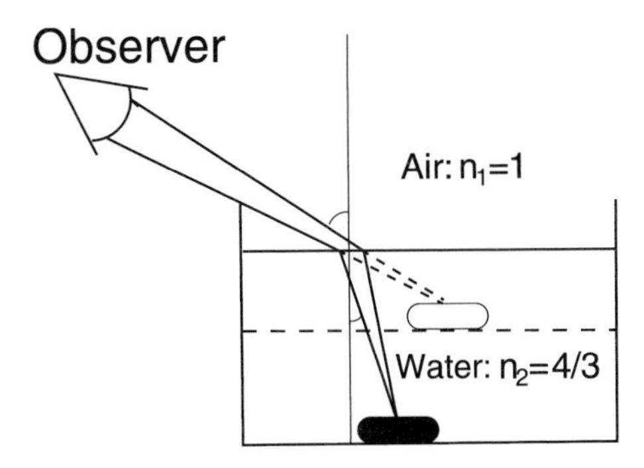

Fig. 5.2. An illustration of Snell's law. The apparent position of a coin (*light shape*), as it appears to the observer due to refraction in the water. The observer is positioned in such a way that he or she would not be able to see the coin (*dark*) at the bottom of the container before it is filled with the water

make up the antecedent state of the universe, some will be relevant, others irrelevant to the present behaviour of a particular system. Rather than a grand principle like the 'law of causation' a more specific model of causation is required. For instance, William Herschel discovered Uranus in 1781. When the orbit of Uranus was plotted back to 1690 and compared to the calculated orbit according to Newton's laws slight deviations appeared.[23] Some unknown planet was disturbing the orbit of Uranus. Neptune's existence was inferred from the observed determination of Uranus's orbit. Neptune causes Uranus to deviate from its Newtonian path. Furthermore this deviation can be precisely calculated due the causal influence of Uranus's orbit on Neptune. Here more is said than that the orbit of Neptune is determined and that its past and future trajectories can be calculated from appropriate laws. A specific causal dependence is invoked to account for the deviation, which Neptune's orbit suffers in the presence of Uranus.

In Kepler's laws and Snell's law we find examples of deterministic relations, in classical physics, without any causal component. What is more – a further point against the functional view – these are not even differential laws. We can also reverse the situation. Then we get causal relations without deterministic components. To see this, consider spectral analysis.

Spectral Analysis was established by Gustav Kirchhoff and Robert Bunsen in 1859. Spectral analysis promised to be a method with tremendous potential. It allowed inferences from the optical spectrum of chemical elements to the chemical composition of the sun – and even the age and distribution of geological formations – which could not be based on direct observation. The two researchers found that chemical elements produced characteristic spectral lines – bright lines of light in particular colours or wavelengths – when held in the flame of the Bunsen burner. Sodium (Na), for instance, produces two particularly bright yellow lines, which can be detected in a spectroscope. Each chemical element produces a *characteristic*

[23] Harwit, *Cosmic Discovery* (1981), 65; Zeilik, *Astronomy* (1988), 219–23

pattern of colour lines, which it shares with no other. The spectral lines of chemical elements are passports to their identity. From the observation of the particular bands of bright lines in a spectroscope it is possible to identify the chemical element, which produces it. Such bright lines are called *emission lines*. When common salt is put into a flame, the spectrum of sodium shows two characteristic bright yellow lines.

Kirchhoff and Bunsen rightly suspected that spectral analysis would make it possible to 'go beyond the limits of the earth, even the solar system.'[24] When the spectrum of sunlight is analysed by these methods a sequence of colours from short wavelengths in the blue region to long wavelengths in the red region appears. However, this spectrum is marked by characteristic *dark* lines, which appear in the blue, green and yellow region of the solar spectrum. Such dark lines are called *absorption lines*, because the light seems to have been removed from these narrow bands. Kirchhoff and Bunsen were particularly interested in two dark lines (*D* lines) in the solar spectrum, which appear in the yellow region. They coincide with the two yellow lines of sodium. In their experiments they were able to produce the *D* lines of the solar spectrum artificially in the spectrum of an element, in which these lines did not normally occur. They came to a startling conclusion:

> We may assume that the bright lines corresponding with the *D* lines in the spectrum of a flame always arise from the presence of sodium; the dark *D* lines in the solar spectrum permit us to conclude that sodium is present in the sun's atmosphere.[25]

From the point of view of causation, this was a puzzling conclusion. What caused the absorption of the *D* lines in the solar spectrum? Why were the *D* lines in the solar spectrum dark? Today, an explanation in terms of Bohr's model of the atom, with its discrete energy levels could be given. But Kirchhoff could not operate at such a deep level of causal explanation. Kirchhoff explained the appearance of dark *D* lines in the solar spectrum by the difference in temperature between the hot interior of the sun and the cooler gases of its surrounding atmosphere. The continuous spectrum of the sun (with no breaks in it) travels through the cooler transparent gases surrounding it. The cooler transparent gases cause the absorption lines in the solar spectrum.[26] Kirchhoff was therefore in a position to affirm that the famous *D* lines in the solar spectrum were produced by the presence of sodium in the atmospheric gases surrounding the interior of the sun. More generally, Kirchhoff was aware of a causal dependence of the spectrum on the chemical elements. The positions, which the coloured lines occupy in the spectrum, he writes,

> determine [i.e. identify] a chemical property, which is of the same unchangeable and fundamental nature as the atomic weight of materials, and they can therefore be determined with an almost astronomical accuracy.[27]

[24] Kirchhoff/Bunsen, 'Chemische Analyse durch Spectralbeobachtungen' (1860), 187

[25] Kirchhoff, 'The Fraunhofer Lines' (1859/1935), 355

[26] Kirchhoff, 'Emission and Absorption' (1859/1935), 357; see Zeilik, *Astronomy* (1988), 78

[27] Kirchhoff/Bunsen, 'Chemische Analyse' (1860), 185; see the qualifying statement in Pais, *Inward Bound* (1986), 168; for a general analysis of determinism in classical physics, see Earman, *Primer* (1986), Chaps. I–III

The chemical elements cause characteristic spectral lines. Kirchhoff and Bunsen clearly tell a causal story. It carries a certain degree of determination – the determination of the wavelengths emitted and absorbed by chemical elements. Yet it lacks the precise determination required by determinism. There is no appeal to a grand principle of causation. Causal stories are open-ended. On one level of causal explanation it is possible to stipulate that the presence of sodium in the cooler gases surrounding the sun's interior cause the appearance of *D* lines in the sun's spectrum. But on another level it will be asked why chemical elements produce spectra in the first place. In 1860 such further questions were beyond the reach of physics. The atom's structure had to be unveiled. Quantum mechanics had to be born before such deeper lying causal questions could be asked and answered.

So as with all forms of explanation there are different levels of causal explanations. On one level it is sufficient to say that hot water causes a sugar lump to dissolve. On another level this is an unsatisfactory answer for it remains an outstanding question why sugar but not oil is dissolved in water. Such questions must be answered on the molecular level.[28] There is a hierarchy of causal explanations. Causal stories abound in science and everyday life in the absence of deeper levels. And when in the 20th century the level of atomic and subatomic phenomena was reached, Laplace's hope that even the motion of the lightest atoms could be predicted with certainty was finally shattered. So was the idea of the causal concatenation of all events. With the discovery of atomic decay, Mill's universal law of causation came under a cloud of suspicion. At this moment, the bond between the notions of determinism and causation weakened. But in the heyday of classical mechanism Laplacean hopes of causal determinism could at least be upheld. So even the existence of causal stories without precise determination – say, spectral analysis, as well as Harvey's explanation of the circulation of blood, Guericke's experiments on the mechanical force of a vacuum, Hooke's explanation of planetary orbits and Newton's explanation of colours – held the promise that the precise determination was still to follow. This was the nature of the argument employed by d'Holbach and Laplace. From a postulation of a universal causal chain of events they inferred the possibility of precise predictability.

Soon new discoveries would show that such a programme could not be carried out. Evidence of causal situations transpired, which lacked the degree of determination required by determinism. But even from the precise determination of the states of systems it does not follow that a causal determination is involved. The new discoveries, which preceded the emergence of quantum mechanics, led to a questioning of both Laplace's major premise and his conclusion. Quantum mechanics confirmed that the axiom of a causal concatenation of all events was doubtful and that the demand for a precise determination of the succeeding states of a system, in terms of individual parameters, cannot be satisfied. But this left room for a conditional view of causation in the face of indeterminism.

---

[28] See, for example Gerstein/Levitt, 'Simulating Water' (1998); Eberhart, 'Why Things Break' (1999); Mill, *A System of Logic* (1843), Bk. III, Chap. IX, §3 on the cause of dew. Causal stories are incomplete and open-ended, see Lewis, 'Causal Explanation' (1986)

5.2 New Discoveries – New Ideas

Often fundamental presuppositions in science are not questioned till they clash with well-established facts. Or they disturb the coherence of a theory. Such assumptions frequently reveal themselves in the use scientists make of fundamental notions. Often they are inherited from philosophy. This was the case with the Kantian notion of time and it is the case with the Laplacean identification of causation and determinism.

New discoveries may eventually lead to a questioning of the fundamental notions. This happened with the notions of causation and determinism. The 19th century saw the development of thermodynamics, electromagnetism, the Darwinian revolution and the discovery of atomic structure. These confirmed the cosmos-view of Nature. The new discoveries, which led to the physics of the atom, had a decisive influence on the reformulation of the notions of causation and determinism by scientists in the early part of the 20th century. This was due to the need to understand the new discoveries. The established fundamental notions caved in under the cumulative impact of the new discoveries.

5.2.1 Planck's Constant

Max Planck (1858–1947)

A particularly emminent contribution, with far-reaching implications for the understanding of the concept of physical law and the understanding of causation and determinism was Planck's introduction of a new fundamental physical constant. The *Planck constant*, h, was originally introduced as an *ad hoc* device to save a mismatch between theory and experience. Although Planck calls h a natural constant in his original publication,[29] its value at the time (1900) was not independently established. Rather Planck formulated an empirical law for energy distribution in blackbody radiation, then inserted known data from other researchers into the new formula to work out the value of h. Planck at that time was not sure whether h was just a fictitious unit or whether it incorporated a statement about the material constitution of the physical world. Indirect and direct measurements soon confirmed that h was a new fundamental physical constant. The *ad hoc* assumption had turned into one of the most

[29] Planck, 'Zur Theorie des Gesetzes der Energieverteilung im Normalspectrum' (1900/1972), 8. Planck already uses the letter h. Given the *ad hoc* nature of Planck's introduction of h, the surprising agreement of the magnitude of this value at which Planck arrived (6.55×10^{-27} erg. s) with the modern value (6.626×10^{-27} erg. s) speaks in favour of Planck's hunch that he had indeed found a new fundamental constant. Planck's original doubts as to the status of h are described in his acceptance speech for the Nobel prize in physics: 'Die Entstehung und bisherige Entwicklung der Quantentheorie' (1920), 131

important physical discoveries in the 20th century. It engendered a revision of the physical worldview. The Greek metaphor that nature does not make jumps (*natura non facit saltus*), which had taken the form of continuity of functions in classical mechanics, gave way to the fundamental idea of discontinuity – expressed in the Planck constant – which characterises the precise mathematical analysis of atomic processes. With the introduction of the idea of *discontinuity*, Planck became the founding father of quantum mechanics. Everyday experience instils in us a belief in the continuity of physical processes. A car can be driven at a continuous velocity scale from zero to some maximum speed. A body can have a continuous mass from zero (the rest mass of the photon) to some maximum value. But an atom cannot acquire continuous values for its various parameters. For instance, it has discrete energy levels and discrete orbits.

Planck was aware that his discovery of discontinuity in microphysical processes would have important philosophical consequences. Unlike Kirchhoff, Thomson and Rutherford he considered these philosophical consequences during his most creative years as a physicist. To introduce the problem situation and appreciate the revolutionary nature of the solution, which included the postulation of h, the constant of discontinuity, Planck suggested a helpful analogy.[30] Imagine a lake in which the water has been whipped up by the wind. Waves will move from one shore to the other. When the wind dies down the waves will continue for a while. Slowly, however, a transformation will take place from ordered waves to an unordered calm lake, from ordered molar energy to unordered molecular energy. In the ordered visible waves many molecules have a similar velocity. Once the waves have calmed, the molecules will be in an unordered state with many different velocities. Now imagine a cavity, in which light and radiation are reflected between the walls. Such laboratory cavities, which became experimentally available at the end of the 19th century[31], are often called *black bodies* because radiation is allowed to enter from outside through a hole and then has little chance of escape. As with the water waves, it may be suspected that radiant energy in a cavity will also experience a slow transformation from ordered long infrared waves to unordered short ultraviolet waves. Infrared waves have longer wavelengths than visible light; ultraviolet wavelengths are shorter than visible light. On the analogy with the water waves the infrared waves are expected to disappear and be transformed into ultraviolet waves. This expectation would be in accordance with classical theory. But observation of the actual behaviour of radiation inside a black cavity did not agree with the expectation. The radiation distribution in the cavity reaches a maximum of intensity near the visible region of the spectrum, then falls off sharply through the visible region and tapers off towards the infrared on the other side of the curve (Fig. 5.3). It is the sharp peak of the curve towards the ultraviolet region, which is totally unexpected from classical theory (Fig. 5.4). Because the

[30] Planck, 'Neue Bahnen der physikalischen Erkenntnis' (1913), 74–5; for a more fancyful analogy see Greene, *Elegant Universe* (1999), 91–3

[31] Otto Lummer and Ernst Pringsheim were the first physicists to use such cavities for the study of radiation, see Kuhn, *Black-Body Theory* (1978), 11

data at short wavelengths do not fit the expected curve from classical theory, this failure has become known as the *ultraviolet catastrophe*. To make the theory fit the curve, Planck introduced the hypothesis that energy can only be absorbed and re-emitted in discrete bundles of energy or *quanta*. This discreteness was expressed by the constant h, which, as Planck put it, 'was at first totally hanging in the air.'[32] Furthermore, the energy of each quanta, ε, is determined by its frequency, ν. In Planck's notation the element of energy of the quantum is written in the form $\varepsilon = h\nu$. This reasoning leads to an explanation of the fit between theory and experimental data brought about by Planck's formula. The atoms in the wall of the cavity absorb and re-emit the radiation in the cavity. But they do so according to the pattern expressed in Figs. 5.3, 5.4. This can be understood from Planck's formula: $\varepsilon = h\nu$. Expressed in terms of wavelengths (λ), this reads $\varepsilon = \frac{hc}{\lambda}$, where c is the velocity of light and h is Planck's constant. As c and h are constants, the energy becomes proportional to the wavelength: $\varepsilon \propto \frac{1}{\lambda}$. According to the Rutherford-Bohr model of the atom, atoms have discrete energy levels. In the ultraviolet region, the frequency of radiation is high (10^{15}–10^{16} Hz) and the wavelengths short (10^{-7}–10^{-8} m). The atoms cannot easily absorb and re-emit the radiation. For fixed amounts of energy, ε, there is a shortest wavelength, which can be excited. This explains the sharp drop in the curve in the ultraviolet region. In the infrared region, where the frequency of radiation is lower (10^{12}–10^{14} Hz) and the wavelengths are longer (10^{-4}–10^{-6} m), the problem is reversed. Here the atoms can lose energy only to higher-frequency

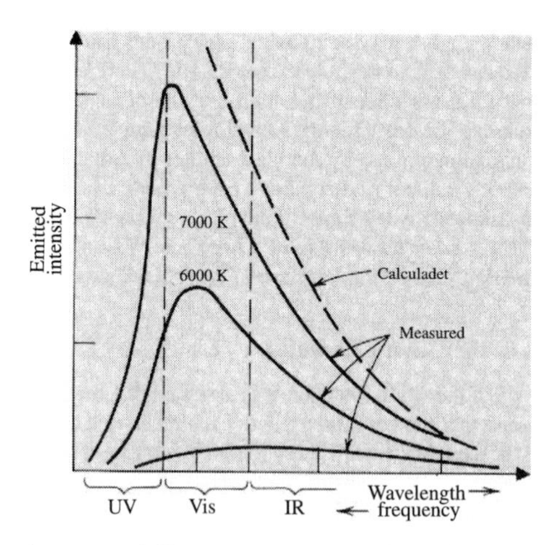

Fig. 5.3. Spectral distribution at different temperatures found in experiments. At very short wavelengths, in the ultraviolet region (UV), there is no emitted intensity. At very long wavelengths in the infrared region (IR), the radiation intensity again falls off. The failure of classical theory to explain the peak in the visible region (Vis) of the spectrum is called the *ultraviolet catastrophe*. Source: Spielberg/Anderson, *Seven Ideas* (1995), 266, by permission of John Wiley & Sons

[32] Planck, 'Zur Geschichte der Auffindung des physikalischen Wirkungsquantums' (1913), 27; author's own translation

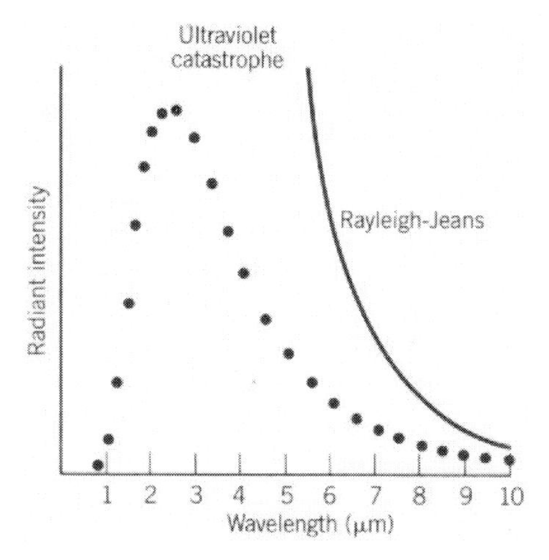

Fig. 5.4. Spectral distribution or radiant intensity as a function of wavelength. According to the classical Rayleigh-Jeans view the spectral distribution goes off to infinity, leading to the ultraviolet catastrophe. Planck's curve (dotted line) fits the experimental results. Source: Krane, *Modern Physics* (1983), 64, by permission of the author

atoms, if they can emit the right amount of energy. Their energy is low (wavelength is inversely proportional to their energy), so that higher energy atoms are less 'ready' to accept these amounts of energy. There will be some longest wavelength, which can be emitted. There is a very low probability that radiation at either very short or very long wavelengths will be excited in the cavity. This leaves, as the curve shows, a most probable radiation distribution, where the atoms can absorb and re-emit energy.[33]

This discussion already anticipates the essentially probabilistic nature of explanation, which was to become the hallmark of Planck's revolution in science. The radiation distribution in the cavity is a probabilistic statement. To speak of the emission and absorption of discrete quanta by the atoms in the cavity wall is to speak about what they are most likely to do.

As early as 1908 Planck was aware of the philosophical consequences of the new discoveries. He saw a disturbing split in the unity of the physical worldview – between the old dynamic laws and the new statistical laws, between apparently deterministic behaviour in the macro-world and apparently indeterministic behaviour in the micro-world. Only statistical statements could be made concerning the behaviour of radiation in the black body. If only average values of parameters can be calculated, nothing can be known about the individual behaviour of the elements. More seriously, two different types of causal connection between physical states had to be admitted: an absolutely necessary link and a merely probabilistic

[33] Planck, 'Neue Bahnen' (1913), 75; 'Entstehung' (1920), 129; see Cushing, *Philosophical Concepts* (1998), 274, 278; Spielberg/Anderson, *Seven Ideas* (1995), 263–69; Krane, *Modern Physics* (1983), 61–6

link between cause and effect.[34] In 1908 Planck remained a lone voice till some other decisive discoveries completed the case against the classical view.

5.2.2 Radioactive Decay

The *radioactive decay law* was the next step on the road to a questioning of the notions of causation and determinism. The formulation of this law in 1902 was the work of Ernest Rutherford and Frederick Soddy. It was preceded by some feverish activity in the physics community, totally unanticipated by von Jolly in 1874. In November 1895, Röntgen discovered X-rays (light of very short wavelength). In 1896, Becquerel discovered radioactivity and in 1897 J.J. Thomson discovered the electron. These important discoveries were milestones in the passage of physics from the classical period to the modern period. They paved the way for the formulation of quantum mechanics – the physics of the atom. But these discoveries did not yet give any reason to doubt the validity of the classical notions of determinism and causation. For in these early years of the dawn of the new physics, researchers would consistently try to understand the new phenomena in terms of the models of classical physics. But the Rutherford-Soddy theory of radioactivity and the formulation of their decay law were a different matter. Here was an apparently causal account – Rutherford and Soddy called their paper 'The Cause and Nature of Radioactivity' – which failed to offer any deterministic specifications of a causal mechanism. And the decay law was a statistical law. The theory explained that radioactivity was an atomic phenomenon, which was accompanied by chemical changes. By emitting charged particles – beta particles (electrons) or alpha particles (helium nuclei) – one chemical element was transformed into another. These changes therefore had their seat within the atom and not on the molecular level. The change in the chemical elements was caused by the emission of these subatomic particles. This theory could also explain the observable series of changes in which chemical elements cascaded from longer-lived into shorter-lived elements. But what in turn caused the emission of the particles itself? The radioactive elements, the theory continued, 'must be undergoing *spontaneous* transformation.'[35] The transformation theory did not stipulate a particular cause inside the atom, which would be responsible for the emission of, say, an alpha particle at a particular time and thus bring about a chemical change. Like Kirchhoff's

Ernest Rutherford (1871–1937)

[34] Planck, 'Einheit des physikalischen Weltbildes (1908), 41; 'Dynamische und Statistische Gesetzmäßigkeit' (1914), 81–4; 'Physikalische Gesetzlichkeit' (1926), 195. Schrödinger, *What is Life* (1967), 92–4 makes probabilistic laws the fundamental laws of nature.

[35] Rutherford/Soddy, 'The Cause and Nature of Radioactivity I' (1902), 493; italics added.

spectral law, it gives a causal explanation at one level but fails to specify a causal mechanism at a deeper level. In terms of the classical notion of determinism the transformation theory did not give rise to the precise prediction of emission events. By contrast the theory proposed a formula, which stated the probability value of an alpha- or beta particle being emitted from a certain element after a certain amount of time. This statistical formula has become known as the *decay law*. In the original words of Rutherford and Soddy, 'if I_0 represent the initial activity and I_t the activity after time t,

$$\frac{I_t}{I_0} = \mathrm{e}^{-\lambda t}$$

where λ is a constant and e the base of natural logarithm.'[36] The statistical nature of this law can be illustrated, using Rutherford's original data, as in Fig. 5.5.

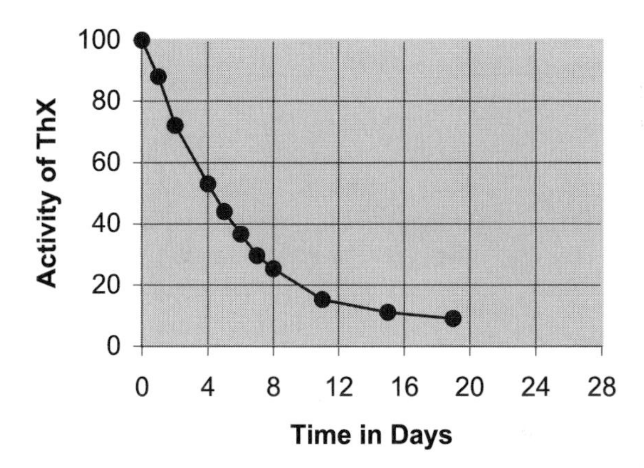

Fig. 5.5. Rutherford's Curve

Unlike the next generation of physicists, which was to provide the theoretical foundations to these new discoveries, Rutherford did not draw any philosophical consequences, which may have followed from the statistical nature of the decay law. Like Planck's radiation law, the decay law introduced a new type of law into physics. But it had no immediate philosophical impact. Nevertheless, the evidence that the new discoveries would have philosophical consequences was mounting.

5.2.3 Spontaneous Emissions

Spontaneous emissions are one last scientific discovery to be discussed before we can consider how the scientists reacted philosophically to the threat hanging over the notion of causation. In 1917 Einstein showed that Planck's quantum hypothesis

[36] Rutherford/Soddy, 'The Cause and Nature I' (1902), 482; for a modern discussion, see Weinberg, *The Discovery of Subatomic Particles* (1993), 109–21

and Rutherford's law of radioactive decay could be united into one model. It is this contribution, which implanted in Einstein's mind the worry about the seemingly *acausal* nature of quantum-mechanical processes. And it left him permanently dissatisfied with quantum mechanics. As so often during these decades of feverish scientific activities, the rigorous mathematical treatment of a topic is interwoven with brief philosophical flurries about the fundamental notions. This was sufficient to plant a grain of doubt and, as we shall see in the next section, did not grant the great scientists any peace of mind. Formally, Einstein's contribution was concerned with a new, more elegant derivation of the Planck radiation law. From this derivation, Einstein hoped to throw some new light on the poorly understood mechanism by which matter could absorb and emit radiation. Such questions had not been addressed in a new model of the atom: the Rutherford-Bohr nucleus model. But Einstein's work also went beyond the habitual concern with the exchange of energy in transition processes, which is central to the nucleus model. There was an additional question to be addressed: 'Does a molecule suffer an impulse when it emits or absorbs energy ε?'[37] Just as a shooter experiences a recoil from firing a gun the question was whether an atomic particle, which emits energy, experiences a recoil. Einstein's answer was that the particle, when it emits or absorbs radiation in the transition from one stationary state to another, experiences a recoil momentum of magnitude ε/c (where $\varepsilon = h\nu$ and c is the velocity of light). The central point is that such elementary processes as the emission and absorption of radiation can only be consistently described by the quantum theory, if they are considered 'to be perfectly directional'. That means that the emission of radiation from the 'molecule' is not to be thought of, as classical theory would suggest, as a spherical wave going out in all directions. Einstein distinguished two processes: *spontaneous emissions* in which the atom, without excitation from an external source, suddenly emits radiation of energy equal to $\varepsilon_m - \varepsilon_n$ with frequency ν; and *induced absorption and emission*, which occur when the atom is placed in a radiation field such that the work done on it by the radiation field induces transitions from a higher energy level, ε_m, to a lower energy level, ε_n, (emission) or from ε_n to ε_m (absorption). In the latter case the momentum recoil is determined by the direction from which the radiation beam strikes the atom. There is a causal component. The worry occurs in the case of spontaneous emissions, in which the atom emits radiation without the presence of an external cause. The emission process is directional but how is this direction determined? Einstein's answer became the source of the doubt about causation looming large in discussions on the foundation of quantum mechanics.

Niels Bohr
(1885–1962)

[37] Einstein, 'On the Quantum Theory of Radiation' (1917), 65

> Outgoing radiation in the form of spherical waves does not exist. During the elementary process of radiative loss, the molecule suffers a recoil of magnitude $h\nu/c$ in a direction which is only determined by 'chance', according to the present state of the theory.[38]

Einstein saw it as a weakness of the theory that 'it leaves the duration and direction of the elementary processes to "chance".' As is clear from his correspondence with Born, Einstein resisted the apparently acausal view of the nature of atomic processes. He declares himself unwilling to give up the notion of strict causation.[39] It should not be forgotten, however, that the identification of determinism and causation, due to Laplace and d'Holbach, constituted a powerful paradigm, which befitted the hopes and aspirations of the mechanical worldview. There is the possibility that this Laplacean identification may have misled Einstein, along with many others, into thinking that quantum mechanics heralded the abandonment of the cherished notion of causation (understood, implicitly, as causal determinism). Even physicists like Bohr and Heisenberg, who wholeheartedly embraced the demise of the notion of causation in quantum mechanics, fell victim to the same mistaken identification. This will be shown in Sect. 5.3.2.

5.2.4 Awakening Doubts and the Rutherford-Bohr Model of the Atom

> It is impossible to trap modern physics into predicting anything with perfect determinism because it deals with probabilities from the outset. A. Eddington, *New Pathways in Science* (1935), 105

The worry about causation began to make itself felt. Curiously, it was Rutherford, whose philosophical instinct had remained idle when the introduction of the decay law could have awoken it, who raised the issue of causation in a letter to Bohr (1913). Reflecting on Bohr's nucleus model of the atom, Rutherford wrote to Bohr:

> There appears to me one grave difficulty in your hypothesis which I have no doubt you fully realise, namely, how does an electron decide with what frequency it is going to vibrate at when it passes from one stationary state to another? It seems to me that you would have to assume that the electron knows beforehand where it is going to stop.[40]

This was not the first awakening of the causation issue. In 1908 Planck had already expressed concern about the applicability of classical notions to the atomic realm. The issue was not to remain dormant for long. The quantum jumps of atoms were to inspire Sommerfeld and others to propose a new *teleological* conception of causation. It briefly preoccupied physicists. Then other philosophical models followed. Before we turn to the philosophical consequences, which the scientists themselves drew from these discoveries, let us briefly consider the main features of the Rutherford-Bohr model.

---

[38] Einstein, 'On the Quantum Theory of Radiation' (1917), 76

[39] Einstein-Born *Briefwechsel* (1969), 44, 117

[40] This passage is quoted and discussed in Pais, *Inward Bound* (1986), 212 and *Niels Bohr's Times* (1991), 153, as well as in Cushing, 'Background Essay' (1989), 2

Starting in 1909 Geiger and Marsden in Manchester carried out scattering experiments on gold atoms. A high-velocity projectile like an α-particle (a doubly ionised helium atom) is fired at a thin sheet of gold. Geiger and Marsden found that 1 in 8000 projectiles were scattered back at angles greater than 90°. (See Figs. 5.11, 5.12 in Sect. 5.3.4 below). These were unexpected, amazing results because the common assumption at that time was that atoms consisted of rings of electrons in a sphere of positive electricity with no nucleus. This model of the atom was due to Thomson and is often referred to as the *plum pudding model*. It was the genius of Rutherford to realise that the scattering experiments of his colleagues Geiger and Marsden could not be explained on the plum pudding model. If the atom is *like* a plum pudding and the α-particles are *like* bullets, then the analogy illustrates that the plum pudding could not return any bullet in the direction of its source. If we shot a great number of bullets at a plum pudding and miraculously some of them were scattered at angles greater than 90° we would suspect that there was something buried inside the plum pudding, which was causally responsible for the behaviour of the bullets. This is precisely the argument, which Rutherford used. There must be a nucleus inside the atom, which has properties able to scatter the α-particles in the manner observed by Geiger and Marsden. Rutherford's proposal of a *nucleus model* (1911) was a major step forward in the understanding of atomic structure. Bohr (1913) developed this model mathematically. By making some fundamental assumptions about the nature of the atom, which were in radical opposition to established views in the physical sciences, he was able to show that puzzling observable phenomena, like the radiation distribution in black bodies and the scattering experiments, as well as others, could be explained mathematically from the Rutherford-Bohr model. As he developed his theory, Bohr made a number of assumptions, two of which are important for the understanding of the causation issue.[41]

I. The first assumption is that atomic systems can only exist in stable energy states, in which no emission or absorption of energy takes place. Each state is characterised by a certain value of energy. The energy values of different states are discontinuous. Such states of the atomic systems are called *stationary states*.
II. Atomic systems can 'jump' between different stationary states by emitting or absorbing energy in terms of radiation. These jumps are called *transitions* between stationary states. When radiation is emitted the system sinks to a lower discrete energy state ε_n. When radiation is absorbed during a transition the systems jumps to a higher discrete energy state ε_m. The radiation which passes through these transitions is homogeneous ('unifrequentic') and possesses a frequency ν, which is determined by the relation $\varepsilon_m - \varepsilon_n = h\nu$. That is, the radiation must correspond to the energy difference between the allowed states of the system.

Bohr's model of the hydrogen atom combines Planck's discovery of the universal constant, h, the bearer of discreteness, with Rutherford's inference to the existence of

[41] Rather than quoting Bohr verbatim, I have reconstructed these two assumptions from the following two Bohr papers: 'On the Quantum Theory of Radiation and the Structure of the Atom' (1915), 395–6; 'On the Quantum Theory of Line-Spectra' (1918), 97–8

a nucleus from the scattering data. Although in the late 1920s Bohr would become the staunchest defender of the *acausal* understanding of the physics of the atom, in these early scientific papers there is only a 'general acknowledgement of the inadequacy of the classical electrodynamics in describing the behaviour of systems of atomic size.'[42] There is as yet no recommendation that the classical principle of causation had to be abandoned and replaced by the principle of complementarity. Between 1913, when Bohr proposed his first quantized atom model, and 1930 scientists began to think very actively about the philosophical consequences of the new discoveries for the fundamental notions of causation and determinism. This philosophical productivity is accompanied by immense scientific progress in the understanding of atomic phenomena.

5.3 Scientists Draw Philosophical Consequences

> Indispensable est la philosophie.
>
> B. d'Espagnat, *Incertaine Réalité* (1985), 1

In his scientific biography of Einstein, Abraham Pais makes the claim that Einstein was the first to understand that the traditional notion of causation was under threat from the new developments in atomic physics.[43] We have seen that Planck, in fact, deserves the credit for having first raised the question of the philosophical consequences of the new discoveries. Although Einstein does not mention the notion of causation in his 1917 contribution, his critical observation that the moment and direction of emission is left purely to 'chance' is seen as a gap in the quantum theory of radiation. As quantum theory could state no cause for the direction of emission, he concluded that it was incomplete. After his groundbreaking contributions to relativity theory (1905, 1916), Einstein continued to provide proof that he was philosophically alert to the new developments in physics. But from Einstein's ambiguous philosophical attitude emerged, what we called 'Einstein's problem'. Why was it right to abandon the Newtonian notion of absolute time, but wrong to give up the notion of causal determinism against all the evidence?

Einstein's unwillingness to abandon such a fundamental and time-honoured notion as causation was not unique. Especially the older generation of physicists found it difficult to jettison the notion of causation. Max Planck was one of them. Like many of the great physicists, who laid the foundations of the physics of the atom, Planck was keenly interested in philosophical questions. During his scientific career he wrote many papers combining physical and philosophical considerations, which reflect the emerging adaptation of the physical worldview to the new discoveries. One concern was the defence of *realism* against Mach's positivism. In modern terms, realism means the belief that scientific theories capture, by degrees of idealisation and approximation, the structure of a material world. The degrees of idealization

[42] Bohr, 'On the Constitution of Atoms and Molecules' (1913), 2
[43] Pais, *Subtle is the Lord* (1982), 5–6

and approximation are captured in the models, which physicists construct on the basis of observational and experimental phenomena. The scientific theories are at best true of models, not of the real world. The nature of the real world is revealed through controlled interaction with it. The mathematical structure of the theory displays, through robustness of the tests, the ontological structure of the world. This world is taken to exist independently of human consciousness. If the natural world, according to the classical view, is a chain of causes and effects, then a closely related concern is the question of *causation*. One complaint Planck had against the statistical treatment of the behaviour of gases was that it led to two different forms of the causal link.[44] The familiar macro-processes of Newtonian mechanics and Laplacean astronomy obeyed *dynamic laws*, which were strictly causal. The new phenomena, like the conduction of heat, friction, diffusion, were subject to *statistical laws*, which gave only rise to probabilistic predictions. But underlying the statistical behaviour of a collection of molecules was a microscopic structure, which was held to be causally responsible for the collective behaviour. At the micro-level, atoms and molecules behaved in a perfectly deterministic manner. At least, this was the accepted wisdom.[45]

There is a considerable difference between the statistical nature of the collective behaviour of gases and molecules and the statistical nature of quantum-mechanical phenomena. The statistical nature of thermodynamics is compatible with the assumption of strict deterministic behaviour of the single molecules in the ensemble. The statistical laws reflect only our ignorance of the properties of the individual molecules (velocity, position), which would in principle permit a precise determination of their individual paths. The statistical nature of quantum mechanical phenomena is a much more complicated affair. It is more complex, because it depends on the interpretation, which is attached to the theory. If we adopt the *orthodox* interpretation, we are encouraged to hold that a) quantum systems do not possess particular properties like position and momentum before they are measured, b) the indeterministic nature of quantum phenomena is due to the 'collapse' of the ordered, deterministic evolution of the quantum system, according to the Schrödinger equation, in the act of measurement. The act of measurement forces the wave function into the particular, measured state. We shall see later that these assumptions have become questionable. In 1914 a well-developed theory of the atom had not yet been constructed. Only the Bohr model of the hydrogen atom was available. Planck therefore naturally assumes that causal determinism was an indispensable assumption of the scientific enterprise. And he expects that the statistical laws will in the end be reducible to strict dynamic (causal) laws.

But like Einstein a few years later, he is strangely aware that atomic phenomena may pose special difficulties. Ignoring his own discovery of the quantum of action, he uses the decay law of Rutherford and Soddy to express his concern. What induces

---

[44] See Planck, 'Die Einheit des physikalisches Weltbildes' (1908), 41f and 'Dynamische und statistische Gesetzmäßigkeit' (1914), 81ff

[45] For a discussion of the foundational problems of thermodynamics and statistical mechanics, see Sklar, *Physics and Chance* (1993)

one uranium atom to disintegrate all of a sudden while a second uranium atom in its immediate vicinity continues to persist in its state?

> Indeed, even to meddle here with the assumption of a causal determination by the dynamic law seems hopeless at the moment especially as all attempts through external interference, for instance the increase or decrease of temperature, to gain an influence on the course of the radioactive phenomena have led to no result.[46]

In 1908 Planck had raised the issue of causation with respect to statistical laws; then once again, in 1914, with respect to the apparently *a*causal manner in which decay processes seemed to occur. From this moment on and due to the new discoveries, the adequacy of a fundamental notion was in doubt. The empirical evidence posed a threat to the status of the fundamental notions on the conceptual map. Many physicists participated in the discussion, with many different proposals. We will sketch the main solutions. The responses of physicists and philosophers can be divided into three types.

The *conservative* response was to retain the notion of causation and its identification with determinism. This was the attitude of Planck, Einstein and von Laue. But it demanded a heavy conceptual price. The notion of causation lost its effective role in the understanding of quantum mechanical phenomena. It evaporated to a heuristic principle of research. There was, however, an epistemological gain. Quantum mechanics could be treated as an incomplete theory, leaving room for alternative formulations.

The *radical* response was to renounce the notion of causation. This was the attitude of Bohr, Heisenberg and Pauli. This approach reasons from the indeterminism of quantum events to their acausal character. A long list of scientists and philosophers has followed their lead. This radical response has penetrated public awareness. But the well-publicised renunciation of causation in quantum mechanics is based on the mistaken identification of causation and determinism. The conceptual price to be paid is the relinquishment of two of the most fundamental concepts in the history of physical thinking.

The *philosophical* response was that quantum mechanics did not relinquish the notion of causation. Rather, it points the way to a finer conceptual distinction between causation and indeterminism. This main insight was worked out by Born and de Broglie. For a brief period, Sommerfeld's proposal to enlarge the notion of causation gained some currency. But the main point of the philosophical response was the recognition that the Laplacean identification of causation and determinism had to be abandoned. It was incompatible with the evidence. Quantum mechanics suggested a notion of causation *without* determinism. Like no other scientific theory quantum mechanics is beset by the problem of interpretation. While the mathematics is clear, the conceptual issues are confused. The physicists' philosophical reactions are cast in terms of their conceptual understanding of this unusual situation. Many of the technical notions, which were once common currency in the understanding of quantum mechanics – especially those associated with the orthodox interpretation, on which the radical response is based – have lost their value

[46] Planck, 'Dynamische und statistische Gesetzmäßigkeit' (1914), 83; author's own translation

in the eyes of physicists. Current research has retreated from key concepts like the collapse of the wave function, the complementarity principle and the indeterminacy relations to the more reliable notions of entanglement and decoherence. They are more reliable because they are grounded in experimental results. As the story unfolds, Interludes will serve as stopovers to reformulate the physicists' concern in the newer language.

5.3.1 Conservative Response

The conservative response was the most hostile to the budding theory of quantum mechanics. It never questioned the evidence, only the interpretation of the evidence in terms of the Bohr-Rutherford structural model of the atom. This response simply identifies causation with determinism and, like Laplace's demon, shifts the focus from causal to predictive determinism. In his Guthrie Lecture, held before the Physical Society of London in 1932, Planck proposed to define the meaning of causation as a 'lawful connection in the temporal course of events.'[47] Predictability served as the *criterion* of causal determination. 'An event is causally determined when it can be predicted with certainty.'[48] This is basically the functional view of causation.

[47] Planck, 'Kausalität in der Natur' (1932), 250 and 'Kausalgesetz und Willensfreiheit' (1923), 143

[48] Planck, 'Kausalität in der Natur' (1932), 252. Amongst modern physicists and philosophers Planck was not the only one to associate causation with predictability. Schlick, 'Kausalität' (1931), §§1–8 defines physical causation as the lawful dependence between events and uses confirmation of predictions as the criterion of causation. The use of predictability as a criterion of causation was quickly criticised. Although he uses different examples, M. Strauss, 'Komplementarität und Kausalität' (1936), 336 shows that predictability is neither a sufficient nor a necessary criterion of causation. As was pointed out above, knowledge of the orbital period of a planet and of Kepler's third law allows a determination of its average distance from the sun. Predictions can be made from temporally related events. Details of the ticking of a clock on earth and knowledge of the equations of relativistic physics allow us to determine precisely the amount by which a clock in a satellite, orbiting the earth, will differ from the clock on earth. Yet there is no causal connection between these two clocks. Therefore predictability is not a sufficient condition for causation. Nor is it a necessary condition. If a beam of atoms is sent through a certain type of magnet, as in the Stern-Gerlach experiment (Fig. 5.13), the beam will be split in half. This splitting can be demonstrated by mounting a photographic plate behind the magnet, on which the atoms will leave recorded marks (blackening). The magnet is the cause of the splitting but there is no precise predictability: there is a 50% chance that an individual atom will impinge on the upper part of the plate and a 50% chance that it will impinge on the lower part. For similar considerations see Jordan, 'Kausalität' (1927), 105, 'Philosophical Foundations' (1927), 566–7 and Bunge, *Causality* (1979), §3.3.2. As we shall see this approach is very different from that of de Broglie, *Continu et Discontinu* (1941), 59 who defines physical determinism, with Laplace, as rigorous predictability ('prévisibilité rigoureuse') of the phenomena (present initial conditions & laws → future initial conditions). But de Broglie distinguishes determinism from causation in quantum mechanics. It would be a mistake to think that the identification of causation with determinism is a thing of the past, see the literature cited in footnote 4.

Planck realised of course that due to measurement errors no physical event could ever be precisely predicted. But amongst the two conclusions, which presented themselves – either to take statistical laws or dynamical laws as ultimate – he opted for determinism. The consequence is that causation ($\approx$ predictive determinism) can only be upheld as a model-like idealisation. Causation becomes a category of thought, of transcendental nature: a presupposition in scientific thinking.[49] Human, macroscopic observers must content themselves with 'chance and probability', only a microscopic observer, who shares the cognitive abilities of the Laplacean demon, can detect everywhere 'certainty and causation'. Planck never abandoned the quest for a causal understanding even of quantum-mechanical systems. He even predicted confidently that the quantum hypothesis would lead to quantum-mechanical equations, which could be regarded as a 'more exact formulation of the causal law.'[50] Planck's epistemological position on causation remained modelled on the statistical nature of phenomena in thermodynamics. The observable statistical properties of gases, of heat and ultimately of subatomic particles could be related to an underlying dynamic causal structure, in which each constituent possessed precise microscopic properties, which determined its behaviour.

The irony of the matter is, however, that Planck was much less conservative in his philosophical positions than Einstein. His response to the threat to causation was thoroughly conservative. But he realised that the quantum-mechanical evidence demanded a *modification* of the conceptual system somewhere. The younger generation was ready to relinquish the notion of causation. Planck relinquished the notion of the Newtonian point particle and developed, already in the 1930s, a position of nonlocal holism, which is reminiscent of current interpretations of quantum mechanics. First, consider the situation in classical mechanics.

> It used to belong to the presuppositions of causal physical thinking that all events in the physical world – by which I mean, as always, the physical worldview, not the real world – could be presented as composed of local events in various single, infinitely small space volumes. And every single one of these elementary events was determined in its law-like course by the local events in the immediate spatial and temporal neighbourhood, without regard for all other elementary processes. Let us illustrate this with a concrete, sufficiently general case. Let the physical object under consideration consist of a system of material points, which move in a conservative force field with constant total energy. According to classical physics every single point is in a particular state at every point in time; that is, it possess a particular position and velocity and its motion can be exactly calculated from its initial state and the local properties of the force field in space, which it passes during its motion.

But all this has changed with the advent of quantum mechanics.

> All this is very different in the new mechanics. (...) A useful presentation of the lawful behaviour is only obtained when the physical object is considered as a whole.

[49] Planck, 'Kausalgesetz und Willensfreiheit' (1923), 154–62; 'Physikalische Gesetzlichkeit' (1926), 184; Meyer, 'Kausalitätsfragen' (1936)

[50] Planck, 'Kausalgesetz und Willensfreiheit' (1923), 154–6; 'Die Physik im Kampf um die Weltanschauung' (1935), 292

> According to the new mechanics every single material point of the system is, so to speak, to be found in a certain sense simultaneously in all points of the space available to the system (...).
>
> You see: what is at stake is nothing less than the notion of the material point, the most elementary notion of classical physics. The previously central significance of this notion must be sacrificed.[51]
>
> (...) the elements of the new worldview are not the material corpuscles but the simple harmonic material waves corresponding to the physical system under consideration. (...) There is no question of indeterminism.[52]

Although Planck is ready to replace the particle view for a field view of Nature, his determinist impulses are too strong for him to want to abandon the principle of determinism – if not on the experimental level, at least on the mathematical level. On the one hand, Planck accepts the inherent indeterminacy in all quantum systems, expressed in the Heisenberg principle. For instance, the blackening of a photographic plate, on which a beam of atoms impinges, does not permit an unambiguous inference to all the details of the process under investigation. But this aspiration of precise determination on the experimental level arises from an adherence to the classical worldview. It is not compatible with the physics of the atom. Nevertheless, measurements on quantum systems show certain effects (blackening on photographic plates), which are conditionally dependent on a given experimental set-up.

On the other hand, the doctrine of determinism can be saved, if we consider mathematically the whole state of a quantum system (not that of its individual constituents). The demand for determinism must then be shifted from the concrete experimental situation to an abstract mathematical formalism – the Schrödinger equation. This equation mathematically describes the deterministic evolution of the quantum-mechanical system. It is specified by a wave function, ψ. This wave function cannot be observed directly, unlike some of the observables in quantum mechanical experiments (like spin and polarization).[53] Nevertheless the wave function and its mathematical properties (like superposition) are needed to describe the motion of atomic systems and predict their behaviour.

The powerful influence of the Laplacean spirit prevented Planck from ever making this distinction explicit. There is a price to be paid for this refusal to relinquish the notion of strict causation. Causation becomes a *heuristic* principle of research, an ideal, whose only hope of realisation resides in the assumption of an ideal Laplacean spirit.[54] There is a perfect analogy between Planck's microscopic

[51] Planck, 'Das Weltbild der neuen Physik' (1929), 214–5; author's own translation; see also Planck, 'Lichtquanten' (1927)

[52] Planck, 'Das Weltbild der neuen Physik' (1929), 223; author's own translation

[53] Planck, 'Das Weltbild der neuen Physik' (1929), 224; 'Kausalität in der Natur' (1932), 261–2; 'Physik im Kampf um die Weltanschauung' (1935), 285–300; 'Determinismus und Indeterminismus' (1937), 334–49; Cushing, *Philosophical Concepts* (1998), 298

[54] Planck, 'Kausalität in der Natur' (1932), 265. In a similar vein von Laue, 'Kausalität' (1932), and 'Heisenbergs Ungenauigkeitsbeziehungen' (1934) argued that from the new

observer of the atomic world and Laplace's macroscopic observer of the universe. Both have to be equipped with the same capacity to encompass all the initial conditions necessary for solving the dynamic causal laws. The consequence is that human ability to emulate these ideals pales to insignificance.

Like Planck and von Laue, Einstein was willing to pay the price for the commitment to determinism. Einstein took issue with Born's rather measured and philosophical response to the conceptual problems. In the margins of Born's book *Natural Philosophy of Cause and Chance* (1949), Einstein jotted down his objections:

> I know very well that with respect to the observable, causality does not exist; I regard this as a definite piece of knowledge. But in my opinion it should not be inferred from this that the theory has to rest on fundamental laws of a statistical nature.[55]

Einstein took causation to be such a powerful motivating principle of research that quantum mechanics should not be made to rest on statistical notions. It is in this context that we see Einstein's more conservative attitude. Planck was willing to concede to quantum mechanics a revision of the conceptual structure of the physical worldview. He pinned his hope on a new ontology. But when it came to quantum mechanics Einstein was unwilling to abandon any classical notions. Not only did he want to retain a classical notion of causal determinism. He was unwilling to concede that quantum mechanical particles may not possess definite properties at definite times, independent of the act of measurement. In quantum mechanics, there are puzzling correlations between spin properties of particles, observed in certain laboratory set-ups. These correlations have been dubbed 'action-at-a-distance', because the particles are so far apart that no detectable causal influence has been observed. If accepted, this leads to a form of non-local interaction. Planck accepted this holism and called for a consideration of the physical object as a whole. Today this phenomenon of correlated properties of widely separated atomic particles is described as *entanglement*. It is experimentally well established but defies our traditional view of causation. As we have seen on the chapter on Nature, Einstein was vehemently opposed to the idea of action-at-a-distance. In his correspondence with Born, he wrote:

> For the relative independence of spatially distant objects (A and B) there is a characteristic idea: an external influence on A has no immediate influence on B; this is known as the 'principle of proximate action', which is only systematically applied in field theory.[56]

---

evidence and the importance of statistical laws in quantum mechanics 'no epistemological renunciation of the causal interpretation of individual events should be inferred'.

[55] Born-Einstein, *Briefwechsel* (1969), 217; author's own translation

[56] *Briefwechsel* (1969), 226; Einstein's famous snide on 'spooky action-at-a-distance' appears on page 210. B. d'Espagnat, *Conceptual Foundations* (1971), 114 gives the following formulation of the locality condition: 'If a physical system remains, during a certain time, mechanically (including electromechanically, etc.) isolated from other systems, then the evolution in time of its properties during the whole time interval cannot be influenced by operations carried out on other systems.' Because direct causal influences between these spin correlations can be excluded, some authors prefer to call the phenomenon

Recent experiments in quantum mechanics do not agree with the two ideas expressed by Einstein in his correspondence: that causal relations are not observable in the quantum realm and that some form of 'action-at-a-distance' between two spatially separated photons should not be incorporated into the physics of the atom. These non-local correlations between spatially separated particles are routinely observed. It is true that they cannot be understood in terms of causal determinism. But once the bond between determinism and causation is broken, quantum mechanical experiments, as we shall see below, can be interpreted causally.

As we have already remarked there are several important features, which are traditionally associated with the notion of causation: the temporal priority of the cause over the effect (*antecedence*); the regular, perhaps law-like *link* between cause and effect (Hume's constant conjunction); and the local proximity of cause and effect (*contiguity*). Sometimes, the traceability of the link between the cause and the effect (*causal mechanism*) is postulated. It is the aspects of contiguity, known as *locality* in discussions of the interpretation of quantum mechanics, and causal mechanism to which Einstein appeals in his correspondence with Born. Einstein had no objection to the essentially statistical character of quantum mechanics as long it could be regarded as a statement about the ensemble of particles. He assumed, however, that the particles themselves obeyed dynamic laws and possessed definite properties.[57] The acceptance of quantum mechanics as a working model implied that it may not be possible for quantum mechanical systems to state with precision and without exception, that the *same* effect always followed the *same* cause. Einstein found this consequence acceptable, in practice. But the nature of the quantum-mechanical evidence *seemed* to require conceptual changes to the physical worldview – non-locality or action-at-a-distance and acausality – which Einstein found distasteful.

The transformation of the principle of causation from an achievable to an unachievable aim of scientific research was the conceptual price, which was demanded by a graft of the quantum mechanical evidence onto the conceptual structure of an Einsteinian field view. There was also a heuristic bonus. The heuristic gain from an adherence to the classical notion of causation, not in the Humean but the Laplacean sense, was an encouragement of causal interpretations of quantum mechanics. David Bohm, for instance, appealed to some feedback thesis (reminiscent of Jammer's thesis) to justify the search for alternative interpretations of quantum mechanics.[58] (Box 5.3)

'passion-at-a-distance'; see Redhead, 'Factorizability' (1989), 145–52 and 'Nature of Reality' (1989b), 440

[57] *Briefwechsel* (1969), 270ff; see also Einstein/Podolsky/Rosen, 'Quantum-Mechanical Description' (1935). It is interesting to note that in Einstein/Tolman/Podolsky, 'Knowledge of Past and Future' (1931), 780 the authors try to show, by the construction of a paradox, that 'the principles of quantum mechanics actually involve an uncertainty in the description of past events which is analogous to the uncertainy in the prediction of future events.'

[58] Bohm, *Causality and Chance* (1957), 97–99

Box 5.3:

Discovery of New Facts $\Longrightarrow$ Refinement of Well-Established Concepts
Fundamental, Rich Concepts $\Longrightarrow$ Discovery of New Facts

We see how the notion of causation plays its role as a thematic presupposition. It led Einstein to a rejection of quantum mechanics as a fundamental theory. It encouraged Bohm to look for an alternative interpretation of quantum mechanics. One side of the *dialectic* is that the discovery of new facts leads to a refinement of well-established concepts. But the concepts should not be made slaves to the facts, the other side of the dialectic should not be forgotten. Adherence to rich concepts, like causation, could eventually lead to the discovery of new interpretations of quantum mechanics and even to the discovery of new facts.

Bohm's appeal to hidden variables was designed to avoid the radical response of Heisenberg and Bohr to the problem of causation. To re-establish causation, Bohm appeals to a quantum-mechanical sublevel in which new kinds of forces and causal laws govern. Their function is to reproduce the quantum-mechanical evidence from numerous experiments, without accepting the orthodox conception of lawless fluctuation at the level of individual quantum events. Bohm's version of quantum mechanics aims at reducing indeterminacy from Heisenberg's ontological level to a mere epistemological level. At a quantum mechanical sublevel the individual processes have well-defined space-time trajectories and properties.[59]

We see that the notion of causation is not disposable baggage in the physical understanding of quantum mechanics. If a functional view of causation is retained, with its associated predictive determinism, quantum mechanics must be seen as incomplete. This encourages a search for deeper levels, at which causal, perhaps even deterministic relations can be found. Einstein would have encouraged Bohm in his endeavour to provide a causal interpretation of quantum mechanics. Bohm, though, falls short of embracing Laplacean determinism, since 'every real causal relationship, which necessarily operates in a finite context, has been found to be subject to contingencies arising outside the context in question'.[60]

There is a conceptual alternative to the conservative response. Let us retain the Laplacean presupposition and identify causation with predictive determinism. Imagine that Planck's microscopic observer fails to determine the initial conditions of atomic particles and to trace dynamic laws for their evolution. We would conclude

[59] Bohm, *Causality and Chance* (1957), Chap. I. Bohm comes very to close to a conception of conditional causation, which will be developed later as the most adequate model of causation in view of quantum-mechanical evidence. On Bohm's version of quantum mechanics, see Cushing, *Philosophical Concepts* (1998), Chaps. 23, 24.3; Toretti, *Philosophy of Physics* (1999), Chap. VI. A newer attempt to provide a causal interpretation of quantum mechanics is given by Vignier/Dewdney/Holland/ Kyprianidis, 'Causal particle trajectories' (1987), 169–204. Although the authors use the term 'causality', they mean 'physicalistically deterministic reality' (1987, 190).

[60] Bohm, *Causality and Chance* (1957), 3, 13

that causation was absent from the microscopic world. This move would reverse the conservative response and turn it into a radical response.

5.3.2 The Radical Response

> The observer is like the comedian with an armful of parcels; each time he picks up one he drops another.
>
> A. Eddington, *New Pathways in Science* (1935), 100

The *radical response*, too, works on the assumption of an identity between causation and classical determinism. But instead of suspecting quantum mechanics of being incomplete, it rejects classical mechanics and the fundamental notions, which are part and parcel of this view. In terms of *Einstein's problem* the conservative response adheres to fundamental notions, like causation and determinism, and refuses to listen to the persuasiveness of the new facts. Proponents of the radical response, like Heisenberg and Bohr, also concentrate on only one side of the dialectic. By contrast, they allow the quantum-mechanical facts to determine the fate of the fundamental notions. In terms of *Einstein's problem* they adhere to the new facts, accumulated since Planck's constant and the radioactive law, and refuse to listen to the echo of the inherited fundamental notions. The new discoveries give rise to *new* concepts, of which the notion of causation is not part. Both attitudes are similarly indebted to the identification of causation and determinism. To the conservative response the 'truth' of causation means the invalidity of the orthodox interpretation of quantum mechanics. To the radical response the indeterminacy of quantum-mechanical relations means the end of causation.

Werner Heisenberg (1901–1976)

The indeterminacy relations, developed by Heisenberg in 1927, provide the cornerstone of the radical response. In this otherwise mathematical paper, Heisenberg concludes with a brief discussion of the philosophical consequences of his interpretation of quantum mechanics. The major consequence is, of course, that quantum mechanics shows 'conclusively' that the principle of causation is untenable. Heisenberg formulates this principle in the statement: 'If we know exactly the determinable properties of a closed system at a given point in time, we can calculate precisely the future behaviour of the properties of this system.'[61] How reminiscent this statement is of Laplace! It is evidence of our claim that the radical response

[61] This statement is constructed out of two similar formulations, which Heisenberg offers in 'Über den anschaulichen Inhalt' (1927), 197, 174–79 and 'Kausalgesetz und Quantenmechanik' (1931), 179–81. Heisenberg's term Unbestimmtheitsrelationen is often translated in physics textbooks as uncertainty relations. Čapek, *Philosophical Impact* (1962), 311, Bohm, *Causality and Chance* (1957), 85–86 and Salmon, *Causality and Explanation*

is indebted to the identification of causation with determinism. If we consider, James Jeans wrote, that '500 million atoms are due to disintegrate in the next second', then it cannot be determined, even with a knowledge of the present conditions of individual atoms or their past histories, which ones will disintegrate first. We seem to have 'an event without a cause.' As a philosophical consequence, Jeans stated that causation seemed to have been removed from our physical world picture. Even a Laplacean spirit would not be able to calculate the disintegration rate of individual atoms.

> From the state of the matter at one instant, it is impossible in principle to discover what the state will be at a future instant.[62]

This, however, is the classical identification of causation with determinism. The Laplacean spirit holds a powerful sway over modern minds. The causation principle becomes untenable in the quantum-mechanical realm because the antecedent conditions, needed to calculate the future evolution of the properties of a system, cannot be satisfied, if the indeterminacy relations hold in the quantum world.

However, the quantum world also forbids the satisfaction of the *consequent*, as Heisenberg stressed soon afterwards.[63] The physical behaviour of anatomic system can in general not be precisely calculated. For instance, even if we can specify an atomic region, Δx, say of length 10^{-10} m, the probability of finding an electron, in its ground state, inside a specified band of size $(0.09 \text{ to } 0.11) \times 10^{-10}$ m is approximately 40 percent. This is an illustration of the often-expressed view that quantum mechanics only leads to statistical statements about the behaviour of quantum systems. Sometimes we get luckier. Consider the Stern-Gerlach experiment: a beam of atoms travels through a non-uniform magnetic field. In the simplest case each atom in the beam has a 50% of being deflected upwards or downwards by the field. There are certain experiments, which can be performed on quantum mechanical systems whose statistical outcome can be determined more exactly.[64]

(1998), 261 have pointed out that the term indeterminacy relations is more appropriate, since orthodox quantum theory takes these relations to be an expression of the indeterminate nature of quantum particles and not just human inability to measure their properties with precision. To readers of modern technical or popularised physics texts, the inclusion of a philosophical discussion, as in Heisenberg's paper, seems incredulous. It was, however, common practice in the first forty years of the 20th century, as we have already observed in the discussion of the block universe. It will be illustrated further in this chapter.

[62] Jeans, *Physics and Philosophy* (1943), 149–50; see also Frank, *Kausalgesetz* (1932), Chap. VII. A contemporary physicist arrives at much the same conclusion as Jeans, namely that the decay law spelt the end of classical causation. Note, however, that his conclusion is based on the understanding of a classical notion of causation as being identical with determinism: where strict prediction fails, causation disappears. See Pais, *Inward Bound* (1986), 121, *Niels Bohr's Times* (1991), 82

[63] Heisenberg, 'Kausalgesetz und Quantenmechanik' (1931), 36–38; 'Die Rolle der Unbestimmtheitsrelationen' (1931), 44–45

[64] In the two articles, referred to in the previous footnote, Heisenberg mentions the Stern-Gerlach experiment and the passage of light quanta through two prisms with a variable angle between them.

The indeterminacy relations were once considered fundamental to quantum mechanics. Today they are often circumvented. In the indeterminacy relations the precision with which paired operators can be measured is inversely related. The indeterminacy relations comprise two situations[65]: (a) Imagine we attempted to determine exactly the location, x, of an electron in an atom. This could be done by shining a light on it. Our attempt will be accompanied by a discontinuous increase in the imprecision with which its momentum, p, can be known. The more precise the location of the electron is, the less precise is its momentum and vice versa. The precision is dependent on the wavelength of the light, with which the atom is observed. The smaller the wavelength of the observing light, the more precise the determination of the location. But the change in the momentum of the electron is the greater, the smaller the wavelength of the observing light.[66] If Δx is the uncertainty with which the location, x, can be measured and Δp is the amount of uncertainty with which the momentum can be stated, then the product of these two operators is always greater than the Planck constant. In symbols:

$$\Delta x \Delta p > h\,.$$

(b) A similar inverse relationship applies to the pair energy, E, and time, t. The precision with which the energy states of a particle can be measured increases with the amount time, t, which is available for the measurement. If an infinite amount of time is available the energy of the system can be measured exactly. Thus, for short-lived particles, the energy cannot be known precisely.[67] But precision in the determination of energy, ΔE, is paid for by the corresponding imprecision in the phase of the atomic motion. This is a measure of the amount of change of the state of the system during time, Δt. It would remain completely unknown if E were known precisely.[68] In symbols:

$$\Delta E \Delta t > h\,.$$

No experiments can ever be performed, which will result in indeterminacy below the limits expressed in these equations.

In more rigorous discussions of these relations, it is pointed out that these pairs of operators, which represent the observable variables, do *not commute*. We have

[65] Two further forms of the indeterminacy principles can be stated: one refers to the indeterminacy of angular momentum and angular position, the other refers to the indeterminacy of the moment of inertia and the angular velocity; see Čapek, *Philosophical Impact* (1962), 290–3.

[66] This can be seen from the de Broglie relation ($\lambda = h/p$ or $p = h/\lambda$), where h is the Planck constant.

[67] See Tipler, *Modern Physics* (1978), 194. Ground state energies are known precisely, because the atom remains in its ground state indefinitely, if no transitions occur. Recall from the end of Chap. 4 that time is not strictly speaking an operator in quantum mechanics. It is a background parameter without observables.

[68] See Heisenberg, 'Schwankungserscheinungen' (1926–27), 502–6; 'Über den anschaulichen Inhalt' (1927), 177; *Physical Principles of Quantum Theory* (1930), 39–46

just seen what this means. Physically, the determination of the pair x and p can not be performed independently of each other, nor can the determination of the pair E and t; mathematically, the product of the spread in the indeterminacy of two non-commuting operators will always be greater than zero.

Bohr agreed with Heisenberg that the indeterminacy relations were responsible for the collapse of the classical ideal of causation in the quantum world. As usual, the classical ideal of causation was seen as the complete determination of a physical system in terms of its well-defined properties. But Bohr went further than Heisenberg. Rather than just renouncing the notion of causation, leaving a vacuum in which the celebrated '*a*causality' of quantum mechanics could nestle, he argued that the notion of *complementarity* was a generalisation of the notion of classical causation.[69] The culprit was the Planck constant h. Complementarity means that quantum mechanics needs to employ both the particle picture and the wave picture to make statements about the behaviour of quantum mechanical systems. But the employment of each picture comes at a price: a loss of information concerning properties associated with the alternative picture. The observer is like Eddington's comedian with an armful of parcels. Energy, E, and momentum, p, are associated with particles and allow space-time co-ordination. Periods of vibration, t, and wavelengths, λ, are associated with waves extended in space. In order to obtain any information at all, there must be an interaction between the system of particles and the measurement instrument. This interaction, however, produces uncontrollable changes in the measured quantum system. As it turns out these uncontrollable kicks are just features of the orthodox interpretation. The following dilemma arises for two alternative experimental situations[70] (Box 5.4):

Box 5.4:

I. Determination of the spatio-temporal location of atomic particles
$\Rightarrow$ Loss of the precise value of the dynamic variables (energy/momentum)

II. Determination of the value of the dynamic variables (energy/momentum)
$\Rightarrow$ Loss of the precise value of the spatio-temporal paths of the particles

The dilemma arises from the indeterminacy relations. The determination of the location of an electron in the atom, Δx, leads to an unavoidable fuzziness in the

[69] Bohr, 'Quantum Postulate' (1927), 'Atomic Theory' (1929), 'Wirkungsquantum' (1929), 'Kausalität and Komplementarität' (1936), 'Quantum Physics and Philosophy' (1958). Support for this interpretation came from Pauli, 'Raum, Zeit und Kausalität' (1936) and 'Physikalische Realität' (1957); Heisenberg, *Principles* (1930) Chap. IV, §3, Heisenberg, 'Atomphysik und Kausalgesetz' (1952). For Pauli, Bohr and Heisenberg the notion of determinism as regards atomic phenomena had altogether lost its sense. Pauli, 'Raum, Zeit' (1936), 743 discusses an example.

[70] Bohr, 'Kausalität' (1936), 295–6; Pauli, 'Raum, Zeit' (1936), 742; see also Heisenberg, *Principles* (1930), Chap. I, §1. It should be noted that in modern *welcher Weg* experiments, discussed below, such uncontrollable changes can be avoided, see Dürr et al., 'Origin' (1998); Scully et al., 'Quantum optical tests' (1991).

value of its momentum, Δp. And this means, according to Bohr, 'a complete rupture in the causal description of its dynamical behaviour.' By contrast, 'the determination of its momentum always implies a gap in the knowledge of its spatial propagation.'[71]

One reason why the radical response rejected determinism was that it simply argued on the basis of the Laplacean identification of causation and determinism. On the analogy with the fate of the notion of absolute time in Einstein's theory of relativity, Heisenberg and Bohr were keen to stress the inadequacy of classical notions in the quantum domains. To show the inadequacy, they basically concentrated on only two kinds of examples, which would illustrate the limited usefulness of classical notions.[72] *Radioactivity* or the natural decay of certain atoms figured prominently in their discussions. The 'disintegration of atomic nuclei without external cause' showed the failure of the notion of causation. No initial conditions can be specified, which could be shown to have a causal effect on the moment and direction of decay. The antecedent conditions of the atom before the moment of decay display no causal priority. The decay law does not specify a conditional dependence of the effect on the cause. Is it possible that the *known* conditions of radioactive decay amount to an incomplete knowledge of the decay process? No, says Heisenberg, from other experiments it is known that the atom possesses no further properties, which could specify the direction and moment of decay.

Even more revealing for the defence of the Copenhagen interpretation of quantum events is the use of the *double-slit experiment*. The double-slit experiment is designed to show the complementary nature of quantum systems. For a complete description of quantum events, both the wave picture and the corpuscular picture are needed. Both convey only partial information. As we have seen, Bohr, Heisenberg and Pauli felt that the complementarity principle was a natural successor to the classical causation principle. There is a complementarity between the space-time description and the causal claims. Quantum mechanics permits the specification of location, Δx, or time, Δt alternatively, but each time the energy components, Δp and ΔE, which could have completed the dynamic or causal picture, becomes imprecise (Box 5.4). The price is a renunciation of a continuous causal, spatio-temporal description of atomic processes.

What is at stake can be shown by a consideration of the double-slit experiment, which for a long time existed only as a *Gedankenexperiment*.[73] Imagine a photon or electron source emitting particles towards a black or metal sheet. There are two slits in the sheet. Behind the sheet a photographic or electron detector is placed. Although the photons or electrons are considered to be particles of which some

[71] Bohr, 'Quantum Postulate' (1928), 584. This is the replacement of the ideal of causation with the notion of complementarity, embraced by Bohr and Pauli.

[72] See Bohr, 'Quantum Postulate' (1928), 589; 'Atomic Theory' (1929), 104, 108; Heisenberg, *Physik und Philosophie* (1959), 68; *Der Teil und das Ganze* (1973), 142–3

[73] Heisenberg, *Physik und Philosophie* (1959), 34–5; *Naturbild* (1955), 29–30. Modern discussions of the double-slit experiment can be found in Feynman, *Physical Law* (1965), Chap. 6; Pagels, *Cosmic Code* (1983), 129–30; Cushing, *Philosophical Concepts* (1998), 299ff; Dürr et al., 'Origin' (1998); Scully et al., 'Quantum optical tests' (1991); Deutsch, *Fabric* (1997), Chap. I.

of the physical properties are known, they leave wavelike traces or interference patterns on the detector (see Fig. 2.7).

When both slits are open, the atomic particles behave just like waves, seemingly co-ordinating their behaviour. Should we attempt to determine through which slit individual photons have travelled, the interference pattern will be destroyed. The waves are extended in space and seem to defy the requirements of a 'causal', spatio-temporal description. Their non-local interaction may lead us to think that the wave nature of quantum systems does indeed violate the notion of causation.[74]

The aim of the radical response was to emphasise the differences between classical and quantum mechanical approaches to the natural world. According to Bohr and Heisenberg the contrast between classical physics and the quantum theory could conveniently be summarised as in Table 5.1.[75]

The classical association of causation with determinism allows the proponents of the radical response to conclude that the notion of causation has lost its place on the conceptual map of quantum mechanics. A causal description of atomic phenomena – the disintegration of atomic nuclei without an external cause, interference

Table 5.1. Classical versus Quantum Theory, according to the Radical Response

Classical Theory	**Quantum Theory**		
	Either		*Or*
– Description of Phenomena in Space and Time – Mathematical Laws – Causation (= Determinism)	Classical Description of Phenomena in Space and Time *But* Indeterminacy Relations ↓ Indeterminism of Quantum World	**Statistical Relations**	Mathematical Equation (i.e. Schrödinger equation) in abstract Hilbert space (Determinism) *But* Classical, Physical Description of Phenomena in Space-Time Impossible (*A*causality)

[74] In 1961 C. Jönsson, 'Electron Diffraction at Multiple Slits' (1974), 4–11 carried out the double-slit experiment on electrons for the first time in the laboratory. This type of experiment has subsequently been refined and repeated. A modern version of this experiment is the so-called welcher-weg or which-way experiment. It allows a determination of the trajectory of particles, without involving the indeterminacy relations, see Dürr et al., 'Origin' (1998); Scully et al., 'Quantum optical tests' (1991). Recall from Chap. 3 that there are several ways of understanding the double-slit experiment.

[75] This table in slightly amended form in Heisenberg, *Principles* (1930), 65 and 'Die Rolle der Unbestimmtheitsrelationen in der modernen Physik' (1931), 46. See Bohr, 'Quantum Postulate' (1928), 580–82. The amendment is significant for it reveals the vacillation of proponents of the radical response between causation and determinism. In his *Principles* Heisenberg uses the expression 'Causal relationship expressed by mathematical laws' instead of the expression 'Mathematical Equation (Determinism).'

patterns in double-slit experiments – had become impossible. Quite generally, the determination of a causal chain, leading from a cause to an effect, was impossible because of the necessary gap in the knowledge of some of the variables. This knowledge, however, was required for a causal account.

Einstein objections to the quantum theory were based on the association of the concept of causation with one of its standard features, contiguity or locality. When Bohr speaks of the 'causal description of the dynamical behaviour' of the system[76], he appeals to another feature of the concept of causation, the causal link. What Bohr deplores is the loss of the traceability of individual causal lines from a cause, *C*, to an effect, *E*. It is the loss of a determinate causal link between a cause and its effect. But Bohr has inherited Laplace's view of causal determinism. So he concludes that an indeterminate link between *C* and *E* must mean a loss of causation in quantum mechanics.

Is this conclusion inevitable? In their later years, Bohr and Heisenberg reveal a shift in their philosophical thinking. If quantum mechanics can be regarded 'as a rational generalisation of the causal space-time description of classical physics',[77] is it not conceivable that an indeterminate, yet causal link exists between *C* and *E*? Shining light of a certain wavelength on the atomic system, for instance, helps to determine the location of the electron, but changes its momentum unpredictably. This interference is a causal process. The causal interference changes the prior state of an atomic system to a posterior state. The effect of the interference on the system is revealed to us by a gain of causal information. That is, a conditional dependence can be established between a set of antecedent conditions (cause) and a set of subsequent conditions (effect) it produces in the system.

It is a testimony to their continued philosophical awareness that both Bohr and Heisenberg, in their later writings, began to separate the notions of causation and determinism. Determinism was an ideal form of the causal relationship. This model of determinism had to be abandoned in quantum mechanics. The principle of complementarity took its place. But there was still a causal account at play for 'the experimental arrangement and the irreversibility of the recordings concerning the atomic objects ensure a sequence of cause and effect.' This conforms, as Bohr says, to 'elementary demands of causation.'[78] Heisenberg, too, came to suggest an understanding of the notion of causation, which was in conformity with the demands of quantum mechanics. This, for Heisenberg, was an example of how scientific findings had an immediate impact on philosophical notions.

> Causation here does not mean that from identical initial conditions identical events must necessarily follow for the future. We know from quantum theory that such a conception of causation is too narrow. Rather the term causation is taken to mean that the relation between cause and effect, the possibility of which already follows from the existence of laws of nature, in the first place expresses a relation

[76] Bohr, 'Quantum Postulate' (1928), 584
[77] Bohr, 'Quantum Postulate' (1928), 589
[78] Bohr, 'Quantum Physics and Philosophy' (1958), 1, 5

> between the events in one spatio-temporal point and succeeding events in another spatio-temporal point, which is in immediate proximity to it.[79]

Heisenberg appeals to features like lawlike dependence of the effect on the cause, locality and temporal succession, often associated with the notion of causation. He implies that some of these features may well be compatible with the empirical findings of quantum mechanics.

We can be fairly certain today that locality is *not* a feature of the quantum world. Numerous experiments have demonstrated that quantum systems are entangled. For instance, curious correlations exist between spin or polarization states of quantum systems, which do not depend on the distance between the systems. This entanglement comes in degrees and is a matter of experimental manipulation. It is not compatible with the requirement of locality. It holds the promise of new technological innovations. The philosophical lessons drawn from such experiments, already pointed out by Planck, is that *non-locality* must be a feature of reality. But we should not conclude, on the strength of a *mistaken* identification of causation with determinism, that the quantum world is acausal. Rather, the experimental evidence must guide our assessment. This approach is in agreement with the philosophical response.

Quantum mechanics had clearly brought about an upheaval in the fundamental notions, with the help of which physicists tried to make sense of the material world. The Laplacean presupposition had led the conservative and radical response in opposite directions. The radical response was more willing to abandon the fundamental notion of causation than the conservative response. Then there was the increasing amount of evidence. The redrawing of the conceptual map after the quantum revolution proved to be much more difficult than a similar redrawing after Einstein's Special theory of relativity. The Special theory of relativity led to an agreement concerning changes in the notion of clock time but left disagreement about the ontological status of time: is it real or is it an illusion? The difficulties in the redrawing of the conceptual map after the quantum revolution was partly due to the lack of conceptual clarity about the understanding of quantum mechanics (including the notion of causation), partly to the bewildering empirical discoveries, which cried out for conceptual understanding. Nevertheless, distinctive features of the notion of causation were considered, if not jointly at least separately. Slowly,

[79] Heisenberg, 'Grundlegende Voraussetzungen' (1959), 253; author's own translation. Heisenberg, who in his early writings had announced the end of the law of causation 'as a definite prediction' (see his 'Erkenntnistheoretische Probleme der modernen Physik' [1928], §II and 'Kausalgesetz und Quantentheorie' [1930]), here considers a revised notion of causation, which is compatible with quantum mechanics. In his book, *Das Naturbild der heutigen Physik* (1955), 24–32 he holds that a narrow conception of causation as causal determinism is incompatible with quantum mechanics. This is essentially the attitude of the philosophical response. Heisenberg often stressed the interrelation between science and philosophy and considered that the results of natural science could be used to decide philosophical problems empirically. See Heisenberg, 'Atomphysik und Kausalgesetz' (1952) and 'Philosophische Probleme' (1967), 421–22

tentative models of causation for quantum mechanics began to emerge. The notion of non-locality or entanglement exercised the physicists' minds.

Interlude D. *Entanglement and the States of a Quantum System.* In the 1920s, several physicists proposed the notion of *non-locality* as a way out of the dilemma, which quantum mechanical phenomena had posed for the notion of causation.[80] Discontinuous absorption and emission processes were the culprits. Einstein had identified the particles, involved in the interaction between matter and radiation as photons or light quanta. To deal with the puzzle of the chance character of the direction of the emitted photons in spontaneous emissions (in transitions from ε_m to ε_n), Einstein had declared the quantum theory incomplete. This attitude permitted him to retain a strong notion of causal determinism, despite the verdict of the experimental results. On the basis of the nature of how atoms absorbed and emitted discrete bundles of energy (quanta) other physicists even suggested a *teleological* model of causation.[81] This would be in conformity with the view that the fundamental notions had to undergo conceptual revision, if the empirical results required it. But which part of the conceptual scheme should the revision target? As we have seen, Planck attributed a central position to the classical notion of causation within the conceptual scheme. If it is to agree with the empirical results, any conceptual adaptations will have to affect other parts of the conceptual scheme. Planck was ready to jettison the Newtonian point particle. The light quanta were not to be modelled as spatially located Newtonian corpuscles but as spatially spread wave packets. This view has an implication for the conception of causation on which at least some of the physicists (Planck, Sommerfeld, Schottky) agreed. The assumption of spatio-temporal locality had to be abandoned. The existence of light quanta and quantum jumps required the introduction of a notion of spatio-temporal *non-locality*. The preference in the classical model of causation for the antecedent conditions, which in conjunction with general laws would determine the

[80] The first seems to have been Schottky, 'Das Kausalgesetz' (1921). He is the discoverer of the Schottky defect. (This describes a defect in crystals, when an ion moves to the surface of the crystal and leaves a vacant position inside.) The same idea was proposed by Sommerfeld, 'Quantentheorie' (1924), 'Atomphysik' (1927) and 'Grundlegende Bemerkungen' (1929) and Planck, 'Lichtquanten' (1927), 'Weltbild der neuen Physik' (1929), 206–27 and 'Die Physik im Kampf' (1935), 285–300. Planck's 'Lichtquanten' (1927), 531 adds that nonlocal, teleological causation would not change the essence of the principle of causation, only its form. The proposal was attacked by Stark, 'Axialität' (1930), 718–22. It received little sympathy from Zilsel, 'Asymmetrie' (1927), 284 and Schlick, 'Kausalität' (1931), 159–60.

The reader should be aware that the philosophy of quantum mechanics has produced a variety of notions of locality: *Einstein locality* $\approx$ no 'spooky' action at a distance; *Bell locality* $\approx$ factorizability of joint probabilities; *Jarrett locality* $\approx$ parameter independence; see Cushing/McMullin eds., *Philosophical Consequences* (1989); Cushing, *Quantum Mechanics* (1994), §4.4; Maudlin, *Quantum Non-Locality* (2002). In the main text we use non-locality in the sense of entanglement between the states of quantum systems in *time-like* or *space-like* separation; with decoherence a different sense of entanglement will be introduced.

[81] Sommerfeld, 'Grundlagen der Quantentheorie' (1924), 1049

consequent conditions, had to be given up. In quantum jumps, light quanta seem to be affected by both the antecedent and the consequent conditions. Recall that Rutherford, in his letter to Bohr, expressed grave concerns about the assumption that the electron knows beforehand, where its jump will lead it. If an atom makes a transition, say, from a higher energy level (ε_m) to a lower level (ε_n) it must emit photons whose energy is equal to the energy difference between ε_m and ε_n. Schottky imagines that the photons must extend 'feelers' so that it 'knows' where it is supposed to go.[82] It seems that the course of a quantum event, like the transition between quantum states, becomes dependent on both the anterior and the posterior states. In order to calculate the probability of a transition between antecedent and consequent states of the atom, both states must be included in the calculation.[83]

> The notion of causation of the 20th century must no longer be restricted to the initial state but must account for the final state as a determining moment. Quantum physics creates a new form of causation, which is different from mechanical compulsion and accounts for the multiplicity of the quantum transitions. The question no longer is, 'Given the initial state in all its details, which one is the final state?' The exact knowledge of the initial state is not sufficient for the posing of this question. The question is, 'Given a certain knowledge of the initial state, i.e. its energy or quantum numbers, and a general knowledge of the possible final states, what is the probability of a transition from that initial state into one of these final states?' The consequent state is therefore no longer necessarily but only conditionally determined due to a certain predictability of the permitted possibilities.[84]

We need not worry, as others have done[85], about the re-introduction of *final* causes into modern physical theory. The decisive aspect of the teleological model is the intimation of two aspects of a probabilistic conception of causation for quantum mechanics:

I. Traditional features of causation, e.g. locality, may be abandoned, without jeopardising the adequacy of the notion of causation in the atomic realm. Non-local interactions are part and parcel of the quantum world. The problem is that the traditional notion of causation leaves no room for non-locality. This *entanglement* expresses a correlation between, say, two quantum systems. Standard examples are spin or polarization states of such systems. Consider an experiment, in which a photon pair is prepared in a common source, after which the two photons fly in opposite directions along the arms of the interferometer. The objective of the experiment is to measure the polarization states of the photons. The French physicist Alain Aspect carried out such experiments in 1982. The novel feature of Aspect's experiments was that the photons were

[82] Schottky, 'Kausalproblem' (1921), 509

[83] Sommerfeld, 'Atomphysik' (1927), 234; Hund, *Geschichte* Vol. I (1978), 10, Vol. II (1978), 147–8, 160

[84] Sommerfeld, 'Bemerkungen' (1929), 868; author's own translation

[85] Frank, *Kausalgesetz* (1932), §26; Stark, 'Axialität' (1930), 718–22. For a general discussion of teleological ideas and action principles in the history of physics and cosmology, see Barrow/Tipler, *Anthropic Cosmological Principle* (1986)

space-like separated. The experimenters could therefore be certain that no familiar causal message had been exchanged between the photons. When the measurements are made along the same direction, strictly correlated outcomes are measured (either both photons pass the filter or are absorbed by it). When the polarization states are measured by analysers at different angles, α, the relative probabilistic distribution of outcomes will be:

$$
\begin{aligned}
p(1|1) = p(0|0) &= \tfrac{1}{2}\sin^2\alpha \\
p(1|0) = p(0|1) &= \tfrac{1}{2}\cos^2\alpha
\end{aligned}
\tag{5.1}
$$

Schrödinger called this type of correlation 'entanglement of our predictions or of our knowledge' concerning the quantum states of the photon pair.[86] This kind of entanglement is now such a common feature of experiments around the world that we should note several of its properties, which go well beyond Schrödinger's original conception.

(a) It is *kinematic*: it is an observable fact of the quantum world, which can be expressed in quantitative relations like (5.1). Its dynamic *cause* is still unknown.

(b) It is not restricted to two quantum systems, or to subatomic particles. The next step was the entanglement of three photons in so-called *GHZ* entangled states (after Daniel Greenberger, Michael Horne and Anton Zeilinger). But nowadays, small collections of atoms can be entangled. The entanglement of molecules is only a question of time.

(c) It can be *edited*: the polarization states of photons can be entangled by degrees. Imagine that the two arms of the interferometer are equipped with polarizers, which can be set at different angles. Then different kinds of interference patterns will be recorded. A dip can be transformed into a peak. Or intermediate forms of interference patterns can be manufactured. The interference patterns, which are the physical manifestation of quantum entanglement, can be suppressed and restored, according to the experimentalists' desire. In this way entanglement, as the discussion below of *which-way* experiments will demonstrate, can even be used to circumvent the Heisenberg indeterminacy relations. It is important to note that the degree of entanglement does not depend on human observation.

(d) It no longer serves as the prime example of the paradoxical nature of the quantum world. Rather, entanglement is a *physical resource* which can already be technologically exploited in quantum computation, quantum cryptography and quantum teleportation.[87]

---

[86] See Schrödinger, 'Die gegenwärtige Situation' (1935), 827. Schrödinger speaks of a *Verschränkung der Voraussagen*. He regarded entanglement as *the* decisive feature of quantum mechanics. The notation (1|1), (0|0) etc. generally refers to excitation, polarization or spin states.

[87] A vast amount of literature on entanglement and its aspects is now available, see for instance Aczel, *Entanglement* (2001); DiVincenzo/Terhal, 'Decoherence' (1998); Nielsen, 'Simple Rules' (2002); Polzik, 'Atomic entanglement' (2002); Tegmark/Wheeler, '100 years'

(e) The entanglement between correlated states of quantum systems – Schrödinger's *Verschränkung* – should be distinguished from the entanglement of quantum systems with their environment. This second sense of entanglement is central to the programme of decoherence – our next stopover. The distinguishing mark of entanglement in the first sense is the existence of interference term; that of the second sense is the irreversible loss of interference terms to the environment.

II. Indeterminism may be compatible with a *probabilistic* notion of causation. A particular consequent state of the quantum system does not follow, by necessity, from an initial state. Rather, a probabilistic distribution of consequent states follows from the potential initial states of the quantum system. Consequent conditions are related to antecedent conditions by a probabilistic form of conditional dependency. This does not apply to individual particles but to quantum mechanical systems. The fact that 'a probabilistic distribution of consequent states follows from the potential initial states of the quantum system' is due to the interaction of the quantum system with the measurement apparatus. This statement enshrines all the problems of the interpretation of quantum mechanics. To see this, let us investigate the *states* of a quantum system.

We have already discussed that the modern view of Nature involves systems, rather than point particles. The atom is a system as Thomson and Rutherford discovered. But it is not a system in the traditional sense. According to the Rutherford-Bohr model, the atom is a quantum system with peculiar idiosyncrasies. To see how the notion of probabilistic causation emerged from the experimental work and to take a step towards the philosophical response, three different cases of the 'states' of a quantum mechanical system, like an atom, need to be distinguished.[88]

In the *first case*, the system is undisturbed. The evolution of its potential states is described mathematically by the Schrödinger equation. The system will develop in a deterministic and continuous manner in accordance with this equation (or some mathematically equivalent representation). But it would be a mistake to regard this equation as a statement about the differential change of real physical values, which the system possesses at all times. Rather, it is taken to represent the allowed *potential* states open to the system. The point is that the deterministic development of the system only exists in an abstract mathematical Hilbert space. The Schrödinger equation describes the evolution of quantum systems in this abstract configuration space. It is a differential equation about the temporal evolution of the *potential* states of the quantum variables in a quantum system. These variables are represented by

---

(2001); Terhal/Wolff/Doherty, 'Quantum Entanglement' (2003); Zeilinger, 'Fundamentals' (2001), 'Quantum Teleportation' (2002), *Einsteins Schleier* (2003)

[88] See Heisenberg, *Physical Principles* (1930), Chap. IV, §2; Strauss, 'Quantentheorie und Philosophie' (1960), 170–1; Maxwell, *Comprehensibility* (1998), 267; Earman, *Primer* (1986), Chap. XI. Precise treatments are given, for instance, by Heisenberg, *Physical Principles* (1930), Chap. IV, §1, 2; Hughes, *Quantum Mechanics* (1989), Pt. I, §§2,3 and d'Espagnat, *Conceptual Foundations* (1971), Chap. 6; van Fraassen, *Quantum Mechanics* (1991); Unruh, 'Time' (1995); Joos, 'Decoherence' (1996), §§2.3, 3.3; Mainzer, *Symmetries* (1996), §5.21

operators. There is a special class of Hermitian operators, whose function it is to represent physical observables. The equation is deterministic by specifying the rate of change of the dynamic variables of the quantum systems with respect to time. No actual values can be known about it until some interaction has taken place. It is the determinism of a mathematical scheme. Born expressed this succinctly:

> The motion of particles follows the laws of probability, but the probability itself propagates in accordance with the causal law. That is, such that the knowledge of the state in all respects in a particular moment determines the distribution of the state in all later times.[89]

This quote reveals the thinking of the early Born, when he still accepted the radical response. Later, as we shall see in a moment, he contributed to the philosophical response. Born's statement reflects Table 5.1. In the world of experiment, we can only have stochastic knowledge about the motion of particles. But in an abstract mathematical space, the Schrödinger equation is the causal (*read*: deterministic) law, which determines the Laplacean evolution of the potential quantum states. As Dirac pointed out, a particularly important property of quantum systems in a pure case[90] is the superposition of their component states. In the first instance the *super-*

[89] 'Quantenmechanik der Stoßvorgänge' (1926), 804. This characterisation is repeated almost verbatim in Born, *Natural Philosophy* (1949), 103. J. Jeans, *Philosophy and Physics* (1943), Chap. VI makes a very similar point when he stated that 'causality disappears from the events themselves to reappear in the knowledge of events.' See also Eddington, *Philosophy of Physical Science* (1939), Chap. VI

[90] Hermann Weyl introduced the distinction between *pure case* (*reiner Fall*) and *mixed case* (*Gemenge*). Mathematically, a pure case is represented by a vector or wave function. The representation of a mixed case requires a density operator. Today a uniform representation of pure and mixed cases is achieved through the use of statistical operators.

Physically a *pure case* occurs when some operator of a quantum system possesses an eigenvalue with certainty. For instance, once the inhomogeneous magnet in the Stern-Gerlach experiment has split the atom beam into two parts, it is certain that atoms in the upper beam possess, say, the value $+1$, while those in the lower beam have the value -1. The most important characteristic in the present context is that pure cases lead to *coherent superpositions*, with their unmistakable interference fringes. When interference patterns are observed, say between photons, their paths are indistinguishable. The experimenter cannot know *in principle* through which arm of the interferometer the respective photons have travelled. The *degree of indistinguishability* is directly related to the *degree of coherence.*

Mixed cases are sometimes called *incoherent superpositions*. That is they do not display interference patterns so that the paths of their respective components are distinguishable. The experimenter can therefore tell through which arm a particular photon travelled. It is the distinguishability, which destroys the interference. Mixed states can be formed out of any other states; for instance two electron beams, pointed in slightly different x-directions, or a two-system entangled spin state, measured with polarizers tipped at different angles. It is the distinguishing mark of mixed states that their component states are only known probabilistically. If we adopt Dirac's symbol for a quantum state, $|\ \rangle$, and $|\varphi_1\rangle$ and $|\varphi_2\rangle$ are two different states of an ensemble of electrons, then there is a state $|\Psi\rangle = a|\varphi_1\rangle + b|\varphi_2\rangle$, where a, b are probability coefficients, with $|a|^2 + |b|^2 = 1$. See Weyl, 'Quantenmechanik'

position principle is due to a mathematical property of quantum mechanics, called *linearity*. If a quantum system can exist in possible states, say, $|\Phi\rangle$ and $|\Theta\rangle$, then it can also exist in a superposed new state $|\Psi\rangle = \alpha_1|\Phi\rangle + \alpha_2|\Theta\rangle$, formed from its components (where α_1, α_2 are proportionality constants). Many new states can be formed in this way. Let a system exist in a state $|spin\text{-}up\uparrow\rangle$ and a state $|spin\text{-}down\downarrow\rangle$, then it can also exist in a superposition of spin states $|S\rangle = \alpha_1|\uparrow\rangle + \alpha_2|\downarrow\rangle$. This is a general property of quantum systems. So a quantum system can also be found in a superposition of energy or polarization states. Any polarization state can be formed from a superposition of horizontal and vertical polarization states. The superposition of such a two-state system can be extended to the superposition of entangled systems. But the superposition principle is not just a mathematical property of the linearity of quantum mechanics. It has *predictable* consequences. Consider again an atom beam, which is sent through an interferometer and split into two parts. No attempt is made to measure the result. Instead the split beams are recombined. The recombined atoms will interfere constructively and display the familiar interference effects. In order to explain this result, it has to be assumed that the atoms, during their journey, exist in a superposition of translational states. Schrödinger[91] argued that the linearity of quantum mechanics allowed the construction of 'burlesque' cases. For instance, we should in principle expect macroscopic superpositions of the form

$$|\text{cat}_{\text{here}}\rangle + |\text{cat}_{\text{there}}\rangle$$
$$|\text{cat}_{\text{alive}}\rangle + |\text{cat}_{\text{dead}}\rangle \ .$$

In the macroscopic world of human experience, such superposed macroscopic effects are never observed. Their correlations leak into the environment. The aim of the programme of decoherence is to explain how this happens.

In the *second case*, the system is disturbed but the disturbance remains unobserved. This happens whenever atoms decay, collide in nature, travel through an interferometer, or when they interact with their environment. This interaction can

(1927); Hughes, *Structure* (1989); van Fraassen, *Quantum Mechanics* (1991); Toretti, *Philosophy of Physics* (1999). D'Espagnat, *Foundations* (1971), Chap. 6 distinguishes *mixed states* further into *proper* and *improper* mixtures. An improper mixture is a superposition of *pure* states, which occur with a certain proportionality, which is not known prior to measurement. A *proper* mixture is a statement about the observer's state of knowledge of the quantum system, with attendant probability values.

[91] The example refers to the famous cat paradox, which Schrödinger introduced in his paper 'Die gegenwärtige Situation in der Quantenmechanik' (1935), 812. A cat is trapped in a sealed steel chamber. A 'hellish device' threatens its well-being. The device consists of a small radioactive substance, mechanically linked to a vial of cyanide. In the course of an hour there is an equal chance that one of the atoms or none will decay. If an atom decays, it will trigger the release of the cyanide and kill the cat. But the observer cannot know what has happened before the steel chamber is opened for examination. Before this act of measurement the cat exists in a superposition of being alive and dead. It is often forgotten that Schrödinger employed this paradox as an argument against a realistic interpretation of the wave function.

be regarded as a form of measurement. It is again the whole *state* of the quantum system that is determined by its initial conditions and the type of disturbance that has taken place. The disturbance flips the state vector into a different direction. This second case is sometimes called the *pre-measurement* stage.[92] The quantum system is coupled to the measurement apparatus but no reading of the measurement result occurs. This is the situation envisaged in Schrödinger's cat paradox. As long as *no* observer opens the steel chamber to check whether the cat is dead or alive, the wave function of the whole system is in a coherent superposition of

$$\frac{1}{\sqrt{2}}\left|\text{atom}_{\text{undecayed}}\right\rangle \left|\text{cat}_{\text{alive}}\right\rangle + \frac{1}{\sqrt{2}}\left|\text{atom}_{\text{decayed}}\right\rangle \left|\text{cat}_{\text{dead}}\right\rangle \ .$$

The quantum system, here represented by the cat, is entangled with the rest of the environment, represented by the atom and the cyanide.

A more realistic illustration of entanglement, the pre-measurement as well as the measurement stage exist in the so-called *welcher Weg* (or *which-way*) experiments. These are sophisticated versions of the double-slit experiment. They have been performed in laboratories since the 1990s.

Which-way experiments offer a way of finding out about the passage of particles without involving the indeterminacy relations. The idea of a *which-way* experiment can be introduced in two steps.

Consider *first* the atom interferometer without the storing of *which-way* information. A beam of atoms, A, is passed through a standing light wave, which splits the beam into a transmitted beam, C, and a reflected beam, B (Fig. 5.6). The first standing light wave is switched off: Beams B and C each move horizontally (B to the left d/2, C to the right d/2). After a time interval t_{sep}, they enter a second standing light wave, which again splits the beams ($B \to D, F$; $C \to E, G$). On a recording screen, the pairs of overlapping beams form a classic spatial interference pattern (Fig. 5.7).

Now consider the storage of *which-way* information in this experimental arrangement (Fig. 5.8). This is done by interfering with the internal electronic states of the atoms, using microwaves. These hyperfine states experience phase shift as the result of the application of the microwaves. The transmitted and reflected beams are converted to different states, $|2\rangle$, $|3\rangle$ respectively, which reveal the paths (B or C), which the atoms have travelled. The result of storing the *which-way* information is significant: the interference patterns are destroyed (Fig. 5.9). In the classic double-slit experiment nothing could be known about the passage of the particles so as to preserve the interference patterns. The *which-way* experiments carry information about the internal states of the atoms. These internal electronic states are then used as *which-way* detectors or *particle*-like information. (Alternatively, polarisation states of photons can be used as *which-way* detectors.) The result is always the same: When the *which-way* information is stored in the internal atomic state, the interference patterns disappear (Fig. 5.9). There is more: When the *which-way*

[92] Mittelstaedt, *Philosophische Probleme* (1972), Chap. III, 'The Problem of Decoherence' (2000), 154–7

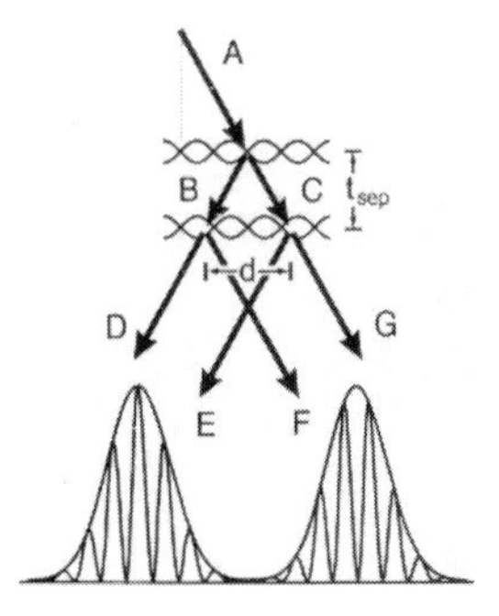

Fig. 5.6. Scheme of an atomic interferometer. Source: Dürr et al., 'Origin' (1998) 34, by permission of the authors and *Nature*

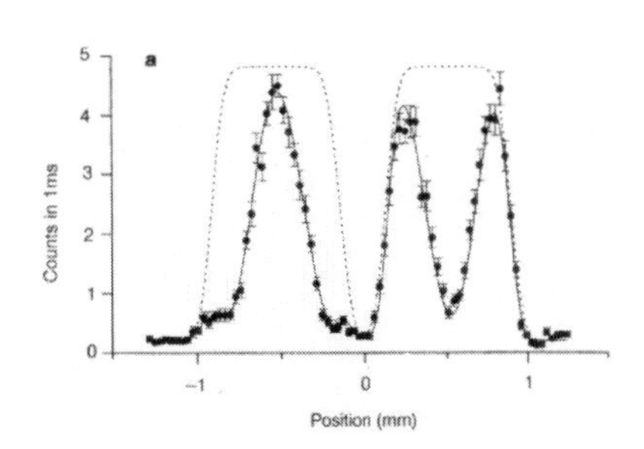

Fig. 5.7. Spatial interference pattern. Source: Dürr et al., 'Origin' (1998) 34, by permission of the authors and *Nature*

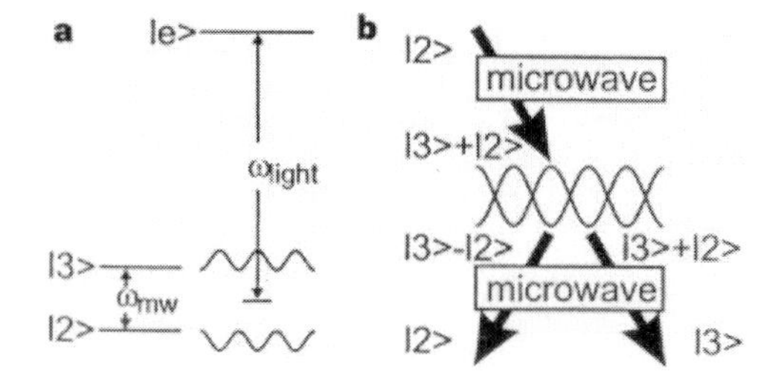

Fig. 5.8. Storage of *which-way* information. Source: Dürr et al., 'Origin' (1998) 35, by permission of the authors and *Nature*

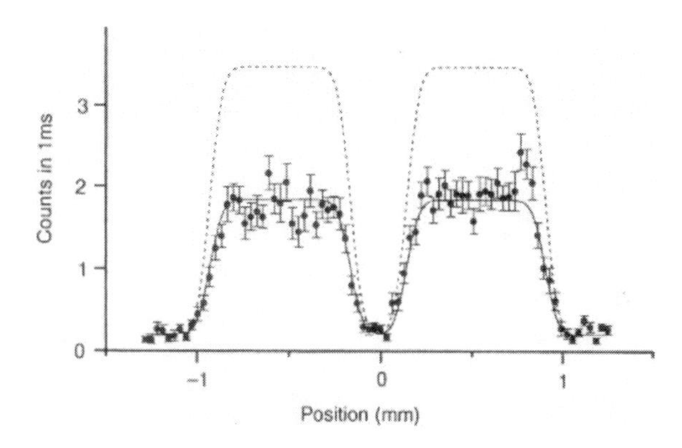

Fig. 5.9. Disappearance of interference fringes as a result of storing *which-way* information in the internal atomic state. Source: Dürr *et al.*, 'Origin' (1998) 35, by permission of the authors and *Nature*

information is erased again, the interference patterns reappear. The crucial point of the experiment is that the indeterminacy relations cannot be held responsible for the disappearance of the interference pattern. The momentum kick from the microwaves is far too small to account for the disappearance of the fringes. And the atom remains 'de-localised' during its whole passage through the interaction.

The correlations between the *which-way* detector (the internal atomic states) and the atomic motion, which destroy the interference, are a form of *entanglement.* It is not the entanglement of the microwave fields with the atoms, which causes the loss of interference. The microwave fields do not 'transfer enough momentum to the atom to wash out the fringes'. What 'causes' the loss of interference is the *entanglement* between the *which-way* detectors and the atomic motion.[93] So it is environmental entanglement with a microscopic environment. This is a form of decoherence but it is reversible. The physical manifestation of this entanglement is the physical *distinguishability* of alternative particle paths. As soon as the possibility of distinguishability exists – when *which-way* information is stored in the atoms or photons – the interference effects are destroyed. However, when the pathways of the photons are indistinguishable the interference effects reappear. This is achieved through a quantum eraser.[94] A quantum eraser is a device, which destroys the *which-way* information just before the atoms hit the recording screen. This erasure of path information restores the interference patterns. The quantum eraser reverses this type of decoherence. Both the storage and the erasure of path information have a strong causal nature. It is a question of quantum editing. The appearance of the interference effects and the form of the interference pattern are dependent on the way and the degree to which the pathway information is erased.[95] To achieve this, two polarizers are placed in front of the detectors, but after they have left the output port of the interferometer. This restoring of the interference pattern (after the erasure of *which-way* information) displays the same 'spooky' non-local cooperation between the particles as in the famous *EPR*-type experiments. Quantum editing and therefore partial interference patterns occur when the two polarizers in front of the detectors are set at different angles.

In the study of *which-way* experiments the notion of entanglement[96] replaces the indeterminacy relations. The relationship between *which-way* information and the appearance and disappearance of interference patterns is clearly a *causal* relation. There is a *conditional priority* of antecedent over subsequent conditions: the storing of *which-way* information is prior to the loss of interference and the erasing of *which-way* information is prior to the restoring of the interference effects. There is a *physical dependence* of the effects (appearance or disappearance of fringes) on the

[93] Dürr/Nonn/Rempe, 'Origin', (1998), 33–7. A good description of this experiment, with useful illustrations, is given by Buchanan, 'End' (1999), 25–8; see also Zeilinger, *Einsteins Schleier* (2003). Note that mathematically the interference is represented by the cross terms '$\Psi_1^*\Psi_2 + \Psi_2^*\Psi_1$' in the state vector. It is the disappearance of this term, once *which-way* information has been stored, which indicates the loss of interference.

[94] See Kwiat/Steinberg/Chiao, 'Observation' (1992), 7729–7739; Davies, *About Time* (1995), 171; Seager, 'A Note' (1996); Zeh, 'Programme of Decoherence' (1996), 23

[95] Kwiat *et al.*, 'Observation' (1992), 45

[96] Zou/Wang/ Mandel, 'Induced Coherence' (1991), 321. The notion of entanglement is used frequently in experimental contexts to give a causal account of the observed phenomena, see Kwiat *et al.*, 'Observation' (1992), 7729–7739

causal conditions: the editing of interference fringes is dependent on the degree of *which-way* information.[97] There is a clear correlation between the appearance and disappearance of the interference patterns on the one hand, and the entanglement between internal and external degrees of freedom of the particles (or between the measuring apparatus with the system wave function) on the other. Note first that there is no particular role for the observer in this conditional dependence. It is the physical process of storing the *which-way* information in the internal particle states and not the actual reading out of this information by the experimenters, which destroys the interference effects. Second, the interference patterns can be retrieved by erasing the *which-way* information in the particles on two spatially separated arms of the interferometer. Significantly, *welcher Weg* information must be erased in both spatially separated particles. If it is erased only on one arm of the interferometer, the interference effects will not re-appear because one of the photons still retains *welcher Weg* information.

There is a *third case*, which causes many problems.[98] The third case takes the form of a measured interaction with the system and the recording of the result. This situation involves the much-discussed *measurement problem* in quantum mechanics. Von Neumann distinguished between the unitary evolution of the potential quantum states according to the Schrödinger equation (Case 1) and the registering of an actual recorded value in the act of measurement (Case 3). Relying on the double-slit experiment for elucidation of the problem, Heisenberg had proposed a *Gedankenexperiment* in which a γ-ray microscope is used to determine the pathway of an electron through the slits. It is assumed that this position measurement forces the system to transit abruptly from a coherent superposition of the spatial wave packets to a mixed state of two definite position values on the screen without interference effects. This has often been dubbed the *collapse* of the wave function. This is a problematic assumption for several reasons: (a) If a collapse of the wave function does take place, quantum mechanics requires two different forms of dynamics – one for the deterministic evolution of the quantum system in accordance with the Schrödinger equation and one for the collapse of the wave function; (b) if there is 'collapse', it happens for unknown dynamic reasons; (c) this remains a problem whether the wave function is regarded as a mere calculation device – a *Gedankending*, as Schrödinger concluded from the cat paradox - or as a mathematical representation of the structure of quantum systems. For in both cases the 'jump' from potential to actual values calls for an explanation. If the wave function is understood in an instrumentalist sense, it is difficult to explain dynamically *how* one actual value of the observable is chosen in the measurement process. If the wave function is understood in a realist sense, a strong demand arises to provide an equally realist dynamic account of the choice of one particular value.

[97] Kwiat *et al.*, 'Observation' (1992) stress that interference can be lost and regained in degrees, which is dependent on the state of the entanglement!

[98] Sklar, *Philosophy of Physics* (1992), 170 Fig. 4.4; Rae, *Quantum Physics* (1986), Chap. 2; McCall, *Model* (1994), Chap. 4; Mittelstaedt, *Philosophische Probleme* (1972), Chap. III; Mittelstaedt, *Interpretation of Quantum Mechanics* (1998); Toretti, *Philosophy of Physics* (1999)

At this juncture, the famous *indeterminism* of quantum mechanics enters the scene. According to the radical response it is due to uncontrollable 'kicks', which the quantum system receives from its interaction with the measurement device. This leads to Bohr's attempt to replace the traditional notion of causation with his concept of complementarity. To learn about the position of an electron at the slits Heisenberg's γ-ray microscope must kick it with a bundle of photons. Its momentum spread will increase. So its classical trajectory cannot be determined. As the *which-way* experiments show, however, the indeterminacy relations can be circumvented. Decoherence offers an alternative interpretation. The wave function does not collapse. Environmental entanglement reveals the pathway information. Quantum indeterminism is not to be lodged in quantum jumps.

From the point of view of the notion of causation, the third case is problematic, at least if we adopt the Laplacean notion of causation. No precise knowledge of the initial conditions exists, so that the subsequent conditions cannot be predicted deterministically. The motion of particles, as Born said, follows probabilistic laws. But do we conclude from this sense of indeterminacy that these processes are *acausal*? Not according to the *philosophical* response. We could allow for a probabilistic notion of causation, which would be incorporated into a conditional model. Let us bracket the fundamental level of quantum indeterminism till we discuss the programme of decoherence. For the purpose of a causal description of the famous experiments of quantum mechanics, the macroscopic level of probabilistic causation is sufficient.

5.3.3 Philosophical Response

> Philosophy without science is empty. Science without philosophy is muddled.
>
> Paraphrasing A. Einstein, 'Reply to Criticisms' (1949), 684

The philosophical response carefully separates determinism and causation and holds that quantum mechanics is compatible with a refined notion of causation. Max Born and Louis de Broglie, who both made fundamental contributions to quantum mechanics, are the principal authors of the philosophical response. As in the scientific writings of Planck and Einstein in the 1910s, and the scientific publications of Bohr and especially Heisenberg from the 1920s onwards, so Born's scientific articles of that period freely move from mathematical calculations and physical interpretations to philosophical deliberations. In a number of publications Louis de Broglie tried to show that there was a mutual dependence between the progress of physics and the conceptions of natural philosophy. This deserves emphasis to dispel the usual impression that scientists 'wax philosophical' when their creative periods are over.

Born, at first, followed the line of argument pursued by Heisenberg and Bohr. Considering the collision of a free electron with an atom, he comes to the conclusion that this interaction cannot be construed as a causal relation. Heisenberg's consequent is not satisfied. The effect of the impact cannot be precisely determined. Quantum mechanics gives no answer to the question, 'In what state is the system after the collision?' It provides a probabilistic answer to the question, How probable is a given effect?' This raises the problem of determinism. Born recommends its renunciation in the quantum world, because quantum mechanics says 'nothing about the course of individual events.'[99] However, we often make causal statements, which contain no information about the course of individual events. Kirchhoff affirmed that the famous *D* lines in the solar spectrum were caused by the presence of sodium in the atmospheric gases surrounding the interior of the sun. Darwin affirmed that favourable modifications in a group of organisms were the cause of an increased survival rate of members of that group. Gibbon affirmed that barbarism and religion were the cause of the decline and fall of the Roman Empire. Such statements contain causal stories but at different levels of specification. Some of the features, with which causation is traditionally characterised, are discernible in these examples. The locality (contiguity) condition is satisfied by the presence of sodium in the atmospheric gases of the sun and by the Darwinian stipulation that modifications are favourable in local environments. The causal priority (antecedence) condition is equally satisfied, since the observation of the absorption line follows the presence of sodium and the improved survival rate follows the presence of favourable modifications. There is in these cases also a lawful regularity between cause and effect. The demand for a causal mechanism, however, is not satisfied. Kirchhoff was not in a position to spell out in detail how the presence of sodium could be causally responsible for the absorption lines. Nor was Darwin able to specify how favourable modifications in organisms were preserved from generation to generation through inheritance. Historians and social scientists can only describe a set of most likely factors as the cause of an event. Still even though these causal stories remain incomplete, as they contain no precise specification of causal mechanisms ('the course of individual events'), they refer at least to some kind of dependence of the effect upon the cause. Kirchhoff was aware of a *causal dependence* of the spectrum on the chemical elements. Darwin frankly admits 'our' ignorance of the cause of mutations in individual organisms. Yet, without the benefit of knowledge of Mendel's laws, he suspected that 'disturbances in the reproductive system' of the parents chiefly contributed to the 'plastic condition of the offspring'.[100] Although this hypothesis is vague, the essential point is that Darwin suspects some *causal dependence* between the preservation of favourable modifications in the children and their possession by their parents. Much has been learnt about the causal dependencies of absorption and emission lines in the electromagnetic spectrum and the inheritance of favourable and unfavourable modifications in organisms. In the case of Kirchhoff's spectral lines, a rather exhaustive causal account can be given. It essentially involves Bohr's account of the energy levels of atoms and Einstein's account of the emission and absorption of photons at specific wavelengths. What can

[99] Born, 'Zur Quantenmechanik der Stoßvorgänge', (1926), 866; 'Quantenmechanik der Stoßvorgänge (1926), 826; 'Physical Aspects of Quantum Mechanics' (1927), 355. Modern authors who have stressed the compatibility of causation and quantum mechanics include Bunge, *Causality* (1979), 14–5; Salmon, *Causality* (1998), 116; McMullin, 'Shaping of Scientific Rationality' (1988), 37; Falkenburg, *Teilchenmetaphysik* (1995), 74–5

[100] Darwin, *Origin* (1859), 173

be known about the course of individual quantum events may remain inherently limited. However, some form of causal dependence may emerge even in quantum systems.

As we have seen in the last section, quantum systems are very different from more familiar physical systems. As long as no interference in terms of measurements is attempted, the quantum systems evolve according to the deterministic Schrödinger equation. The causal aspect comes to light when deliberate measurements are made on the systems. We are dealing with a collection of particles. The causal interference produces events, which happen with degrees of probability.

To get causal stories in quantum mechanics, the notions of determinism and causation must be separated. Once this separation is effected, the notion of causation will need to be modified. This is the line of argument, proposed by the philosophical response.

Consider a phenomenon, A, such as the firing of an electron gun at a crystal, which is always succeeded by one of several phenomena, $B_1, B_2, B_3, \ldots B_n$. These may be scintillation effects at different points on the surface of a screen erected near the crystal. Furthermore, none of the phenomena, $B_1, B_2, B_3, \ldots B_n$ will be recorded if A is absent (Fig. 5.10).

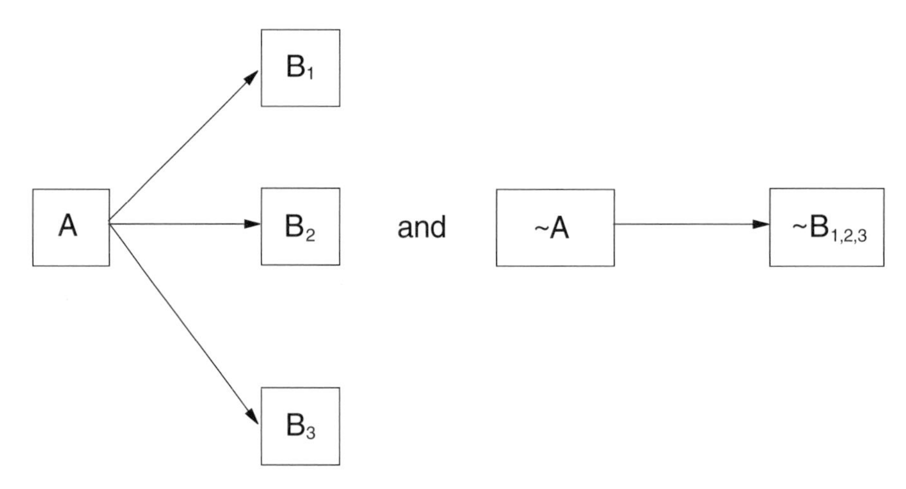

Fig. 5.10. De Broglie's causal thought experiment

In this situation, we would consider that A must be the cause of B. It is a case of causation without determinism because quantum mechanics cannot predict which of the phenomena, B_1, B_2, B_3, $\ldots B_n$ will actually occur at which place and time on the surface of the crystal.[101] There is a causal dependence. It cannot take the

[101] The example, which leads to the distinction between causation and determinism, is taken from de Broglie, *Continu et Discontinu* (1941), 64–66. This experiment was already experimentally realized in 1927, when Davisson and Germer measured the wavelengths of electrons scattered off the surface of nickel crystals. When the detector was set at

form of a spatio-temporal causal mechanism, because the situation is subject to the statistical laws of quantum mechanics and, in particular, the restrictions imposed by the indeterminacy relations. While there are regions of the screen where the probability of impact is greater than in other regions, it is impossible to make any statements about the path and impact area of individual electrons. But there is a causal dependence of the scintillation effects on the firing of the electron gun. It is reminiscent of the causal accounts, without precise determinations, encountered in the pre-quantum era. As we shall see below, there are a number of well-known experiments in quantum mechanics which satisfy the details of de Broglie's thought experiment.

De Broglie and Born state clearly that physical dependence, without 'infinite sharpness', gives rise to causal accounts in quantum mechanics. It is a step towards meeting Bohr's challenge. Quantum mechanical discoveries, from Kirchhoff's spectral analysis to Einstein's spontaneous emission, have driven a wedge between the notions of determinism and causation. The quantum-mechanical experiments of the 1920s served to bring into focus a notion of causation compatible with atomic phenomena.

Thus it would seem that quantum mechanics does not relinquish the notion of causation. It only eliminates the traditional interpretation of causation in terms of determinism. If causation means *physical dependence* of one set of observable parameters, *A*, on another set, *B*, then quantum physics has not abandoned causation. This general notion of physical dependence must be specified in terms of laws, which lay down how the occurrence of one set of events, *A*, is dependent on the occurrence of another set of events, *B*. As Born states: 'the objects of observation for which a dependence is claimed (...) are the probabilities of elementary events, not those single events themselves' (as we may expect from classical physics and everyday experience).[102] This gives rise to a characterization of causation, compatible with quantum mechanical evidence, which lays the basis for a conditional model of causation. What is new is that Born, de Broglie, and other observers of the scene, like Frank and Cassirer, take the empirical discoveries of quantum mechanics as empirical constraints on an adequate notion of causation. Empirical discoveries have an impact on fundamental philosophical notions because these notions belong to the larger categorical framework, within which the empirical phenomena are interpreted. An appropriate understanding of such phenomena involves the employment of fundamental philosophical notions.

It may appear that with the separation of the notions of causation and determinism, we pay an unacceptable price: the loss of a causal, traceable link between cause and effect. In de Broglie's thought experiment some of the traditional features of causation are still present: *antecedence* (the cause is prior or simultaneous with

specific angles, which differed for different atoms, the intensity of the reflected beam reached a maximum. It should be added that de Broglie went through various phases of deterministic and indeterministic convictions; the above thought experiment belongs to his indeterministic phase; on de Broglie's conversions and reconversions see Čapek, *Philosophical Impact* (1962), 317, 320–1

[102] Born, *Natural Philosophy* (1949), 102, cf. 76, 124

the effect) and *contiguity* (cause and effect must be in spatial contact or connected by a chain of events, which are in spatial contact with each other). But we know from the double slit experiment that nonlocal interactions are observed in quantum mechanics. Equally alarming seems to be the lack of a traceable causal mechanism. However incompleteness of causal accounts occurs both in the classical era (the planet's orbit, the *D* lines in the solar spectrum) and the quantum era (the electron gun and the experimental realisation of the de Broglie thought experiment). De Broglie and Born stress that causal relationships deal with observable, measurable quantities. Admittedly, the interaction with quantum systems in the laboratory only leads to statistical laws about their observable behaviour.[103] But this does not deprive us of the notion of causation. Causal stories are possible in quantum mechanics.

5.3.4 Some Causal Stories in Quantum Mechanics

The discovery of radioactive decay (1901) and the phenomenon of spontaneous photon emissions during the transitions of atoms from higher to lower energy states (1917) baffled physicists because of their indeterministic and acausal character. Only a statistical determination could be achieved in the sense of calculating the probability of jumps between permissible quantum states. Physicists tended to concentrate on only one of the features traditionally associated with the notion of causation – the lack of locality (Einstein) or of a traceable causal mechanism (Bohr, Heisenberg, Pauli) – to either regret the incompleteness of the theory due to its failure to specify the causal mechanism or to reject the adequacy of the notion for quantum mechanics. It is to the credit of de Broglie and Born to have sketched a philosophical response, in which the notions of causation, determinism and indeterminism could be accommodated. Determinism is confined to the evolution of quantum systems as described by the Schrödinger equation in an abstract mathematical Hilbert space (*Case 1: the undisturbed system*). Indeterminism enters at the level of the interaction between experimental devices and quantum systems in the sense specified by Born – the probabilistic motion of the particles (*Case 3: measurement of quantum system*). Indeterminism means that the antecedent conditions (cause) do not specify a unique trajectory for the particles, which would give rise to deterministic causation. Even given a set of causal factors, the consequent

[103] Frank, *Kausalgesetz* (1932), 199; Born, 'Aspects of Quantum Mechanics' (1927), 355–6; 'Quantenmechanik und Statistik' (1927), 240–42. Already in 1927 Jordan argued that the notion of physical causation was not identical with the notion of deterministic lawfulness. If many identical observations or experiments are performed on a few quantum systems or many identical quantum systems are subjected to a few similar experiments, quantum mechanics is able to predict the mean value of these experimental results. Hence quantum mechanics preserves a notion of probabilistic causation. See Jordan, 'Kausalität und Statistik' (1927), 105–106. It is a testimony to the powerful sway of the Laplacean spirit over occidental scientists that R. Oppenheimer, in the English version of Jordan's article entitled 'Philosophical Foundations of Quantum Theory' (*Nature* 1927), 566–9 translates Jordan's term 'Kausalität/causality' with 'determinism.'

conditions (effect) can only be calculated probabilistically. Thus indeterminism does not mean complete randomness. A statistical prediction of the behaviour of quantum systems under measurement is still possible. As the following examples show, this statistical determination can be developed into causal accounts. They are analogous to causal stories, embedded in the classical domain (Newton, Kirchhoff, Darwin). Causal explanations come at different levels of specification. There is not just a functional dependence between parameters, expressed in a differential equation and giving rise to predictability. The other causal features - the conditional priority of the cause, a general conditional dependence of the effect on the cause, even in some cases the locality condition - are also satisfied. The incompleteness of the causal information is not such that the existence of causal relations must be doubted. Sufficient empirical constraints are available to infer, by eliminative induction, that the effect is due to the conditions at hand. A closer scrutiny of some of the famous experiments in the area of quantum mechanics reveals that the de Broglie thought experiment has become a reality.

According to Bohr and Heisenberg, the wave-particle duality of quantum systems renders the notion of causation inapplicable to the quantum domain. It will therefore be important to scrutinise causal aspects of experiments, which require the *particle* picture and the *wave* picture alternatively. We start with the particle picture.

The Particle Picture

Scattering Experiments. In Chap. 2, we discussed the scattering experiments and the Stern-Gerlach experiments to highlight the role of modelling in physical understanding and the peculiar nature of quantum systems. We can now see that these experiments give rise to a causal interpretation. In the first chapter we introduced Rutherford's model of the trajectory of an α-particle as an illustration of a structural model, in which both a topologic and an algebraic structure combined to account for the observable behaviour of the hydrogen atom. The scattering experiments, which led to the construction of the nucleus model of the atom, reveal causal components. They amount to causal stories in quantum mechanics, which are close to de Broglie's *Gedankenexperiment*. Recall the basic structure of the scattering event, in which highly energetic alpha particles are fired at a gold foil, F. The deflexions of the particles are recorded by the scintillation method, which counts the flashes produced by the deflected particles on a zinc sulfide screen, S. The screen, S, is placed behind a lead plate, P, so that no α-particles from the source can strike the screen directly. The following diagram (Fig. 5.11) is similar to that used by Geiger and Marsden (Geiger 1909–1910, 496).

The atoms in the gold foil cause the α-particles to scatter in various directions, which depends on the impact parameter b. The positively charged nucleus exerts a repulsive force on the positively charged α-particles. The closer the particle gets to the nucleus the greater the scattering angle, ϕ, will be (Fig. 5.12). The causal aspects of this situation can be discerned by looking at what Geiger and Marsden found. The most surprising effect was, of course, that 1 in 8000 particles were deflected

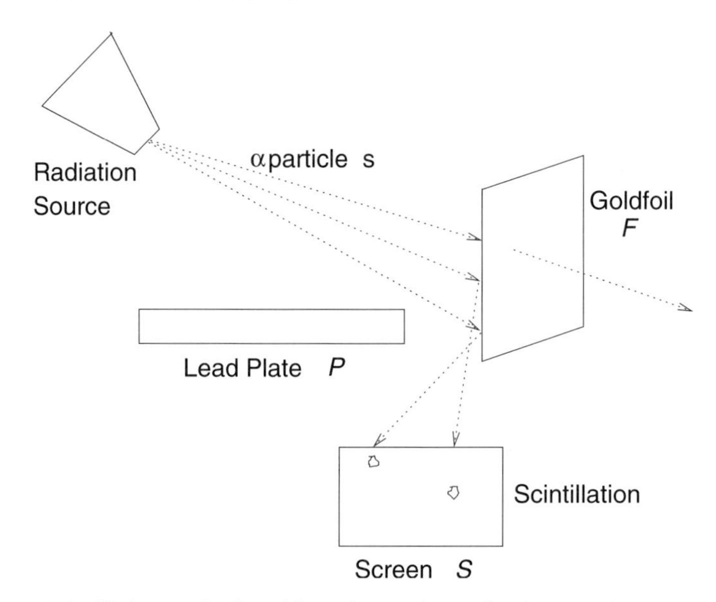

Fig. 5.11. Scintillation method used by Geiger and Marsden in scattering experiments

by more than 90°. To account for this result, Rutherford assumed the existence of a nucleus at the centre of the atom. The most probable angles of scattering through a given thickness of matter were 2.8° and 3°. The most probable angle of scattering through a layer of gold equivalent to 1 cm of air was 2.1°. The atoms in the gold foil clearly cause the scattering of the particles. The distance of the particle as it approaches the nucleus also counts among the causal factors, since the smaller the value for b, the greater the scattering angle ϕ.

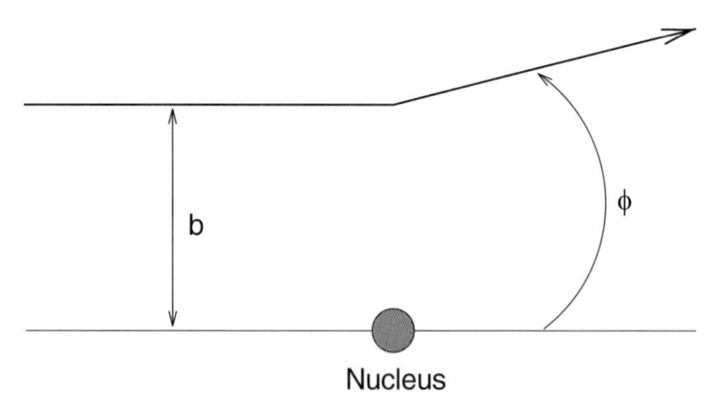

Fig. 5.12. Closeness of approach and scattering angle

The *Stern-Gerlach* experiments were carried out in the early 1920s. They established experimentally the so-called quantization of angular momentum.[104] This means that electrons orbit the nucleus only in certain permitted planes. The experiments demonstrated, for the first time, the idea, proposed by Sommerfeld, of the quantization of the orbital planes of the electron in the atom. The orbital planes of electrons do not only possess discrete sizes and shapes. These orbital planes must also be inclined in certain ways. They must have discrete spatial orientations in relation to a co-ordinate system like an external magnetic field. The size, shape and orientation of the orbital planes are indicated by quantum numbers (n, l, m_l, m_s). These quantum numbers specify the state of the atoms in an atom beam. When a beam of atoms is sent through a non-uniform magnetic field, this discrete spatial orientation will be revealed on a screen mounted behind the magnet. Otto Stern and Walter Gerlach ran these experiments between 1921 and 1925 on beams of silver atoms. Concentrating on the causal aspect of these experiments, two scenarios can be distinguished:

1. The beam of silver atoms is sent through the magnet but the magnet is switched off. A screen mounted behind the magnet will record the impact of the atoms. When the magnet is switched off, one central dot will be recorded after the passage of the atom beam (dashed line and white central spot indicated on the screen in Fig. 5.13).
2. The magnet is now switched on when the beam of atoms is sent through. Depending on the precise state of the atom beam, specified by its quantum numbers, and assuming the simplest case, two dots will appear on the screen. The *effect* of the magnet will be an intensity shift. When the magnet was switched off the intensity maximum was in the centre of the screen. But with the magnet switched on, this central intensity maximum will become a minimum. The central dot will disappear and two clearly separated dots will appear, deflected upwards and downwards respectively (Fig. 5.13). With the magnet switched on, the magnet will *cause* the atom beam to split exactly into two halves.

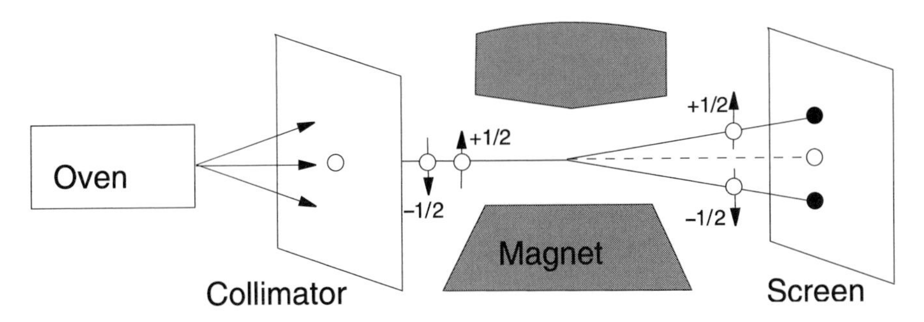

Fig. 5.13. The Stern-Gerlach Experiment

[104] Hughes, *Quantum Mechanics* (1989), 1–8; Weinert, 'Wrong Theory' (1995), 75–86

It is not difficult to apply Mill's 'method of difference', a form of eliminative induction, to this situation to establish its causal nature. The only difference between otherwise two identical situations, including the preparation of the atoms in identical atomic states, specified by the quantum numbers, lies in the behaviour of the magnet. If it is not switched on and there is no magnetic field, one central dot appears; if it is switched on and a magnetic field is applied to the passing atoms, two dots appear in the simplest case. The set of causal conditions is closed. There are no other interfering factors to be considered. We are therefore justified in concluding that the creation of the non-uniform magnetic field is the cause, given the initial state of the atoms, of the splitting of the atomic beam into two parts. As is customary in quantum mechanics, no claim is made about the behaviour of the individual atoms making up the beam. Since their initial orientation is random it is not possible to predict, which way they will turn under the influence of the magnet. But statistical predictions can be made about the behaviour of the whole beam. The rules of quantum mechanics specify how atom beams in different states behave. For instance, under certain conditions three dots will be seen, under others five and so forth.

These experiments are concrete illustrations of de Broglie's thought experiment. They confirm the thesis embodied in the philosophical response: causal relations exist in quantum mechanics. And they disconfirm once more the adequacy of classical determinism in the quantum mechanical realm. There is causation without classical determination. It is probabilistic causation.

To show how causal accounts can be obtained from quantum mechanical systems, we have concentrated on two experiments, which essentially involve the particle nature of quantum systems. However, Heisenberg and Bohr would argue that this notion of probabilistic causation does not apply to the wave aspect of quantum systems. For instance, in the double slit experiment, an attempt to determine through which slit the photons pass will annihilate the interference pattern and produce simple, not superimposed maxima. This wave aspect defies a notion of causation, inspired by classical determinism, with its heavy emphasis on local, causal mechanisms between cause and effect.

Are there experiments, involving the wave nature of quantum systems, for which more detailed causal accounts could be given? The philosophical account would be strengthened if causal relations, involving the wave nature of quantum systems, could be found.

The Wave Picture

Compton scattering involves such a system, which makes the point about probabilistic causation equally well. It had been observed that the wavelength of X-rays is increased when they are scattered off matter. Compton showed that this behaviour could be explained by assuming that the X-rays were photons. When photons are scattered off electrons, part of their energy is transferred to the electrons. The loss of energy is translated into a reduction of frequency, which in turn leads to a lengthening of the wavelength of the scattered photons. This happens because the

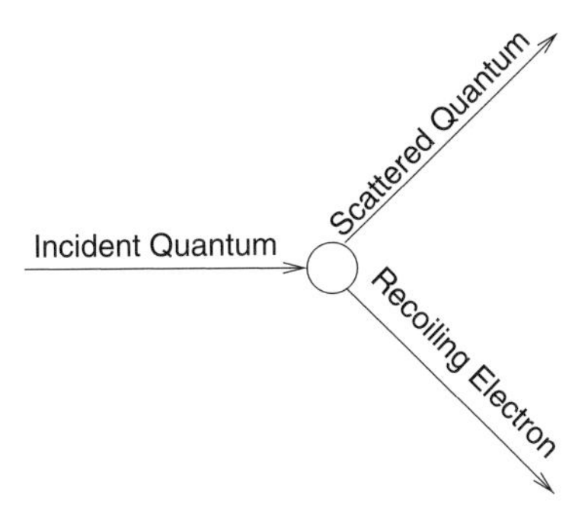

Fig. 5.14. Compton's Model of the Scattering process

relation $E = h\nu = hc/\lambda$ holds. In this experiment, first carried out in 1923 by Arthur Compton, the scattering of X-rays is treated as a collision of photons with electrons (Fig. 5.14).

The wavelength of the scattered photon, λ, can be related to its *initial* wavelength, λ_0, to the electron mass, m, and the scattering angle, θ, by the relation $\lambda - \lambda_0 = h/mc(1 - \cos\theta)$. Compton's description of his theory conveys the flavour of a perfectly causal explanation.

> From the point of view of the quantum theory, we may suppose that any particular quantum of X-rays is not scattered by all the electrons in the radiator, but spends all of its energy upon some particular electron. This electron will in turn scatter the ray in some definite direction, at an angle with the incident beam. This bending of the path of the quantum of radiation results in a change in its momentum. As a consequence, the scattering electron will recoil with a momentum equal to the change in momentum of the X-ray. The energy in the scattered ray will be equal to that in the incident ray minus the kinetic energy of the recoil of the scattering electron; and since the scattered ray must be a complete quantum, the frequency will be reduced in the same ratio as is the energy. Thus on the quantum theory we should expect the wave-length of the scattered X-rays to be greater than that of the incident rays.[105]

There is thus a causal dependence of the increase in wavelength of the photon (the effect), on the antecedent conditions, like the collision of the photon with an electron (the cause).

How is it possible that the causal aspects of important experiments in quantum mechanics can so easily be constructed when such a strong tradition seems to throw its weight behind the apparent *a*causal nature of quantum mechanical events? Part of the answer is that the picture of *a*causality and indeterminism was encouraged by a presupposition: the identification of causation with (classical) determinism. Another reason, which emerged, is related to the famous wave-particle duality of quantum mechanics. Proponents of the radical response to the problem of causa-

---

105 Compton, 'A Quantum Theory of the Scattering of X-Rays by Light Elements' (1923), 485; and 'The Spectrum of Scattered X-Rays' (1923), 410–13

tion in quantum mechanics concentrated on experiments, which reveal the wave nature of quantum systems. But according to Bohr and Heisenberg, both the wave and particle pictures are reciprocal descriptions of the natural processes. Recall that according to the radical response waves are extended in space. They form interference patterns. But no causal description can be given of *wave*-like processes. Their behaviour simply defies the behaviour expected from Laplacean causation: the precise spatio-temporal determination of the trajectories.

But we can turn the emphasis around. If the *particle*- and *wave*-pictures are complementary, the causal aspects of quantum systems should have as much legitimacy as the apparently *a*causal aspects.[106] Besides, as we have seen, both the particle- and wave-picture give rise to causal interpretations. The only unavoidable concession is the demise of deterministic causation. For the causal accounts of the quantum mechanical experiments discussed above do not reveal the trajectories of individual particles. Hence no complete determination can be given. Still it is possible to retain the notion of *probabilistic causation* for quantum mechanical systems.[107]

In the second chapter we encountered Bohr's challenge to philosophy. The significance of physical science for philosophy lies above all in the opportunity it provides for the test of some of the most fundamental concepts. We have considered the features of the notion of causation in the light of a number of experiments. It is a *cluster concept*: if at least a majority of its features are satisfied, then we are entitled to speak of a causal situation. To be philosophically satisfying, the *notion* of *probabilistic* causation, which emerges from observations of experimental interference with quantum systems, should be developed into a philosophical *model* of causation. Ideally, this model should describe causal situations in both the micro- and the macro-world. Just as the equations for micro-systems approximate equations for macro-systems in the limit of large quantum numbers, so an adequate model of causation should apply to both macro- and micro-systems in the limit of available causal conditions.[108]

The functional model of causation governed much of the thinking of scientists before the advent of quantum mechanics. In the conservative response it lingered on. With the separation of the notions of causation and determinism a conceptual space was created in which a new philosophical model of causation could take root.

---

[106] The argument of this section was inspired by Sommerfeld, 'Bemerkungen' (1929), 870

[107] As far as I can see Exner, *Vorlesungen* (1919), §§88–95 was the first to propose this notion of probabilistic causation as a reaction to the quantum mechanical evidence. Schrödinger wholeheartedly embraced this notion, see 'Was ist ein Naturgesetz' (1922), 14–17 and 'Indeterminism in Physics' (1931). See also Nernst, 'Gültigkeitsbereich' (1922), 494–5; Jordan, 'Kausalität' (1927), 105–10, 106; Schlick, 'Kausalität' (1931), §10. In view of later development I will take up Salmon's suggestion to speak generically of probabilistic causation. Salmon subdivides this notion further into statistical causation, emphasising constant conjunctions and statistical regularities and aleatory causation, emphasising causal mechanisms; see Salmon, *Causality* (1998), 207

[108] Weinert, 'Correspondence Principle' (1994), 303–23

This *conditional model of causation* emerges in philosophical reflections on the new situation in physics.

5.3.5 The Conditional Model of Causation

Einstein's problem is clearly before us. The evidence from the quantum-mechanical experiments tells us that there are causal relations in the atomic realm. It is a new notion of causation, probabilistic in essence, which is required. Yet our traditional models, often shared by scientists, imply that there can be no causation in the quantum realm. For instance, any philosophical model in terms of 'spatio-temporal continuous causal processes', satisfying a locality condition, will have to draw this conclusion.[109] Do we cling to cherished notions of causation? Do we listen to the evidence? In this section we propose to tailor our philosophical model of causation to the available evidence.

The notion of causation has had a long, if not altogether distinguished career in philosophical and scientific thinking, which goes back to antiquity. Many different notions of causation have been in vogue. To mention but a few, Aristotle distinguished four notions of causation (efficient, formal, final and material). The transition from the organismic to the mechanistic worldview reduced this matrix of causal notions to efficient causation. Leibniz's 'Principle of Sufficient Reason' was interpreted by Laplace as affirming the universal statement that 'every event has a unique cause', a version of the principle of causation to which J.S. Mill subscribed. The 16th century discovery of mathematical laws of science, first introduced by Kepler with his three laws of planetary motion and the 17th century development of differential equations promulgated a functional view of causation.[110] The mathematization of the physical sciences, the invention of the calculus and differential equations, strongly encouraged the identification of the notion of causation with that of predictive determinism. This approach favours a functional model of causation. But the Laplacean notion was not the only presupposition, on which scientists could draw. Kant's notion of causation as an a priori category of the mind exerted a strong influence on some continental scientists. Helmholtz, for instance, one of the co-discoverers of the principle of conservation of energy, stated that

> the law of causation, by virtue of which we infer the cause from the effect, has to be considered also as being a law of our thinking which is prior to all experience.[111]

The long arm of Kantianism was also be felt by Bohr, one of the main proponents of the radical response. Insisting on the need for interference to obtain information

---

[109] Salmon, *Causality* (1998), 325, 280 accepts that his insistence on causal mechanisms will limit the applicability of his model.

[110] See Burtt, *Metaphysical Foundations* (1924); Skyrms, 'EPR: Lessons for Metaphysics' (1984) for a discussion of seven views of causation with respect to the EPR correlations.

[111] Quoted in Warren/Warren, *Helmholtz on Perception* (1968), 201, 228. The views expressed by von Helmholtz in his Introduction to *Über die Erhaltung der Kraft* (1847), are also strongly Kantian in flavour. Recall however that in a Footnote, added in 1881, von Helmholtz distanced himself from this earlier Kantian influence and equated the principle of causation with lawfulness (i.e. determinism).

about atomic processes, and not shirking from an invocation of subjective elements in quantum mechanics, Bohr speculated in 1929 that 'causality may be considered as a mode of perception by which we reduce our sense impressions to order.'[112] But both Heisenberg and Planck insisted that the quantum mechanical discoveries had shown that causation was not a necessary category of thought. These discoveries achieved more than the 'refutation' of causation as a Kantian category. As we have already seen, they brought about a separation of the notions of causation and determinism.

Let us approach new philosophical models of causation, not from the point of view of the philosopher but from that of the scientist who wants to *understand* the results of empirical experiments. As we know from Chap. II, understanding in science is achieved by the availability of appropriate concepts and the construction of representational models. In that chapter we regarded philosophical 'theories' as conceptual models. Our scientist may therefore wish to supplement his or her understanding by appeal to an appropriate philosophical model of causation. As we have seen, this procedure was adopted by the founding fathers of quantum mechanics. But progress was hampered by the identification of causation with determinism. How would a scientist fare in this endeavour today?

Recall the situation as it presents itself after a careful analysis of some of the great experiments performed in quantum mechanics – the scattering experiments, the Stern-Gerlach experiments and Compton scattering. In all these situations, causal conditions could be specified, which would produce the specific effects observed in the experiments. Under these specifiable conditions there was a lawlike dependence of the effects on the antecedent conditions. This will give rise to a probabilistic notion of causation because no specific predictions can be made about the path of individual electrons or other subatomic particles. The suggestion could be made to analyse these situations by reference to a number of *causal conditions* or variables, which bring about the effect. Of these conditions some may be *necessary* and others *sufficient*. Of the variables some may be *dependent*, others *independent*. Suggestions to analyse the notion of probabilistic causation in terms of causal conditions were indeed made already in the 1930s.

It appeared to those who sought to analyse a notion of probabilistic causation, which was to be compatible with the new experimental discoveries, that a suitable account may be developed from a consideration of the causal conditions involved in the generation of some phenomenon. The idea goes back to J.S. Mill for whom philosophically, the cause is the sum total of the conditions positive and negative (= absence of counteracting causes) taken together.'[113]

What may be termed the *conditional model of causation* underwent several refinements. According to Exner, already Goethe proposed to replace the notion

[112] Bohr, 'Atomic Theory' (1929), 116; cf. Meyer, 'Kausalitätsfragen' (1934)

[113] Mill, *A System of Logic* (1843), Bk. III, Chap. V, §3. Mill did not lay much store by the terms 'antecedent' and 'consequent', declaring that "cause" may be defined as the assemblage of phenomena, which occurring, some other phenomena invariably commence.' (1843, §7)

of cause by the notion of conditions.[114] Causation and chance are compatible, if the law of causation is taken as a law of averages.[115] However, Tendeloo adopts the interesting approach, which looks at causal relations in terms of energy. In a causal situation, there is an energy exchange between cause and effect. The causal relation itself can be analysed in terms of conditions. The causal law states that every effect is determined by the cause and the constellation of conditions.[116] Reichenbach was one of the first philosophers who attempted a conceptual model of causation, which would be compatible with the new quantum mechanics.[117] Reichenbach sought to achieve this compatibility by associating the notion of causation with that of probability. The first step in any physical situation, in which a causal connection may be suspected between the antecedent and the consequent, is the recognition that the antecedent (or cause) consists of a number of factors. Of these factors some will be measured parameters, others will be unmeasured rest factors. A closer analysis may turn some of these rest factors into measured parameters but the rest factors can never be exhausted. Causation is concerned with the relation between *individual* measured parameters. Probability has to do with the distribution of the rest factors. (Thus Reichenbach adopts the view that singular causation is primary.) According to Reichenbach, the principle of causation cannot be formulated without the principle of a statistical distribution. Causal claims take the form of an implication ('If *C*, then *E*'). But we know that *C* consists of observable measured parameters and unknown rest factors, which may equally have an influence over *E*.

> What we know of *C*, can only be expressed in terms of a statistical statement: we know that subsequent situations, *with great probability*, differ little from *C*. (...) We predict *E* only with probability, not certainty. *Every causal statement, applied to the prediction of a natural event, has the form of a statistical statement. (...) If an event is described by a finite list of parameters, the future evolution of the event can be predicted with probability. This probability tends towards* 1, *the more parameters are taken into account.*[118]

This notion of probabilistic causation, which is implied in Reichenbach's combination of probability and causation, can then be regarded as a generalisation

[114] Exner, *Vorlesungen* (1919), 663

[115] Exner, *Vorlesungen* (1919), 666

[116] Tendeloo, 'Bestimmung' (1913), 154; for a recent discussion of this idea, see Salmon, *Causality* (1998), Chap. 16; Dowe, 'Process Theory' (1992) and *Physical Causation* 2000

[117] Reichenbach, 'Philosophische Kritik' (1920), 'Kausalproblem' (1931). For a modern interpretation of determinism in terms of probabilism, see Omnès, *Understanding* (1999), Chap. 16

[118] Reichenbach, 'Kausalproblem' (1931), 715–6; italics in original. Author's own translation: Reichenbach's letters *A*, *B* have been exchanged for *C*, *E*. Zilsel, 'Asymmetrie' (1927) makes a distinction between dependent (effect) and independent variables (cause) in his discussion of the asymmetry of causation. The independent variable is sometimes accessible to human manipulation. Zilsel's discussion is reminiscent of Mackie's much later account of causal priority in terms of fixity, which is related to the possibility of intervention. See Pfeiffer, 'Zwei Auffassungen des Kausalbegriffs' (1960); von Wright, *Explanation and Understanding* (1971); Mackie, *Cement* (1980), 180

of the classical notion of causation. Only this new notion is applicable to quantum mechanics. The Heisenberg indeterminacy relations forbid a convergence of the predictability of particular, singular quantum events towards 1. Reichenbach's interprets this as an extension, not an abolition of the notion of causation. The analysis of the quantum mechanical experiments confirms the adequacy of this interpretation. A proviso must be entered: for a judgement of causal relations, these experiments can be regarded as closed sets of causal conditions, since the probability of the rest factors' influence on the effect can often be calculated. Furthermore, the cause is probabilistically related to its effect, irrespective of prediction. Experimental results generally strengthen our twofold conclusion: quantum mechanics forces us to jettison the notion of determinism; more importantly it gives rise to a model of probabilistic causation, in which the statistical relevance of the causal factors can be assessed. As the analysis of the famous experiments in quantum mechanics has revealed, the notion of probabilistic causation is not a decorative by-product. It is required for the understanding of the experimental results.

Reichenbach anticipated that a *conditional* view of causation may well be the most adequate answer in the light of the results from quantum mechanics. This is the moment to report that even philosophy was affected by the cold war. In October 1959 an international symposium on philosophy and science was held in Leipzig, then part of the GDR, to celebrate the 550th anniversary of the Karl-Marx-Universität, Leipzig.[119] The only Western scientist who sent a contribution was Max von Laue. In their talks, the scientists and philosophers from what was then known as *the East Bloc* advanced, in an embryonic form, the following theses:

- ☞ *Separation of the notions of determinism and causation*
- ☞ *Recognition that the cause-effect-relation is compatible with the discreteness of the microworld (indeterminacy relations)*
- ☞ *Use of the Davisson-Germer diffraction experiment as an example of a causal relation in quantum mechanics.* (This is another example of the wave picture involving causal relations.)
- ☞ *Insistence on the incomplete knowledge of the conditions involved in causal relationships.*

None of these points were worked out in detail and nobody took notice. Nevertheless the *conditional* view of causation was recorded in the literature and it was not just philosophers who regarded it as a replacement of the functional view. We have already noted that Bohr and Heisenberg, in their later years, began to see a conceptual gap between determinism and causation. For even in quantum mechanical experiments, a cause-effect sequence was observable. But this suggestion needs to be fleshed out in terms of a philosophical model. David Bohm was another physicist, who began to think of the notion of causation in terms of causal relations between physical conditions.[120] Bohm's view is very close to a modern philosophical version

[119] Harig/Scheifstein *eds.*, *Naturwissenschaft und Philosophie* (1960)

[120] Bohm, *Causality* (1957), Chap. I, §4. The conditional model of causation is not the same as the *counterfactual* view, proposed by David Lewis. See Lewis, 'Causation' (1974);

of the conditional view of causation, worked out by John Mackie. He calls it the *INUS* view.

In *The Cement of the Universe* (1974) Mackie made no attempt to measure the adequacy of his *INUS* account against classical physics, let alone quantum mechanics. Rather, Mackie tried to develop a general model of *physical* causation. Causation as it works in the real world. Causation is the cement that holds the universe together. This model stands in the tradition of D. Hume and J.S. Mill. Any questions of the existence of a causal bond in the physical universe between correlated events, over and above their succession, are treated with caution. In the physical world, causation is only regular succession of events. By contrast, W.C. Salmon believes that causal connections between events can be found in causal processes, which establish a physical link between C and E. 'Causal processes are capable of transmitting energy, information, and causal influence from one

---

'Postscripts to Causation' (1986); Butterfield, 'Bell's Theorem' (1992), 51; Butterfield, 'David Lewis meets John Bell' (1992) The conditional view, as present here for the purposes of dealing with quantum mechanical experiments, may be said to be pragmatic. It deals with the physical situation at hand and with the physical parameters, which are considered to be relevant for the execution of the experimental tests. By contrast, David Lewis hopes to analyse causal dependence between events as counterfactual dependence. Counterfactual dependence among events is counterfactual dependence among corresponding propositions $O(c)$, $O(e)$, which state that the events occur. Causal dependence affects the probability with which the effects happen. An event e causally depends on another event c with probability p, if the following two counterfactuals hold:

$$O(c)' \text{ pr } [O(e)] = x$$
$$\neg O(c)' \text{ pr } [O(e)] \ll x$$

(x expresses a probability value between 0 and 1 and $'$ expresses the counterfactual conditional, saying that e would have been much less likely to occur without c). The counterfactual analysis, as it is stated in this way, cannot easily be used in an analysis of causal relations in quantum mechanical experiments. Quantum mechanics is about ensembles, not individual quantum events. It is about atom beams and collisions of electrons with mercury atoms. It is about scattering events. If the above counterfactuals are about individual quantum events, then no sufficiently precise statistical value could be assigned to them. If they are about ensembles, then more than a declaration of counterfactual dependence could be expected. An analysis of the actual physical conditions, which obtain, gives rise to an account of *physical* causation. We would expect an analysis of the cluster of the necessary and sufficient conditions, which enter the experimental situations. The experiments would be placed against a background of the causal field, against which the causal interaction takes place. We are less interested in what would have happened than in what actually happened, since we are interested in establishing the causal conditions and in identifying any possible interfering factors. The counterfactual model may be philosophically deeper – for instance, because it may manage to explain the asymmetric direction of causation, at least this was Lewis's hope – yet the conditional model allows a consideration of the empirical evidence, which has affected models of causation since the birth of quantum mechanics.

part of space-time to another.'[121] This may be true of classical physics. Quantum mechanics poses serious obstacles in the way of establishing such an account. All quantum mechanical experiments at one point fall short of delivering a continuous trace between *C* and *E*, most dramatically in the case of the non-local correlations of entangled quantum systems.

> 'Causation', Mackie holds, 'is not something *between* events in a spatio-temporal sense, but is rather the way in which they follow one another.' [122]

For Mackie, a cause is an *INUS* condition, an *I*nsufficient but *N*on-redundant part of an *U*nnecessary but *S*ufficient condition for some *E*. (Mackie 1980, 62) Thus there is a cluster of factors, making up the cause *C*, which bring about the effect *E*. Unlike Mill, however, who took the cause to be the sum total of the conditions, Mackie makes a distinction between *necessary* and *sufficient* conditions.

If *X* is a *necessary* condition for *Y*, then in the *absence* of *X*, *Y* cannot occur. Anyone who seeks a divorce must possess the marital status, since in the absence of the marital status (*X*), divorce (*Y*) cannot occur. Note that the mere presence of *X* does not mean that *Y* will occur since people get married without getting a divorce. It is just that, in Mackie's words, 'whenever an event of type *Y* occurs, an event of type *X* also occurs.' (1980, 62)

If *X* is a *sufficient* condition for *Y*, then in the *presence* of *X*, *Y* will occur. Rain is a sufficient condition for the street to get wet, since in its presence the streets get wet. But rain is clearly not a necessary condition for the streets to get wet, since in its absence, flooding or sprinklers could achieve the same effect.

Given that Mackie analyses causation in terms of a cluster of conditions, an *INUS* condition is a *partial* cause. The whole cause may consist of a cluster of conditions *ABC*, which combined are sufficient but not necessary for *E*, since the presence of the clusters *DGH* or *JKL* may also bring about *E*. If the cause is the cluster of conditions *ABC*, none of these individual factors is redundant. Consider some of the previous examples. We have encountered two ways of producing traces on a screen: the two-slit experiments and the Stern-Gerlach experiments. In a two-slit experiment, two traces can be produced when photons are scattered at slit 1 (with slit 2 closed) and at slit 2 (with slit 1 closed). An interference pattern emerges when both slits are open. In a Stern-Gerlach experiment, for just two traces to appear on the screen the atom beam must be in an appropriate quantum state and the magnet must create a non-uniform magnetic field. The cluster of conditions, say *MQ* (an appropriate *m*agnetic field, and an appropriately prepared *q*uantum state), will be sufficient to produce two traces on the screen. Each one of those *M*, *Q* will be non-redundant or necessary parts of the conditional state *MQ*. Different atoms can be chosen to produce the traces, but in each case *M* and *Q* must be present. The atom beam is split when it crosses the magnetic field. How it splits physically depends on *MQ*.

One obvious disadvantage of the minimal conditional view of causation presented so far is that it makes no distinction between conditions, whether necessary

[121] Salmon, *Causality* (1998), 71, 16

[122] Mackie, *Cement* (1980), 296; italics in original

or sufficient, which are physically operative in the production of the effect and non-operative conditions, which are merely in the vicinity of the cause-effect relationship. This is sometimes expressed by making the distinction between the 'cause' and 'causal conditions' or by speaking of the 'cause' and 'contributing conditions.' Mill regards all conditions as constituting the cause. But consider the individual atoms, which make up the magnet in the Stern-Gerlach experiment. They play no causal role in this experiment. (Their interactions would be invoked, in a separate causal account, to explain the production of a magnetic field.) To circumvent this difficulty, Mackie introduces the concept of a *causal field*.[123] A causal field comprises the background conditions – Reichenbach's rest factors – which make the normal running of things possible. There are atoms, they interact and lump together to make matter. The antecedent conditions, C, and consequent conditions, E, must make a *difference* within a field to establish a cause-effect relationship. All the quantum mechanical experiments make such a difference. A photon is made to collide with a stationary electron in Compton scattering to cause it to change its wavelength. A beam of atoms is guided through a Stern-Gerlach apparatus or a double-slit arrangement to cause it to leave traces on a screen. Conditions are statistically relevant if their inclusion in the causal account affect the probability of the outcome, otherwise they are statistically irrelevant. In the Stern-Gerlach experiment it is statistically relevant whether the magnetic field is non-uniform but the motion of atoms in the magnet itself are statistically irrelevant. Whether a condition belongs to the causal field (the statistically irrelevant conditions) can often be measured. For instance an important part of the interpretation of the scattering experiments was to show that the existence of electrons in the target atoms was statistically irrelevant. Recall that in these experiments 1 in 8000 α-particles suffered a deflection of more than 90°. Rutherford argued that this large-angle deflection was due to the existence of a nucleus in the atom. The existence of the nucleus was statistically relevant. Rutherford's probability considerations suggest that the large-angle scattering of α-particles from an encounter with electrons would be so exceedingly rare that Marsden and Geiger would have been very unlikely to make their observations. The existence of electrons was statistically irrelevant to the scattering experiments. This does not mean, of course, that electrons will always remain statistically irrelevant. Conditions, which belong to the causal field in one situation, may become statistically relevant in another. Compton fired X-rays at stationary electrons. This causal encounter led to a lengthening of the wavelengths of the scattered X-rays by circa 10^{-12} m. Compton interpreted the X-rays correctly as photons. In Compton scattering the electron is statistically relevant.

The notion of a causal field introduces context-dependent aspects. The same conditions, regarded independently, may be statistically relevant or irrelevant. It depends on the causal situation. Interactions *between* causal fields are subject to laws of nature. When conditions make a difference to a causal field, they do so in a lawful

[123] Mackie, *Cement* (1980), 35, 63; Bohm, *Causality* (1957), Chap. I, §4 similarly discusses significant causes in a given context; Salmon, *Causality* (1998), 130, 194ff makes a distinction between causal process and causal interaction.

way. The photon moves with momentum $h\nu/c$ towards the stationary electron, which has no momentum. After the collision the scattered photon has a new momentum, with a component in the x-direction ($\cos\theta h\nu'/c$) and a component in the y-direction ($\sin\theta h\nu'/c$). The electron, too, has acquired a momentum. Before the atoms traverse the non-uniform magnetic field in the Stern-Gerlach experiment, their magnetic moments are randomly distributed. After the passage through the magnetic field, their magnetic moments are aligned to produce (in the simplest case) two traces on the screen.

Conditions *within* causal fields are also subject to laws of nature. Before the collision, both photon and electron states constitute causal fields. Laws of nature regulate their normal running.

Conditions belonging to a causal relation derive from the interaction between causal fields. The examples encountered are: the non-uniform magnet *and* the beam of silver atoms (Stern-Gerlach apparatus); the nucleus *and* the α-particle (scattering experiments); the photon *and* the electron (Compton scattering); the electron beam *and* the nickel crystals (Davisson-Germer experiment). A causal relation is brought about by the interaction of lawful behaviour in causal fields, which separately would not lead to the implication: $C \rightarrow E$.

Quantum mechanical systems, if subjected to lawlike behaviour from other systems, will show *probabilistic effects*. Importantly, these degrees of interaction are specifiable. The following table (Table 5.2) summarises the results.

In science idealisations and abstractions abound. We have already noted their importance when we considered the role of models in physical understanding. These also come into effect in the conditional view of causation. Firstly, *abstraction* is made from statistically irrelevant factors. It can often be shown mathematically and experimentally that the statistically irrelevant conditions do not need to be included in the antecedent conditions because their inclusion makes no difference to the probability of the outcome. Thus, Rutherford, Marsden and Geiger neglected the presence of electrons in the gold atoms, since they would have no effect on the scattering of α-particles. Secondly, the statistically relevant conditions often appear in idealised form or in mathematical simplification. This is *idealisation*. In other experiments we have discussed, the atoms are modelled, following Bohr, as quasi-planetary systems albeit with discrete orbits. The electrons and photons are taken as point-like particles with no inner structure. (This turned out to be correct).[124]

The examination of causation in terms of necessary and sufficient conditions, however, leaves room for change in the relevant factors. It may be found that factors, hitherto neglected through abstraction, need to be included to give a fuller causal account. Quantum mechanics is able to include factors, which put Kirchhoff's spectral analysis on much firmer ground. Highly idealised factors can also be made more realistic by the operation of *factualisation*. The Bohr model of the atom and

[124] Bohm, *Causality* (1980), 10 makes a very similar observation: 'Yet, because we can never be sure that we have included *all* of the significant causes in our theory, all causal laws must always be completed by specifying the conditions or background in which we have found that they are applicable.'

Table 5.2. Probabilistic Effects of Famous Experiments

Experiment	*Probabilistic Effects*
Scattering Experiments (Marsden, Geiger and Rutherford, 1909–1911)	The most probable angle of scattering and the fraction of particles scattered at angles greater than θ can be determined.
Double-Slit Experiments (1927, 1960)	When both slits are open, we observe *interference* effects. The atoms in the beam have a probabilistic chance of contributing to the peaks or troughs of the interference patterns.
Stern-Gerlach Experiments (1921–25)	A beam of atoms is split into discrete spatial orientations by a non-uniform magnetic field.
Compton Scattering (1923)	The collision between a beam of photons and stationary electrons causes some of the photons to scatter with reduced frequency and larger wavelength. The photon beam splits into a modified and an unmodified part. Thus only part of the total energy of the photons is absorbed by recoiling electrons, the rest reappears as scattered radiation at nearly the original wavelength.
Davisson-Germer Experiment (1927)	Electrons of specific energies are scattered by atomic planes in crystals such that the electrons reach an intensity maximum when the detector is set at a particular angle ϕ. This is very similar to the de Broglie *gedanken* experiment.

the mathematization of the knowledge of lawlike behaviour of atomic particles underwent many modifications between 1913 and 1926. The most recent case of factualization occurred with the discovery of decoherence.

5.3.6 Causal Explanations

From his analysis of causation between *particular* events, Mackie proceeds to argue the case for *general* causal explanations. His reason is plausible. If an important aspect of scientific advance consists in the progressive localisation of causal conditions in causal fields, this advance would be severely hampered 'if only concrete occurrences were recognised as causes.'[125] If causal explanations are needed, we

[125] Mackie, *Cement* (1980), 258–60. In *Causal Structure* (1984), 185–90 Salmon raises two objections against the *INUS* view of causation, which centre on the notion of strict determination. But Mackie (1980), 321 explicitly states that 'a study of causation need not presuppose causal determinism' and that there is no need to require that 'all events have sufficient causes'. Mackie's account is not affected by Salmon's criticism. In fact, as Mackie also uses the notion of a *causally relevant feature*, it seems that it could be made compati-

take facts (a set of relevant conditions) rather than events (concrete occurrences in spatio-temporal regions) as causes. This leads to

> the concept of a *minimally complete causal account* , one which mentions all the features that were actually causally relevant to some event. This is, of course, an ideal: we do not in general expect to be able to give such an account. (...) Now these relevant features are all *general:* what is vital is not *this* bolt's giving way but the giving way of *a* bolt at this place.[126]

The conditional model of causation emerges as the most adequate account of *physical* causation. It can accommodate the quantum-mechanical evidence. It is also flexible enough to be adaptable to causal accounts in both the macro- and the micro-world. This conditional view of causation arises from two central ideas:

☞ Reichenbach's association of causally relevant factors with probability considerations

☞ Mackie's division of relevant factors into necessary and sufficient conditions and the separation of causal relations from causal fields.

It should be emphasised that the conditional view does not necessarily dispense with the notion of a causal mechanism or a spatio-temporal, causal link between cause and effect. But it does not require that such a link be observable before we can speak of causation. If the detection of a causal mechanism is included as an essential feature in a conditional model of causation, the price is the abandonment of causal accounts in quantum mechanics. Salmon is ready to pay this price. But this renunciation involves us in a dilemma. As we have seen many quantum-mechanical experiments demonstrate a causal dependence of the effect on some conditionally prior causal factors. No precise spatio-temporal determination can be given. The evidence suggests that there are causal relations in quantum mechancis. Our philosophical models and presuppositions should not force us to ignore the empirical evidence. In attempting to solve Einstein's problem, our response to empirical evidence must be differential. In the case of the block universe, we found that the evidence from the Special theory of relativity did not compel us to abandon the view that there are temporal relations in the universe, irrespective of the observer. In the case of causation, we find that the empirical evidence from quantum mechanical experiments is sufficient to compel us to abandon the functional model

---

ble with Salmon's account of causation. In *Causality and Explanation* (1998), 75–6 Salmon makes the distinction between fine-grained causal explanations, and coarse-grained functional explanations. Fine-grained causal explanations zoom in on the relevant conditions and establish a traceable connection between cause and effect. This, however, is the decisive difference between the *INUS* account, as we have interpreted it, and Salmon's account: Salmon requires a causal link between cause and effect, which makes quantum mechanical events appear *a*causal. On the conditional view of causation no such link is required.

[126] Mackie, *Cement* (1980), 260; italics in original. There is a more general way of speaking, based on the idea 'that every physical quantity must be describable by a geometric object, and that the laws of physics must all be expressible as geometric relationships between these geometric objects' (i.e. vectors, tensor fields), see Misner/Thorne/Wheeler, *Gravitation* (1973), 48 and Salmon, *Causality* (1998), 286–7

of causation. The evidence also suggests a conditional model of causation, which comprises the following features:

- ☞ a cluster of causally relevant factors being conditionally prior to the effect
- ☞ a cluster of observable probabilistic effects (see Table 5.2), such that the effects and their probabilistic occurrence are conditionally dependent on the cluster of causal factors
- ☞ the activation of these factors through a lawlike interaction between causal fields, where the interaction may be local or nonlocal
- ☞ a separation of determinism and causation. The interplay of idealisations, abstractions and factualisation can give rise to indeterministic causation. The set of relevant conditions is closed, as in the experiments discussed, but no spatio-temporal link can be detected between individual causal factors and individual effects (Table 5.3)
- ☞ an interplay of idealisations, abstractions and factualisations such the conditional model can be used to describe both the macro- and micro-world. For instance, if the set of relevant conditions can be closed, and a spatio-temporal link can be discerned from the antecedent to the consequent conditions, then the Laplacean ideal can be realised for special systems in a limited spatio-temporal region (i.e. planetary systems).

When these factors are found in an experimental situation, they will constitute a causal explanation.

Table 5.3. A Comparison between Causation and Determinism

Causation	***Determinism***
Antecedentand Consequent Conditions – *Time Irreversible*	Prior and Posterior States – *Time Reversible*
Dependence: *Conditional*	Dependence: *Temporal*
Relation: *Physical*	Relation: *Functional*
Law Statement: *A conjunction of several types of laws* Range: *Local & Non-local*	Law Statement: *Differential equations* Range: *Local*
Statement: 'If C happens, then (and only then) E is always produced by it.'[127] But in view of the probabilistic character of quantum mechanics, this should be amended to 'If C happens, then (and only then), E_1, E_2, E_3 are always produced with different degrees of probability.'	Statement: 'A system is said to be "deterministic" when, given certain data, $e_1, e_2, \ldots, e_n$ at times $t_1, t_2, \ldots, t_n$ respectively, concerning this system, if E_t is the state of the system at any time t, there is a functional relation of the form $E_t = f(e_1, t_1, e_2, t_2, \ldots, e_n, t_n, t)$.'[128]

[127] Bunge, *Causality* (1979), 48

[128] Russell, 'On the notion of cause' (1912/1965), 178; see van Fraassen, *Laws* (1989), 250–8

In both deterministic predictions and causal explanations, *inferences* are involved. In causal explanations they take us from the antecedent to the consequent conditions (Fig. 5.15a,b). In deterministic predictions, they takes us from the prior to the posterior states (Fig. 5.16b). In both cases, the inferences can go in the opposite direction. Imagine we observe a particular effect, like the large-angle scattering of α-particles off gold atoms. We may want to causally explain what brings about this effect, E_1, as Rutherford did. We can consider various causal conditions, C_1, C_2, C_3, in our causal explanations. Some of these explanations may turn out to be unsatisfactory, as for instance, the plum pudding model of the atom. Without the assumption of a nucleus in the atom, the large-angle scattering cannot be satisfactorily explained. By testing the causal explanation against the evidence and coherence considerations, Rutherford was able to eliminate certain causal candidates. This procedure is called *eliminative induction* because we make an inductive step from the evidence and general theoretical assumptions to the credibility of our explanatory accounts (Fig. 5.16a).[129]

Inferences in perfectly deterministic systems are temporally symmetric, as we know from Laplace's demon (Fig. 5.16b). Using universal laws of motion and knowledge of the present state of the universe, the demon either infers the future state or the past state of the universe. According to the *philosophy of being*, any objective distinction between past, present and future disappears. By contrast, the temporal asymmetry of the cause-effect relationship – the antecedence of the cause

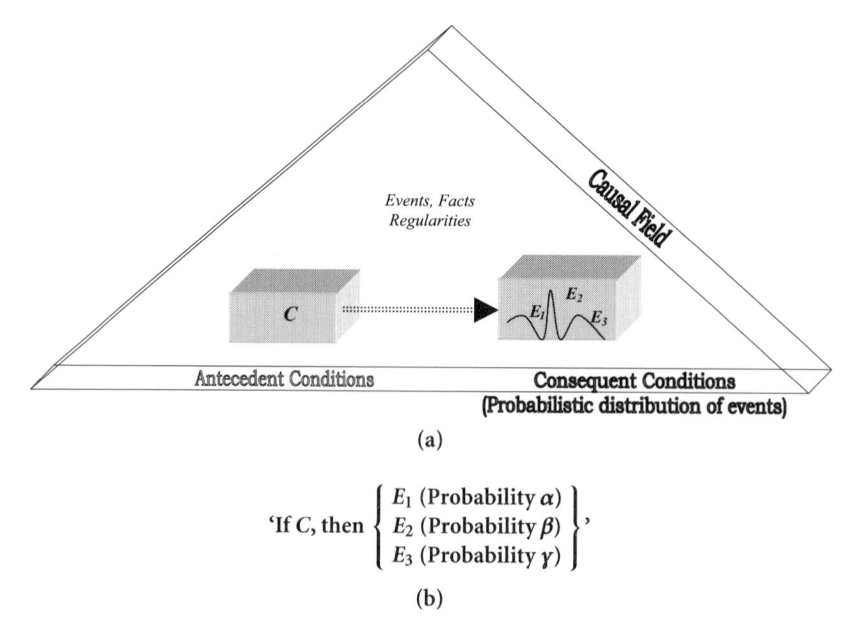

$$\text{'If C, then} \left\{ \begin{array}{l} E_1 \text{ (Probability } \alpha) \\ E_2 \text{ (Probability } \beta) \\ E_3 \text{ (Probability } \gamma) \end{array} \right\} \text{'}$$

(b)

Fig. 5.15. (**a**) Inferences in Causal Fields. (**b**) Inferences from Causal Conditions to Probabilistic Effects

[129] Weinert, 'Construction of Atom Models' (2000)

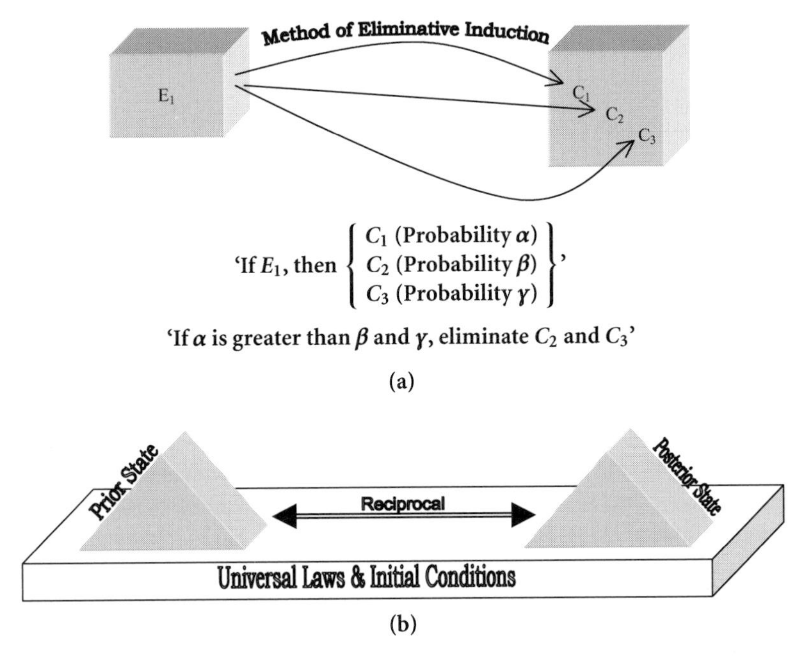

Fig. 5.16. (**a**) The Method of Eliminative Induction. (**b**) Temporally Symmetric Inferences in Deterministic Systems

in *time-like* connected events – is vital for a dynamic interpretation of space-time. It leads to a *philosophy of becoming*.

Causal relations arise when the regularities of causal fields become subject to interference from other fields. The causal behaviour of the quantum systems is due to lawlike interactions between either two quantum systems or a quantum system and a macroscopic device. The quantum system is causally affected by these interactions. If these causal relations are probabilistic in nature, the characterization of probabilistic causation is clear: the causal conditions raise the probability of the effect.[130] We can understand very clearly how this notion arises from the conditional view of causation. Causal relations are characterized in terms of the conditional priority of antecedent conditions over posterior conditions. We can see from this characterization, if we exclude deterministic inklings, that the antecedent conditions may relate to the posterior conditions in two ways: (a) they completely determine the posterior conditions, which are conditionally dependent on the anterior conditions; this is a case of *deterministic causation*; (b) they raise the probability of the occurrence of the posterior conditions, i.e. they determine

[130] See Cartwright, *Laws of Physics* (1983), Essay 1; *Nature's Capacities* (1989), *Dappled World* (1999), Chap. 5; Salmon, *Scientific Explanation* (1984); Eels, *Probabilistic Causality* (1991); Salmon, *Causality and Explanation* (1998), Pt. III; Hughes, *Quantum Mechanics* (1989), Chap. 8

the probability distribution of the cluster of events; this is a case of *probabilistic causation.*

Barring the additional assumption that the anterior conditions are always the same and always produce the same posterior conditions – in which case we have *causal determinism* – the notion of probabilistic causation is the most general notion of causation, as Reichenbach already stressed. It is simply a form of causal relations, in which the occurrence of the consequent conditions is a matter of probability.

Interlude E. *The Programme of Decoherence.*

> Classical concepts and their histories become excess baggage. H.D. Zeh, *Direction of Time* (1992), 94–5

We have already alluded to decoherence in connection with the emergence of classical space-time, the 'collapse' of the wave function (Case 3) and the notion of entanglement. Physicists have hailed the discovery of decoherence as 'the most important advance in the foundation of quantum theory since Bell's inequalities.'[131] This praise is based on good reasons: (a) decoherence offers new ways of understanding quantum mechanics for it embodies a critique of entrenched terminology ('collapse' of the wave function, complementarity); (b) it is testable since it appeals to physical processes; (c) it offers an insight into a much more fundamental level of quantum indeterminacy. Decoherence has been defined as the 'delocalizations of phase relations'; this expresses the 'irreversible formation of quantum correlations of a system with its environment.'[132] This is the second sense of entanglement. But unlike the reversible decoherence in *which-way* experiments due to the quantum eraser, the entanglement with the macroscopic environment cannot be reversed. Due to the dynamic process of decoherence, quantum states degrade. They are subject to numerous scatteringprocesses from their respective environments. As a result, new properties and new forms of behaviour emerge. The programme of decoherence assumes that all objects in the universe interact with each other to

[131] Stamp, 'Time, decoherence' (1995), 108; Tegmark/Wheeler, 'Quantum Mysteries' (2001), 61; Omnès, *Quantum Mechanics* (1999), 74. Major sources for information on decoherence are Giulini *et al.*, *Decoherence* (1996); Blanchard *et al.*, *Decoherence* (2000); Omnès, *Quantum Mechanics* (1999), 'Decoherence' (2000); see also Terhal *et al.*, 'Quantum Entanglement' (2003). Decoherence has been experimentally tested, see Myatt *et al.* 'Decoherence' (2000). Decoherence is a programme rather than an established theory. It still has to overcome many problems; see Bub, *Quantum World* (1997); Omnès, 'Decoherence' (2000). It is *not* considered to be a solution to the measurement problem, precisely because of the continued presence of the interference terms.

[132] Zeh, 'Meaning of Decoherence' (2000), 19; Joos, 'Elements' (2000), 1; DiVincenzo/Terhal: 'Decoherence' (1998); Zeh, 'Interpretation' (1970) was amongst the first to point out the importance of the environment in considerations of quantum systems. The objective of this Interlude is to point to the importance of decoherence in newer interpretations of quantum mechanics. There is no room here to discuss decoherence in the context of Everett's 'many-worlds' interpretation of quantum mechanics. The interested reader may turn to Barrett, *Quantum Mechanics* (1999), Chap. 8; Zeh, 'Interpretation' (1970) and 'Programme of Decoherence' (1996), §2.3

a greater or lesser degree.[133] The classical features of the macroscopic world are not fundamental. Due to the ubiquitous interrelatedness of Nature through quantum correlations, classical particles and their properties only emerge through a process of 'continuous measurement' by the environment. At the most fundamental level Nature is governed by a universal wave function: the Wheeler-de Witt equation, in which there is no room for classical concepts. Quantum mechanics is the universal, unifying theory. At the root of Nature lies a kinematical non-locality, a universal entanglement of all quantum objects in the universe, which makes the universe the only true closed system. The classical world only appears as a consequence of the leaking of quantum correlations into the environment. This is not a true loss, however, since quantum correlations never truly disappear. The *which-way* experiments illustrate the underlying availability of the quantum correlations. If *which-way* information is stored in the atomic systems, their quantum correlations 'decohere' through entanglement with the atomic motion. This results in path distinguishability. On erasure of the path information, the interference patterns are restored. Paths become indistinguishable. The system has 'recohered'. But in *most* cases decoherence is an irreversible process. We therefore never observe macroscopic superpositions. The (Schrödinger) entanglement between the quantum systems is transformed into an entanglement with the environment. In the case of classical properties, the quantum correlations become observationally unobtainable. Their entanglement with the environment makes classical systems look robust. This *robustness* is the effect of a 'continuous measurement' on the part of the environment.

In the analysis of decoherence, relevance concepts are of major importance. One of the most familiar ones is *locality*. In the macroscopic world, objects and their properties are localized in space and time. Humans are so imbued with the experience of locality that philosophers and scientists have elevated it to a cluster feature of causation. The appearance of localization is only approximate. The quantum entanglement is ever present but becomes negligible for all practical purposes. The orthodox interpretation of quantum mechanics omitted the environment from the analysis.

How does the continuous measurement through the environment – the irreversible loss of phase relations – explain some of the quantum mysteries? The quantum correlations in the case of Schrödinger's cat decohere so quickly that the cat is in one state or the other well before the observer opens the steel chamber. The coherent superpositions in the double-slit experiment disappear when the wave function of the electron gets entangled with the γ-ray microscope. The path information is physically transferred to the recording instrument. The measurement process of a quantum system in a spin singlet state

$$S = |\uparrow\downarrow\rangle_{\text{system}_1} + |\downarrow\uparrow\rangle_{\text{system}_2}$$

[133] Joos, 'Decoherence through Interaction' (1996), 35; Zeh, 'Interpretation' (1970), §3; Zeh, *Direction of Time* (1992), 128

produces, in the simplest case, the mixed state 'spin up' for system_1 and 'spin down' for system_2, irrespective of the spatial distance between them. Decoherence explains that the environment acts dynamically and simultaneously on the distant systems. This environmental action results in a degradation of the original quantum correlations between the systems into the degrees of freedom of the environment. The environment acts as a common cause, which flips the state vector in opposite directions.[134]

A comparison of classical and quantum descriptions shows how the interference terms linger on in the wave function.[135] Consider the pure state of a quantum system after it has been coupled to a measurement device (Case 2):

$$|\Psi\rangle = \sum_n \sqrt{p_n}\, |\varphi_n\rangle^{\text{system}}\, |\Phi_n\rangle^{\text{measurement-device}} \,. \tag{5.2}$$

Although Φ_n are generally regarded as 'pointer states' on the measurement device, they can be considered here as general environmental states, which are able to discriminate system states. After pre-measurement, when system and apparatus are considered separately, this leads to 'a ("classical") ensemble of correlated states', described by the density matrix

$$\varrho_{\text{classical}} = \sum_n p_n\, |\varphi_n\rangle\langle\varphi_n| \otimes |\Phi_n\rangle\,\langle\Phi_n| \,. \tag{5.3}$$

The original pure state (5.2) still contains quantum correlations

$$\varrho = |\Psi\rangle\,\langle\Psi| \,, \tag{5.4}$$

so that the density matrix of the ensemble has not lost its quantum correlations:

$$\varrho = \varrho_{\text{classical}} + \sum_{n \neq m} \sqrt{p_n p_m}\, |\varphi_n\rangle\langle\varphi_m| \otimes |\Phi_n\rangle\,\langle\Phi_m| \,. \tag{5.5}$$

The interference terms do not simply vanish, even for macro-systems like Schrödinger cats. When an observer finally reads the measurement result (Case 3), information is gained with a probabilistic distribution of the results. The recording

[134] See Mittelstaedt, 'The Problem of Decoherence' (2000), 157–8; Omnès, *Quantum Mechanics* (1999), 250–5; Zeh, 'Interpretation' (1970). We should note that for the environment to act as a common cause it has to be assumed that the environmental conditions at the separated systems are sufficiently similar to cause the spin states to flip into anti-parallel positions. In view of the indeterministic perturbations involved in the process of decoherence, this assumption needs further justification. Even amongst the proponents of the programme of decoherence, it is a disputed issue whether it solves the measurement problem. Some clearly accept that it does not (E. Joos, H.D. Zeh), since the interference terms never truly disappear. Although unobservable, they continue to exist globally in a total pure state. The measurement apparatus itself contains the total environment. Others admit that the interference terms only disappear *for all practical purposes* (R. Omnès, Ph. Stamp), since they become observationally unobservable. The present writer has some sympathies with this view that this is a pragmatically acceptable solution.

[135] Joos, 'Decoherence through Interaction' (1996), 37–8

does not cause the wave function to collapse. The quantum correlations leak into the environment at a phenomenally fast decoherence time.

The collapse of the wave function is only a pragmatic figure of speech. The radical response located the indeterminism of the measurement process in the quantum kicks. It turns out to be no more than an artefact of this particular interpretation. There *is* quantum indeterminism. It arises from the combined absence of unique consequent and precise initial conditions. If the decoherence programme is correct, it should be an *emergent* property, too. Decoherence is itself an indeterministic, yet mostly irreversible process. The environment performs its continuous measurement processes by way of indeterministic perturbations – scattering processes, heating of vibrational modes, modifications of internal atomic states. These physical perturbations cause quantum indeterminism.

The decay law, $N_t = N_0 e^{-\lambda t}$, served Planck as an alarm bell. He wanted to alert physicists to the threat to the classical worldview. The random, discontinuous behaviour of atoms in radioactive decay needed to be harnessed. Planck did not succeed. Yet if decoherence is correct, no discontinuous jumps are involved in radioactive decay.[136] Rather, the decay status is subject to a continuous measurement. This is still an indeterministic process, but one that settles into causal relations. The observable conditional priority of antecedent conditions emerges from indeterministic perturbations. Once again we have causation without determinism.

5.3.7 Causation Without Determinism

When discussing an appropriate notion of causation in quantum mechanics, the choice of the experimental arrangement to support the argument has turned out to be of vital importance. To support their *radical* response to the experimental situation in quantum mechanics, Bohr and Heisenberg appealed to double-slit experiments. This was then a *Gedankenexperiment* but it illustrated the particle-wave duality, which was given a mathematical expression in the Heisenberg indeterminacy relations. The indeterminacy relations had shown, so it appeared to Bohr and Heisenberg, that no notion of causation could be accommodated in the conceptual space of quantum mechanics. The indeterminacy relations seemed to demonstrate that vital information needed for a classical description of causally deterministic events was missing. Bohr, Heisenberg and Pauli suggested that the complementarity relations were a natural successor to the classical notion of causation. By contrast de Broglie used the *Gedankenexperiment* of atoms impinging on a crystal to draw the conclusion that quantum mechanics could harbour causal relations without determinism. De Broglie was ready to question the traditional Laplacean presuppositions. Years later, Zeh investigated by way of thought experiments which effect the inclusion of the environment would have on the quantum mechanical measurement process. Again we observe that conceptual presuppositions are not idle cogwheels in the machinery of science. Conceptual presuppositions guide, constrain and mislead, yet they are inevitable.

[136] Zeh, *Direction of Time* (1992), §4.3

In this chapter we have argued that essential quantum mechanical experiments, which helped to establish the validity of quantum theory, are not only compatible with but require the notion of probabilistic causation. Clearly, many causal relations were revealed in the history of quantum mechanical interactions, from the scattering to the *which-way* experiments. The behaviour of the quantum systems in these examples is causally regular. There is conditional priority and dependence. At least in some cases it exists under conditions of locality. Yet in all these experiments, there is a clear absence of any spatio-temporal continuity, of any causal link, between C and E. However, decoherence may present us with an indeterministic causal link.

We have seen how the conservative and radical responses were both implicitly committed to the Laplacean identification of causation with determinism. This made quantum mechanics appear acausal. Yet a great number of quantum mechanical experiments satisfy the features of the conditional view of causation. If, however, our philosophical models of causation insist on the satisfaction of features, which traditionally belong to the notion of causation (especially locality and causal mechanisms), then the quantum mechanical evidence will have to be understood in terms of acausality. In this situation the lesson seems obvious to some. Philosophical models of probabilistic causation must be tied to the constraints of quantum mechanical discoveries. But this conclusion is by no means compelling, as Einstein would say. If we adhere to a stricter notion of causation, the physical understanding of quantum mechanics will be very different. The conditional view of causation has the advantage of adaptability. The known conditions permit us to construe causal explanations in quantum mechanics. These are always seen as incomplete and subject to change. In the classical world it could already have been known that we could have spatio-temporal determination without causation. And that we could have causal without deterministic relations. Now we know that we can have probabilistic causation without even entertaining a hope of determinism. There is indeterminism and causation.

6

Conclusion

> When a scientific theory is firmly established and confirmed it changes its character and becomes a part of the metaphysical background of the age: a doctrine is transformed into a dogma.
>
> M. Born, *Natural Philosophy of Cause and Chance* (1949), 47

When we consider the physical view of Nature at the beginning of the 21st century, we must be struck by the conjectural state of scientific knowledge. The understanding of Nature undergoes constant evolution. It engenders a continuous need for *new* concepts. Born was wrong: no scientific dogma is sacrosanct. The electron may be split in two, light pulses may travel faster than the speed of light, which itself may have been variable in the distant past. Such claims have made recent headlines. The conjectural nature of scientific knowledge does not mean that all is pseudo-science. Relativity and quantum theory are well-established scientific theories. Their conceptual innovations drew the fuzzy outlines of a new worldview. Planck divined it. Through recent developments, it has now taken on more definite contours.

- Nature, at the most fundamental level, is a quantum universe. Her genuine discreteness glides into an apparent smoothness of space and time. But space, time and space-time are emergent properties. The precise nature of this *emergence* is still a matter of scientific debate. Philosophy has her own struggle with emergence. What is philosophically most appealing about emergence is the question of its origin. This puzzle has not been clarified in theories of quantum gravity, like string theory or loop quantum theory. Nor in theories of quantum cosmology, like the Wheeler-de Witt equation. Its hallmark is the conspicuous absence of a parameter of time. How can classical space-time be recovered? One answer describes the emergence of space-time in two steps. First, a semi-classical approximation is performed. As the length scales exceed the Planck length, time appears mathematically as an approximate concept. Formidable mathematical obstacles stand in the way of this approximation. Second, decoherence provides the physical mechanism of the emergence. But the action of the environment is central to this process. So one difficulty is to specify which irrelevant subsystems are to act as environment to the relevant ones.

- The quantum nature of the universe compels us to accept a universal *kinematical* non-locality. This entanglement between quantum systems is an experimental reality. A mathematical theory of entanglement has wrapped itself around it. It attempts to quantify notions like *entanglement distillation*, *entanglement cost*, *mixed entanglement* and *bound entangled states*. What is philosophically most enticing about entanglement is the mystery of its dynamic origins. Some physicists and philosophers may want to relegate kinematic entanglement to the status of a brute fact. In the face of the ongoing discoveries and the conjectural state of scientific knowledge, this is as unconvincing as the stipulation of the 'collapse' of the wave function. As entanglement is a feature of the external world – it can be reproduced in experiments and will soon be a central feature of technological innovation – a huge explanatory gap will reside at the heart of modern physics, as long as the causal source of entanglement is not identified. Ironically, Einstein may be right after all: quantum mechanics is still incomplete. And if the cause of entanglement is found, our conceptual models of causation will be revolutionized.
- A familiar everyday friend, the classical particle with all its properties, figures 'only' as an emergent entity. Another acquaintance, the *wave-particle* duality, may have to go into well-deserved retirement if it turns out to be the case that quantum mechanics governs supreme. This does not rule out the physicists' talk of 'particles' and 'waves', though not in the classical sense. If the programme of decoherence and attempts to merge General relativity and Quantum theory are successful, the fundamental oscillations of quantized loops or strings will give rise to particles. The seemingly irreducible wave-particle duality will turn out to be an interpretative ploy in the face of the perplexing nature of quantum systems in a classical world.

The reason for the everlasting interaction between science and philosophy transpires clearly. The human mind musters an admirable ability to think up equations for physical systems. But equations need to be interpreted in terms of physical models and mechanisms. Science requires conceptual understanding. This understanding employs fundamental philosophical notions. Quantum mechanics is only the most conspicuous illustration of the need for physical understanding. The current models of quantum gravity and the new interpretations of quantum mechanics underline the never-ending nature of this process. It will not cease as long as science interprets the world.

Attempts at understanding quantum mechanics, the arrow of time and the nature of space-time indicate the inherent philosophical dimensions of these areas of research. The findings of this study testify to the vitality of philosophy and its interaction with science from a different angle. We can see this by reflection onEinstein's problem *Einstein's problem, Bohr's challenge* and the *feedback thesis*.

The *relative* freedom of the fundamental notions from the new discoveries, of which scientists like Einstein made use, is precisely the leeway, which the philosophical presuppositions enjoy in the face of new scientific discoveries. This dialectic between facts and concepts means that *Einstein's problem* is still with us.

We should not regard Einstein's problem as a bane, but as a blessing. Because science cannot be done without philosophical presuppositions, the scientist would be ill-advised to dismiss philosophy. The scientist uses very basic conceptual tools – Nature and Reality, Understanding and Modelling, Time and Space, Causation and Determinism. This list can be extended to include emergence and entanglement, explanation and prediction, representation and realism, law and regularity. Such notions and themata are often used explicitly in scientific discourse, but they may come with an implicit meaning. The implicit understanding of such notions takes the form of unquestioned presuppositions. And presuppositions, as we have seen, constrain, guide and misguide research. The natural philosopher, in Francis Bacon's apt analogy, must neither act like a fact-gathering ant nor like a thought-weaving spider, but like a fact-concept synthesizing bee. The philosopher, by this analogy, is more like a humble sparrow fluttering around the workplace of the scientist. To some scientists, it seems of little consequence. Yet, its chirping reminds the scientist of the presence of assumptions, which lie at the core of the scientific enterprise. There are real philosophical concerns at the very heart of the scientific activity, pointing, like a signpost away from the purely scientific pursuits to the contemplative rest places on the side.[1]

The scientific enterprise comes with philosophical commitments, whether the scientist likes it or not. The scientist needs philosophical ideas, simply because amongst the experimental and mathematical tools in the toolbox of the scientist there are conceptual tools, like the fundamental notions. The despairing scientist may ask: 'Will we ever get an answer?' The philosopher replies: 'Not *a* definitive answer, but a few tentative answers.' Recall that the philosopher (and the scientist *qua* philosopher) works with conceptual models. At any one time only a few of these models are in circulation. They cannot provide the definitive answers of which the scientist is fond. But this is typical of models even in the natural sciences.

If Einstein's problem is still with us what are we to expect of the dialectic between philosophy and science as we enter the 21st century? Gerald Holton answered the question in his essay title 'Do Scientists Need a Philosophy?' in the affirmative.[2] Their need for a philosophy vanishes from the surface, he argues, when science is untroubled by theoretical or empirical doubts. The scientist's attitude to philosophy becomes nonchalant when the discovery process goes well and no major scientific revolutions threaten the existent world picture. The scientist's presuppositions go underground – they become implicit. But they do not go away. Even implicit presuppositions govern the scientific enterprise. Scientists are committed, in Holton's view, to a few reigning themata, like unification, which remain stable over time. These presuppositions resurface when major obstacles hamper scientific progress. With quantum gravity, decoherence and entanglement a major scientific revolution seems to be in the making. It will revolutionize our conceptions of Nature.

---

[1] Sklar, *Theory and Truth* (2000) for a good overview

[2] Holton, 'Do Scientists Need a Philosophy?' (1986)

This takes us to the *feedback thesis*. Are revolutionary discoveries the only events that challenge the fundamental notions? And do scientific discoveries necessarily lead to a reformulation of the fundamental notions concerned? The *first* point is that, indeed, sometimes a scientific revolution like that of the 17th century leads to a dissatisfaction with an established worldview, like Scholasticism. This motivates the search for a new, more appropriate view of Nature. The resulting world picture, involving philosophical argument and empirical knowledge, was the clockwork universe. The quantum theory was another major conceptual upheaval, which led to a questioning of the notion of causation. It is not easy to say what constitutes a scientific revolution. Some, justifiably, claim that Einstein's Special theory of relativity was the culmination of classical physics. This is due to its adherence to the principle of determinism and its extension of the relativity principle from mechanical to electro-magnetic phenomena. Irrespective of whether or not the Special theory of relativity was a revolutionary theory (Einstein denied that it was), it led to a discussion about the nature of time. In view of the today's research fever in cosmology, quantum mechanics and thermodynamics we should say: an *ongoing* discussion of the nature of time. Thus, scientific revolutions are not necessary for reflections on the fundamental notions. Such reflections may be triggered off simply by new unexpected discoveries – as was the case with radioactive decay and the Planck constant, which early on led to doubts about the notion of causation. Today, decoherence plays a similar role. Or a new view of Nature, like the image of the clockwork universe, inspires philosophers to test its limits. This may then herald an extension or even transformation of the worldview. The cosmos view of Nature was born as a result of such thought experiments.

The *second* point is perhaps the most substantial revision of the feedback thesis: the fundamental notions change *differentially* under the impact of new discoveries. Especially the entrenchment of the notion of determinism and its identification with causation was hard to break. This can be explained by considering the next point.

The *third* point is that there is no straightforward road from the scientific discoveries to the fundamental notions. In this sense much of what scientists have said about the relationship between fact and concept is mistaken. There is no straightforward revision of the concepts in view of new facts. Scientific discoveries present *constraints* for the fundamental notions but not logical compulsions. The philosophical consequences do not follow deductively from the scientific discoveries. While the philosopher should heed these constraints, scientific findings do not uniquely determine the philosophical consequences for the conceptual toolbox of science. This lies in the nature of constraints: they do not determine but channel the consequences. Sometimes the new discoveries confirm the philosophical presuppositions. The philosophical speculations about the interrelatedness of Nature were confirmed by the scientific findings of the 19th century. This now appears in its new guise as a universal entanglement of quantum systems. Sometimes the evidence seems to favour certain conceptions of the fundamental notions. Given the evidence of quantum mechanical experiments, a probabilistic notion of causation is appropriate on the level of empirical adequacy. Even here scientist like David

Bohm and his followers may cling to what they believe is a deterministic notion of causation. Sometimes however the new discoveries do not posses the power to channel the philosophical consequences in a certain direction. It is by no means certain that the passage of time really is just a human illusion. Curiously, this view has become such a commonplace assumption that it has found its way into the pages of the Scientific American.[3]

In spelling out the dialectic relationship between concept and fact, between philosophy and science, we have answered *Bohr's challenge to philosophers.*

How should we expect the *dialectic* between science and philosophy to develop in the future? What will be the role of philosophy in a world of extreme specialization and narrow expertise? To a certain extent it is unavoidable that philosophy, too, undergoes a process of specialization. The philosophy of physics, the philosophy of biology and more recently the philosophy of chemistry testify to this trend. In these areas important foundational work is done, which may well feed back into the scientific domain. But in the light of the dialectic between fact and concept, the role of philosophy should not be confined to narrow specialist concerns. At the beginning of the 20th century Max Weber warned that modern civilization would give birth to 'specialists without spirit, sensualists without heart.' The role of philosophy, I submit, is to put the spirit back into the specialist. Philosophy should act as a *mediator* on several levels:

- Between individual sciences. As we said before, the toolbox of science consists, amongst other things, of fundamental notions – for instance the notion of physical understanding and the role of models in it. Philosophers have studied these tools across several specialized disciplines.
- Between the scientific specialist, the historian and philosopher of science. The task is to construct a model of science as a dynamic enterprise, embedded in a historical and cultural context, entertaining many interrelations with its social and cultural environment. It is these interrelations, ultimately, which render science possible.
- Between the sciences and the educated public. The scientist informs the public about the latest scientific results. The philosopher informs the public about the specific nature of scientific rationality and the role of science in the wider cultural sphere.[4] This is the continuation of the Enlightenment project in a new form. To enlighten the public about the conceptual pillars, on which science is built, the philosopher will need to have some knowledge of the two other mediating roles of philosophy.

Science affects us all, whether we like it or not. It gives us technology and shapes our worldviews. It raises fundamental questions about the way humans interact

[3] See Davies, 'The Mysterious Flow' (2002)

[4] J. Habermas sees the role of philosophy as a mediator between the expert cultures of science and technology, law and morality on the one hand and the everyday life world on the other. The job of philosophy is to work out a theory of communicative rationality. See Habermas, *Nachmetaphysisches Denken* (1988), 35-60; Weinert, 'Habermas, Science and Modernity' (1999); cf. McMullin, 'The Shaping of Scientific Rationality' (1988)

with Nature. As science evolves and transforms the world, every generation must address the foundational and philosophical questions anew. This is just what it means to say that science is dynamic. Its foundations are forever conjectural; its philosophical consequences forever tentative. Philosophers are trained to ask the awkward questions: to show where the foundations are shaky and to evaluate the philosophical consequences of the discoveries.

This book has been one attempt at mediating between philosophy, science and its history. Its finding can be poured into a slogan, reminiscent of Einstein: Science without philosophical presuppositions is impossible. Philosophy without scientific constraint is speculative. And science without philosophical consequences is unimaginable.

Bibliography

Aczel, A.D. (2001): *Entanglement*: The Greatest Mystery in Physics. New York: Four Walls Eight Windows

Ames, J.S. (1920): 'Einstein's Law of Gravitation', *The Physical Review* **XV**, 206–216

Ampère, A.M. (1834): *Essai sur la Philosophie des Sciences* ou Exposition Analytique d'une Classification Naturelle de Toutes les Connaissances Humaines. Paris: Mallet Bachelier

Andrade, E.N. da C. ([3]1927): *The Structure of the Atom*. London: G. Bell & Sons

Andrade, E.N. da C. (1951): 'A Century of Physics', *Nature* **168**, 622–5

Arntzenius, F. (1990): 'Causal Paradoxes in the Special Theory of Relativity', *British Journal for the Philosophy of Science* **41**, 223–43

Aspect, A./J. Dalibard/G. Roger (1982): 'Experimental Tests of Bell's Inequalities Using Time-Varying Analyzers', *Physical Review Letters* **49**, 1804–1807

Atkins, P.W. (1988): 'Time and Dispersal: The Second Law', in Flood, R./M. Lockwood *eds*., (1988), 80–98

Atmanspacher, H./E. Ruhnau *eds*. (1997): *Time, Temporality, Now*. Berlin/New York: Springer

Bacon, F. (1620): *Novum Organum – The New Organon*, ed. by L. Jardine/M. Silverthorne. Cambridge: Cambridge University Press 2000

Baker, L.R. (1975): 'Temporal Becoming: The Argument from Physics', *Philosophical Forum* **6**, 218–36

Balashov, Y. (1999): 'Relativistic Objects', *Noûs* **33**, 644–62

Balazs, N.L. (1929): 'Relativity', *Encyclopaedia Britannica* Volume **19** (14th edition), 95–102

Balazs, N.L. (1929): 'Space-Time, Recent Developments', *Encyclopaedia Britannica* Volume **20** (14th edition), 1073–4

Barbour, J. (1982): 'Relational Concepts of Space and Time', *British Journal for the Philosophy of Science* **33**, 251–74; reprinted in Butterfield, J. *et al*. (1996), 141–164

Barbour, J. (1999): 'The Development of Machian Themes in the Twentieth Century', in J. Butterfield *ed*. (1999), 83–109

Barbour, J. (1999): *The End of Time*. London: Weidenfeld & Nicolson

Barrett, J.A. (1999): *The Quantum Mechanics of Minds and Worlds*. New York: Oxford University Press

Barrett, W. (1968): 'The Flow of Time', in R. Gale *ed.* (1968), 355–77

Barrow, J.D./F.J. Tiper (1986): *The Anthropic Cosmological Principle*. Oxford: Oxford University Press

Bederson, B. *ed.* (1999): *More Things in Heaven and Earth*: A Celebration of Physics at the Millennium. New York/Berlin/Heidelberg: Springer

Beller, M. (1996): 'The Rhetoric of Antirealism and the Copenhagen Spirit', *Philosophy of Science* **63**, 183–204

Beller, M. (1999): *Quantum Dialogue:* The Making of a Revolution. Chicago: Chicago University Press

Belot, G. (1999): 'Rehabilitating Relationism', *International Studies in the Philosophy of Science* **13**, 35–52

Benjamin, C. (1966): 'Ideas of Time in the History of Philosophy', in J.T. Fraser *ed.* (1966), 3–30

Bergmann, H. (1929): *Der Kampf um das Kausalgesetz in der jüngeren Physik*. Braunschweig: Vieweg

Bernard, C. (1865): *Introduction à l'étude de la médecine expérimentale*. Paris: Garnier Flammarion (1966)

Bergson, H. (^{2}1923): *Durée et simultanéité: à propos de la théorie d'Einstein*. Paris: Félix Alcan

Blanchard, Ph./D. Giulini/E. Joos/C. Kiefer/I.-O. Stamatescu *eds.* (2000): *Decoherence:* Theoretical, Experimental and Conceptual Problems. Berlin/Heidelberg/New York: Springer-Verlag

Blumenberg, H. (1975): *Die Genesis der kopernikanischen Welt*. Frankfurt (a. M.): Suhrkamp

Bohm, D. (1957): *Causality and Chance in Modern Physics*. London: Routledge and Kegan Paul

Bohm, D. (1965): *The Special Theory of Relativity*. London/New York: Routledge

Bohr, N. (1913): 'On the Constitution of Atoms and Molecules', *The Philosophical Magazine* **XXVI**, 1–25

Bohr, N. (1915): 'On the Quantum Theory of Radiation and the Structure of the Atom', *The Philosophical Magazine* **XXX**, 394–415

Bohr, N. (1918): 'On the quantum theory of line spectra', reprinted in van der Waerden *ed.* (1968), 95–137

Bohr, N. (1928): 'The Quantum Postulate and the Recent Development of Atomic Theory', Supplement to *Nature* (April 14), 580–90, reprinted in N. Bohr, *Atomic Theory and the Description of Nature*. Cambridge University Press 1934, 52–91

Bohr, N. (1928): 'Das Quantenpostulat und die neuere Entwicklung der Atomistik', *Die Naturwissenschaften* **16**, 245–57

Bohr, N. (1929): 'Wirkungsquantum und Naturbeschreibung', *Die Naturwissenschaften* **17**, 483–6

Bohr, N. (1929): 'Die Atomtheorie und die Prinzipien der Naturbeschreibung', *Die Naturwissenschaften* **18**, 73–8

Bohr, N. (1929): 'The Atomic Theory and the Fundamental Principles underlying the Description of Nature, in N. Bohr', *Atomic Theory and the Description of Nature.* Cambridge University Press 1934, 103–119

Bohr, N. (1936): 'Kausalität und Komplimentarität', *Erkenntnis* **6**, 293–302

Bohr, N. (1958): 'Quantum Physics and Philosophy – Causality and Complementarity', reprinted in N. Bohr, *Essays on Atomic Physics and Human Knowledge, 1958–1962.* New York/London: Interscience Publishers, 1–7

Bohr, N./H.A. Kramers/J.C. Slater (1924): 'Über die Quantentheorie der Strahlung', *Zeitschrift für Physik* **24**, 69–87

Boltzmann, L. (1979): *Populäre Schriften.* Braunschweig: Vieweg

Bondi, H. (1952): 'Relativity and Indeterminacy', *Nature* **169** (April 19), 660

Bondi, H. (1967): *Assumption and Myth in Physical Theory.* Cambridge: Cambridge University Press

Born, M. (1913): 'Zum Relativitätsprinzip', *Die Naturwissenschaften* **1**, 92–4

Born, M. (1926): 'Zur Quantenmechanik der Stoßvorgänge', *Zeitschrift für Physik* **37**, 863–7

Born, M. (1926): 'Quantenmechanik der Stoßvorgänge', *Zeitschrift für Physik* **38**, 803–27

Born, M. (1927): 'Physical Aspects of Quantum Mechanics', *Nature* **119**, 354–57

Born, M. (1927): 'Quantenmechanik und Statistik', *Die Naturwissenschaften* **15**, 238–42

Born, M. (1929): 'Über den Sinn der physikalischen Theorien', *Die Naturwissenschaften* **17**, 109–18

Born, M. (1949): *Natural Philosophy of Cause and Chance.* Oxford: Clarendon

Born, M. (1951): 'Physics in the Last Fifty Years', *Nature* **168**, 625–630

Born, M. (1953): 'Physical Reality', *The Philosophical Quarterly* **3**, 139–49

Born, M. (1962): *Einstein's Theory of Relativity.* New York: Dover

Boyle, R. (1660–63): *Some Considerations touching the Usefulness of Natural Philosophy, Works* Volume I, 423–553; *Works* Volume III, 135–201

Boyle, R. (1666): *The Origin of Forms and Qualities, According to the Corpuscular Philosophy, Works* Volume II, 451–542

Boyle, R. (1679): *The Sceptical Chymist, Works* Volume I, 290–371

Boyle, R. (1686): *A Free Inquiry into the Vulgarly Received Notion of Nature, Works,* Volume IV, 358–424. Modern edition edited by B. Davis/M. Hunter Cambridge: Cambridge University Press 1996

Boyle, R. (1744): *Works in Five Volumes.* London

Brzeziñski, J./F. Coniglione/Th.A.F. Kuipers/L. Nowak *eds.* (1990): *Idealization II:* Forms and Applications. Amsterdam/Atlanta: Rodopi

Brzeziñski, J./L. Nowak *eds.* (1992): *Idealization III:* Approximation and Truth. Amsterdam/Atlanta: Rodopi

Briggs, J./F.D. Peat (1990): *Turbulent Mirror.* New York: Perennial Library

Bub, J. (1997): *Interpreting the Quantum World.* Cambridge: Cambridge University Press

Buchanan, M. (1999): 'An end to uncertainty', *New Scientist* (6 March 1999), 25–8

Brodetsky, S. (1927): 'Some Difficulties in Relativity', *Nature* **120**, 86–9
Bunge, M. (1970): 'Time Asymmetry, Time Reversal and Irreversibility', *Studium Generale* **23**, 562–70; reprinted in J.T. Fraser *et al.* (1972), 122–130
Bunge, M. ([3]1979): *Causality and Modern Science*. New York: Dover
Burtt, E.A. (1924/[2]1980): *The Metaphysical Foundations of Modern Science*. London: Routledge & Kegan Paul
Butterfield, H. ([2]1957): *The Origins of Modern Science, 1300–1800*. London: Bell & Hyman
Butterfield, J. (1987): 'Substantivalism and Determinism', *International Studies in the Philosophy of Science* **2**, 10–31
Butterfield, J. (1992): 'David Lewis Meets John Bell', *Philosophy of Science* **59**, 26–43
Butterfield, J. (1992): 'Bell's Theorem: What it Takes', *British Journal for the Philosophy of Science* **43**, 41–81
Butterfield, J. (2002): 'Critical Notice: Review of J. Barbour, *The End of Time* (1999)', *British Journal for the Philosophy of Science* **53**, 289–330
Butterfield, J./M. Hogarth/G. Belot *eds.* (1996): *Spacetime*. Aldershot/Brookfield: Dartmouth
Butterfield, J./Ch. Isham (1999): 'On the Emergence of Time in Quantum Gravity', in J. Butterfield *ed.* (1999), 111–68
Butterfield, J./Ch. Isham (1999): 'Spacetime and the Philosophical Challenge of Quantum Gravity', in C. Callender/N. Hugget *eds.* (1999), 33–89
Butterfield, J. *ed.* (1999): *The Arguments of Time*. British Academy & Oxford University Press
Callender, C./N. Hugget *eds.* (1999): *Physics Meets Philosophy at the Planck Scale*. Cambridge: Cambridge University Press
Campbell, N.R. (1911): 'The Common Sense of Relativity', *The Philosophical Magazine* **XXI**, 502–17
Campbell, N.R. (1920): *Physics the Elements*. New York: Dover [republished as *Foundations of Science*. New York: Dover 1957]
Campbell, N.R. (1921): 'Theory and Experiment in Relativity', *Nature* **106**, 804–6
Campbell, N.R. (1921): 'Atomic Structure', *Nature* **107**, 170
Campbell, N.R. (1921): 'Metaphysics and Materialism', *Nature* **108**, 399–400
Campbell, N.R. (1921): 'Time and Change', *The Philosophical Magazine* **I**, 1106–17
Campbell, N.R. (1921): *What is Science*? London: Methuen
Čapek, M. (1951): 'The Doctrine of Necessity Re-examined', *Review of Metaphysics* **V**, 11–54
Čapek, M. (1959): 'Towards a Widening of the Notion of Causality', *Diogenes* **28**, 63–90
Čapek, M. (1962): *The Philosophical Impact of Contemporary Physics*. Princeton (N.J.): Van Nostrand
Čapek, M. (1965): 'The Myth of Frozen Passage: the Status of Becoming in the Physical World', *Boston Studies in the Philosophy of Science* **2**: In Honor of Philipp Frank, ed. by R.S. Cohen/M. W. Wartofsky. New York: Humanities Press, 441–62

Čapek, M. (1966): 'Time in Relativity Theory: Arguments for a Philosophy of Becoming', in: J.T. Fraser *ed.* (1966), 434–454
Čapek, M. (1971): 'The Fiction of Instants', *Studium Generale* **24**, 31–43; reprinted in J.T. Fraser *et al.* (1972), 332–44
Čapek, M. (1975): 'Relativity and the Status of Becoming', *Foundations of Physics* **5**, 607–17
Čapek, M. (1976): 'The Inclusion of Becoming in the Physical World', in M. Čapek *ed.*, (1976), 501–24
Čapek, M. *ed.* (1976): *The Concepts of Space and Time* – Their Structure and their Development. *Boston Studies in the Philosophy of Science* **22**. Dordrecht: Reidel
Čapek, M. (1983): 'Time-Space rather than Space-Time', *Diogenes* **123**, 30–49
Čapek, M. (1987): 'The Conflict between the Absolutist and the Relational Theory of Time before Newton', *Journal of the History of Ideas* **48**, 595–608
Carmichael, R.D. (1912): 'On the Theory of Relativity: Analysis of Postulates', *The Physical Review* **XXXV**, 153–76
Carmichael, R.D. (1913): *The Theory of Relativity*. New York: John Wiley
Carmichael, R.D. (1913): 'On the Theory of Relativity: Mass, Force and Energy', *The Physical Review* **I** (2nd series), 161–78
Carmichael, R.D. (1913): 'On The Theory Of Relativity: Philosophical Aspects', *The Physical Review* **I**, 178–97
Carr, H.W. (1920): *The General Principle of Relativity:* In its Philosophical and Historical Aspect. London: Macmillan
Carr, H.W. (1921): 'The Metaphysical Aspects of Relativity', *Nature* **106**, 809–11
Carr, H.W. (1921): 'Metaphysics and Materialism', *Nature* **108**, 247–8, 568–9; reprinted in L.P. Williams *ed.* (1968), 131–3
Carr, H.W. (1921), 'Relativity and Materialism', *Nature* **108**, 467
Carr, H.W. (1922): 'Bergson and Einstein', *Nature* **110**, 503–5
Carr, H.W. (1923): 'Time lived and Time represented' – review of Bergson (1923), *Nature* **112**, 426–8
Carr, H.W. (1924): 'Berson's Theory of Knowledge and Einstein's Theory of Relativity', *The Philosopher* **II**; reprinted in *The Philosopher* **LXXXIX** (2001), 9–10
Carr, H.W. (1924): 'Optical Records and Relativity', *Nature* **114**, 681
Carlson, S. (1999): 'Modeling the Atomic Universe', *Scientific American* (October 1999), 96–7
Cartwright, N. (1983): *How the Laws of Physics Lie*. Oxford: Clarendon Press
Cartwright, N. (1989): *Nature's Capacities and their Measurement*. Oxford: Clarendon Press
Cartwright, N. (1999): *Dappled World*. Cambridge: Cambridge University Press
Cartwright, N. (1999): 'Models and the Limits of Theory: quantum Hamiltonians and the BSC model of superconductivity', in M. Morgan/M. Morrison *eds.* (1999), 241–81
Carus, P. (1913): *The Principle of Relativity in the Light of the Philosophy of Science*. London/Chicago: Open Court
Cassirer, E. (1906/1974): *Das Erkenntnisproblem in der Philosophie und Physik der neueren Zeit*, 4 Bände. Darmstadt: Wissenschaftliche Buchgesellschaft

Cassirer, E. (1921): 'Zur Einsteinschen Relativitätstheorie', in: E. Cassirer, *Zur Modernen Physik*. Darmstadt: Wissenschaftliche Buchgesellschaft (1977), 3–125
Cassirer, E. (1936): 'Determinismus und Indeterminismus in der Modern Physik', in: E. Cassirer, *Zur Modernen Physik*. Darmstadt: Wissenschaftliche Buchgesellschaft (1977), 129–356
Christensen, F. (1974): 'McTaggart's Paradox and the Nature of Time', *Philosophical Quarterly* **24**, 289–99
Christensen, F. (1976): 'The Source of the River of Time', *Ratio* **18**, 131–44
Christensen, F. (1981): 'Special Relativity and Space-like Time', *British Journal for the Philosophy of Science* **32**, 37–53
Clemence, G.M. (1952): 'Time and its Measurement', *American Scientist* **40**, 260–9
Clifford, W. (1955): *The Common Sense of Exact Science*. New York: Dover
Cohen I.B. (1980): *The Newtonian Revolution*. Cambridge: Cambridge University Press
Collingwood, R.G. (1939): *An Autobiography*. Oxford: Oxford University Press. Reprinted with an Introduction by Stephen Toulmin, 1978
Collingwood, R.G. (1940): *An Essay on Metaphysics*. Oxford: Clarendon Press
Collingwood, R.G. (1945): *The Idea of Nature*. Oxford: Clarendon
Colodny, R.G. *ed*. (1966): *Mind and Cosmos*. University of Pittsburgh Press
Compton, A.H. (1923): 'A Quantum Theory of the Scattering of X-Rays By Light Elements', *The Physical Review* **21**, 483–502
Compton, A.H. (1923): 'The Spectrum of Scattered X-Rays', *The Physical Review* **22**, 409–413
Cook, A. (1994): *The Observational Foundations of Physics*. Cambridge: Cambridge University Press
Copi, I.M. (1990): 'The Detective as Scientist', in Burr/Goldfinger *eds*., *Philosophy and Contemporary Issues*. Macmillan 1992, 484–96
Costa de Beauregard, O. (1966): 'Time in Relativity Theory: Arguments for a Philosophy of Being', in J.T. Fraser *ed*., (1966) 417–433
Costa de Beauregard, O. (1971): 'No Paradox in the Theory of Time Anisotropy', *Studium Generale* **24**, 10–8; reprinted in J.T. Fraser *et al*. (1972), 131–9
Costa de Beauregard, O. (1977): 'Two Lectures on the Direction of Time', *Synthese* **35**, 129–54
Costa de Beauregard, O. (1980): 'A Burning Question: Einstein's Paradox of Correlations', *Diogenes* **110**, 83–97
Costa de Beauregard, O. (1987), *Time, The Physical Magnitude*. Dordrecht/Boston: D. Reidel
Craig, W.L. (2000): 'Why Is It Now?', *Ratio* **XIII**, 115–22
Crew, H. (1910): 'The Debt of Physics to Metaphysics', *The Physical Review* **XXXI**, 79–92
Crombie, A.C. (1994): *Styles of Scientific Thinking in the European Tradition*. 3 Volumes. London: Duckworth
Crommelin, A.C.D. (1920): 'The Theory of Relativity', *Nature* **104**, 631–2
Crommelin, A.C.D. (1921): 'Relativity and the Motion of Mercury's Perihelion', *Nature* **106**, 787–9

Cunningham, E. (1914): *The Principle of Relativity*. Cambridge: Cambridge University Press
Cunningham, E. (1914): 'The Principle of Relativity I, II', *Nature* **93**, 378–9, 408–10
Cunningham, E. (1915): *Relativity and the Electron Theory*. New York: Longmans, Green and Co.
Cunningham, E. (1919): 'Einstein's Relativity Theory of Gravitation', *Nature* **104**, 354–6, 374–6, 394–5
Cunningham, E. (1920): 'Relativity and Geometry', *Nature* **105**, 350–1
Cunningham, E. (1921): 'Relativity: The Growth of an Idea', *Nature* **106**, 784–6
Cunningham, E. (1922): 'Prof. Eddington's Romanes Lecture', *Nature* **110**, 568–70
Cunningham, E. (1922): 'The Measurement of Intervals', *Nature* **110**, 698
Cushing, J.T. (1982): 'Models and Methodologies in Current Theoretical High-Energy Physics', *Synthese* **50**, 5–101
Cushing, J.T. (1989): 'A Background Essay', in Cushing/McMullin *eds.* (1989), 1–24
Cushing, J.T. (1991): 'Quantum Theory and Explanatory Discourse: Endgame for Understanding?', *Philosophy of Science* **58**, 337–58
Cushing, J.T. (1994): *Quantum Mechanics*: Historical Contingency and the Copenhagen Hegemony. Chicago: The University of Chicago Press
Cushing, J.T. (1998): *Philosophical Concepts in Physics*. Cambridge: Cambridge University Press
Cushing, J.T./E. McMullin *eds.* (1989): *Philosophical Consequences of Quantum Theory*. Notre Dame (Indiana): University of Notre Dame
Darwin, C.. (1859): *The Origin of Species*, Penguin Books 1968
Daston, L. (1988): *Classical Probability in the Enlightenment*. Princeton: Princeton University Press
Davies, P. (1981): 'Time and Reality', in R. Healey *ed.* (1981), 63–78
Davies, P. (1988): 'Time Asymmetry and Quantum Mechanics', in R. Flood./M. Lockwood *eds.*, (1988), 99–124
Davies, P. (1995): *About Time*: Einstein's Unfinished Revolution. London/New York: Penguin
Davies, P. (2002): 'That Mysterious Flow', in *Scientific American* (September 2002), 40–7
D'Alembert, J.L. (1751): *Discours Préliminaire de l'Encyclopédie*. Hamburg: Felix Meiner 1955
D'Arcy Thompson, N. (1942): *On Growth and Form*. Cambridge: Cambridge University Press
D'Espagnat, B. (1971): *Conceptual Foundations of Quantum Mechanics*. Reading (Mass.): Addison-Wesley-Benjamin
D'Espagnat, B. (1979): 'The Quantum Theory and Reality', *Scientific American* **241**, 128–140
D'Espagnat, B. ([2]1981): *A La Recherche du Réel*. Paris: Gauthier-Villars
D'Espagnat, B. (1985): *Une incertaine réalité*. Paris: Gauthier-Villars
D'Espagnat, B. (1990): *Penser la science*. Paris: Gauthier-Villars

D'Holbach, P.T. (1770): *Système de la Nature.* [English Translation: *The System of Nature*, London 1817; German Translation: *System der Natur*. Frankfurt (a. M.): Suhrkamp, 1978]

De Broglie, L. (1937): *La Physique Nouvelle et les Quanta*. Paris: Flammarion [Engl. Transl. *Revolution in Physics*, 1953]

De Broglie, L. (1939): *Matter and Light* - The New Physics. New York: Dover

De Broglie, L. (1941): *Continu et Discontinu en Physique moderne*. Paris: Albin Michel

De Broglie, L. (1941): *L'Avenir de la Science*. Paris

De Regt, H.W. (1996): 'Philosophy and the Kinetic Theory of Gases', *British Journal for the Philosophy of Science* **47**, 31–62

Denbigh, K.G. (1970): 'In Defence of *the* Direction of Time', *Studium Generale* **23**, 234–44; reprinted in J.T. Fraser *et al.* (1972), 148–58

Denbigh, K.G. (1989): 'Note on Entropy, Disorder and Disorganization', *British Journal for the Philosophy of Science* **40**, 323–331

Denbigh, K.G./J.S. Denbigh (1985): *Entropy in Relation to Incomplete Knowledge*. Cambridge: Cambridge University Press

Denschlag, J./D. Cassettari/J. Schmiedmayer (1999): Guiding Neutral Atoms with a Wire', *Physical Review Letters* **82**, 2014–2017

Descartes, R. (1644): *Les Principes de la Philosophie*, in R. Descartes, *Œuvres et Lettres*. Paris: Gallimard 1953, 553–690

Descartes, R. (1664): *Le Monde et le Traité de l'Homme*, in *Œuvres* XI, ed. C. Adam/ P. Tannery. Paris 1897–1913

Deutsch, D. (1997) : *The Fabric of Reality*. London: Allen Lane The Penguin Press

Diderot (1753): 'De L'Interprétation de la Nature', *Œuvres Philosophiques*. Paris: Garnier 1964, 166–244

Dieks, D. (1988): 'Special Relativity and the Flow of Time', *Philosophy of Science* **55**, 456–460

Dilworth, C. (1996): *The Metaphysics of Science*. An Account of Modern Science in terms of Principles, Laws and Theories. Dordrecht: Kluwer

Dijsterhuis, E.J. (1956): *Die Mechanisierung des Weltbildes*. Berlin/Göttingen/Heidelberg: Springer [English translation: *The Mechanization of the World Picture*. Oxford: Clarendon 1961]

Dingle, H. (1929): 'Relativity: Philosophical Consequences', *Encyclopaedia Britannica* Volume **19** (14th edition), 102–3

Dingle, H. (1937): *Through Science to Philosophy*. Oxford: Clarendon

Dingle, H. (1949): 'Scientific and Philosophical Implications of the Special Theory of Relativity', in P.A. Schilpp *ed.*, Volume II (1949), 557–62

Dingle, H. (1951): 'Philosophy of Physics: 1850–1950', *Nature* **168**, 630–6

Dingle, H. (1966): 'Time in Relativity: Measurement or Coordinate?', in J.T. Fraser *ed.* (1966), 455–78

Dirac, P.A.M. (1930, ^{4}1958): *The Principles of Quantum Mechanics*. Oxford: The Clarendon Press

Dirac, P.A.M. (1963): 'The Evolution of the Physicist's Picture of Nature', *Scientific American* **208**, 45–53; reprinted in J. Mehra *ed.* (1973), 1–14

DiVincenzo, D./ B. Terhal (1998): 'Decoherence: the obstacle to quantum computation', *Physics World* **11** (March 1998), 53–7

Dobbs, H.A.C. (1969): 'The "Present" in Physics', *British Journal for the Philosophy of Science* **19**, 317–24

Dobbs, H.A.C. (1970): 'Reply to Professor Grünbaum', *British Journal for the Philosophy of. Science* **21**, 275–79

Dobbs, H.A.C. (1971): 'The Dimensions of the Sensible Present', *Studium Generale* **24**, 108–26; reprinted in J.T. Fraser *et al.* (1972), 274–92

Dowe, Ph. (1992): 'Wesley Salmon's Process Theory of Causality and the Conserved Quantity Theory', *Philosophy of Science* **59**, 195–216

Dowe, Ph. (2000): *Physical Causation.* Cambridge: Cambidge University Press

Drescher, M. *et al.* (2002): 'Time-resolved atomic inner-shell spectroscopy', *Nature* **419** (24 October 2002), 803–7

Driesch, H. ([2]1930): *Relativitätstheorie und Weltanschauung.* Leipzig: Quelle & Meyer

Du Bois-Reymond, E. (1872): 'Über die Grenzen des Naturerkennens', in Du Bois-Reymond (1886–7), 106–140

Du Bois-Reymond, E. (1882): 'Über die wissenschaftlichen Zustände der Gegenwart', in Du Bois-Reymond (1886–7), 448–64

Du Bois-Reymond, E. (1886–7): *Reden.* Leipzig: Veit & Comp.

Dürr, D./H. Spohn (1998): 'Brownian Motion and Microscopic Chaos', *Nature* **394**, 831–3

Dürr S./T. Nonn/G. Rempe (1998): 'Origin of quantum-mechanical complementarity probed by "which-way" experiments in an atom interferometer', *Nature* **395**, 33–7

Dupré, J. (1995): *The Disorder of Things.* Boston: Harvard University Press

Dyson, F. (1921): 'Relativity and the Eclipse Observations of May, 1919', *Nature* **106**, 786–7

Earman, J. (1967): 'On Going Backward in Time', *Philosophy of Science* **34**, 211–22

Earman, J. (1970): 'Space-Time or How to Solve Philosophical Problems and Dissolve Philosophical Muddles without Really Trying', *Journal of Philosophy* **LXVII**, 259–77; reprinted in J. Butterfield *et al. eds.* (1996), 57–78

Earman, J. (1970): 'Who's Afraid of Absolute Space?', *Journal of Philosophy* **48**, 287–319; reprinted in J. Butterfield *et al. eds.* (1996), 107–40

Earman, J. (1971): 'Kant, Incongruous Counterparts and the Nature of Space and Space-Time', *Ratio* **13**, 1–18

Earman, J. (1972): 'Implications of Causal Propagation Outside the Null Cone', *Australasian Journal of Philosophy* **50**, 222–37

Earman, J. (1986): *A Primer on Determinism.* Dordrecht/Boston: D. Reidel

Earman, J. (1989): *World Enough and Space-Time.* Cambridge (Mass.)/London: MIT Press

Eberhart, M. (1999): 'Why Things Break', *Scientific American* (October 1999), 44–51

Eddington, A.S. (1916): 'Gravitation and the Principle of Relativity', *Nature* **98**, 328–30

Eddington, A.S. (1916): 'Gravitation and the Principle of Relativity', *Nature* **101**, 15–7, 34–6

Eddington, A.S. (1920): 'The Meaning of Matter and the Laws of Nature According to the Theory of Relativity', *Mind* **29**, 145–58

Eddington, A.S. (1920): 'The Philosophical Aspect of the Theory of Relativity', *Mind* **29**, 415–22

Eddington, A.S. (1920): *Space, Time & Gravitation*. Cambridge: Cambridge University Press

Eddington, A.S. (1921): 'The Relativity of Time', *Nature* **106**, 802–4

Eddington, A.S. (1921): 'Space or Æther?', *Nature* **107**, 201

Eddington, A.S. (1922): 'The Measurement of Intervals', *Nature* **110**, 697–8

Eddington, A.S. (1928): *The Nature of the Physical World*. Cambridge: Cambridge University Press

Eddington, A.S. (1935): *New Pathways in Science*. Cambridge: Cambridge University Press

Eddington, A.S.(1939): *The Philosophy of Physical Science*. Cambridge: Cambridge University Press

Eels, E. (1991): *Probabilistic Causality*. Cambridge: Cambridge University Press

Einstein, A. (1905): 'Zur Elektrodynamik bewegter Körper', *Annalen der Physik* **17**; reprinted in H.A. Lorentz *et al.* (1974), 26–50

Einstein, A. (1905): 'Ist die Trägheit eines Körpers von seinem Energiegehalt abhängig?', *Annalen der Physik* **17**; reprinted in H.A. Lorentz *et al.* (1974), 51–3

Einstein, A. (1907): 'Über das Relativitätsprinzip und die aus demselben gezogenen Folgerungen', *Jahrbuch der Radioaktivität und Elektronik* **4**, 411–62

Einstein, A. (1911): 'Über den Einfluß der Schwerkraft auf die Ausbreitung des Lichtes', *Annalen der Physik* **35**; reprinted in H.A. Lorentz *et al.* (1974), 72–80

Einstein, A. (1916): 'Die Grundlage der allgemeinen Relativitätstheorie', *Annalen der Physik* **49**; reprinted in H.A. Lorentz *et al.* (1974), 81–124

Einstein, A. (1916): 'Ernst Mach', *Physikalische Zeitschrift* **7**, 101–104

Einstein, A. (1917): 'Zur Quantentheorie der Strahlung', *Physikalische Zeitschrift* **18**, 121; reprinted in van der Waerden *ed.* (1968), 63–77

Einstein, A. (1918): 'Dialog über die Einwände gegen die Relativitätstheorie', *Die Naturwissenschaften* **6**, 697–702

Einstein, A. (1919): 'Was ist Relativitätstheorie?, in A. Einstein (1977), 127–31

Einstein, A. (1920/^{15}1960): *Relativity*. The Special and the General Theory. London: Methuen

Einstein, A. (1921): 'A Brief Outline of the Development of the Theory of Relativity', *Nature* **106**, 782–4

Einstein, A. (1921): 'Über Relativitätstheorie', in A. Einstein (1977), 131–34

Einstein, A. (1923): 'The Theory of the Affine Field', *Nature* **112**, 448–9

Einstein, A. (1927): 'Newtons Mechanik und ihr Einfluß auf die Gestaltung der theoretischen Physik', *Die Naturwissenschaften* **15**, 273–6

Einstein, A. (1929): 'Space-Time', *Encyclopaedia Britannica* Volume **20** (14th edition), 1070–3

Einstein, A. (1949): 'Autobiographical Notes', in A. Schilpp *ed.* (1949), Voume I, 3–94

Einstein, A. (1949): 'Reply to Criticisms', in A. Schilpp *ed.* (1949), Volume II, 665–88
Einstein, A. (1977): *Mein Weltbild*, hrsg. von Carl Seelig. Frankfurt (a. M.)/Berlin/ Wien: Ullstein
Einstein, A./R.C. Tolman/B. Podolsky (1931): 'Knowledge of Past and Future in Quantum Mechanics', *Physical Review* **37**, 780–1
Einstein, A./B. Podolsky/N. Rosen (1935): 'Can quantum-mechanical description of physical reality be considered complete?', *Physical Review* **47**, 777–80
Einstein, A./L. Infeld (1938): *The Evolution of Physics*. Cambridge: Cambridge University Press
Einstein, A./M. Born (1969): *Briefwechsel 1916–1955*. München: Nymphenburger Verlagshandlung
Elias, N. (1988): *Über die Zeit*. Frankfurt (a. M.): Suhrkamp
Elkana, Y. (1974): *The Discovery of the Conservation of Energy*. London: Hutchinson Educational
Elliot, H. (1921): 'Relativity and Materialism', *Nature* **108**, 432; reprinted in L.P. Williams *ed.* (1968), 133–5
Elsbach, A.C. (1924): *Kant und Einstein*: Untersuchungen über das Verhältnis der modernen Erkenntnistheorie zur Relativitätstheorie. Berlin: de Gruyter
Emden, R. (1926): 'Aberration und Relativitätstheorie', *Die Naturwissenschaften* **14**, 329–35
Enriques, F. (1941): *Causalité et déterminisme dans la philosophie et l'histoire des sciences*. Paris: Hermann
Erkenntnis (1936): 'Das Kausalproblem', Volume **6**
Exner, F. (1919): *Vorlesungen über die physikalischen Grundlagen der Naturwissenschaften*. Wien: Franz Deuticke
Falkenburg, B. (1987): *Die Form der Materie:* Zur Metaphysik der Natur bei Kant und Hegel. Frankfurt (a. M.): Athenäum
Falkenburg, B. ([2]1995): *Teilchenmetaphysik:* Zur Realitätsauffassung in Wissenschaftsphilosophie und Microphysik. Heidelberg: Spektrum [English translation in preparation]
Falkenburg, B. (2000): *Kants Kosmologie:* Die wissenschaftliche Revolution der Naturphilosophie im 18. Jahrhundert. Frankfurt (a. M.): Klostermann
Faraday, M. (1860): *A Course of 6 Lectures on the various forces of matter and their relation to each other*. London/Glasgow: R. Griffin
Faraday, M. (1936): *Faraday's Diary*. London: Bell
Ferré, F. (1970): 'Grünbaum *vs.* Dobbs: The Need for Physical Transiency', *British Journal for the Philosophy of Science* **21**, 278–80
Ferris, T. (1998): *The Whole Shebang*. London: Phoenix
Fetzer, J.H. *ed.* (1988): *Probability and Causality*. Dordrecht/Boston: D. Reidel
Feynman, R.P. (1965): *The Character of Physical Law*. British Broadcasting Cooperation
Feynman, R.P. (1995): *Six Easy Pieces*. Reading (Mass.): Helix Books, Addison Wesley Company
Feynman, R.P. (1997): *Six Not So Easy Pieces*. Cambridge (Mass.): Helix Books, Perseus Books

Fitzgerald, P. (1969): 'The truth about tomorrow's sea flight', *Journal of Philosophy* **66**, 307–29

Fitzgerald, P. (1985): 'Four Kinds of Temporal Being', *Philosophical Topics* **13**, 145–77

Flood, R./M. Lockwood *eds.* (1988): *The Nature of Time*. Oxford: Basil Blackwell

Fleck, L. (1929): 'Zur Krise der "Wirklichkeit"', *Die Naturwissenschaften* **17**, 425–30

Folse, H.J. (1996) 'Ontological constraints and understanding quantum phenomena', *Dialectica* **50**, 121–136

Fokker, A.D. (1915): 'A Summary of Einstein's and Grossmann's Theory of Gravitation', *The Philosophical Magazine* **XXIX**, 77

Frank, J./G. Hertz (1914): 'Über Zusammenstöße zwischen Elektronen und den Molekülen des Quecksilberdampfes und die Ionisierungsspannung desselben', *Verband Deutscher Physikalischer Gesellschaften* **16**, 457–67

Frank, Ph. (1910): 'Das Relativitätsprinzip und die Darstellung der physikalischen Erscheinungen im vierdimensionalen Raum', *Zeitschrift für Physikalische Chemie* 1910, 466–95

Frank, Ph. (1929): 'Was bedeuten die gegenwärtigen physikalischen Theorien für die allgemeine Erkenntnislehre?, *Die Naturwissenschaften* **17**, 971–994

Frank, Ph. (1932): *Das Kausalgesetz und seine Grenzen*. Wien: Julius Springer

Frank, Ph. (1936): 'Philosophische Deutungen und Mißdeutungen der Quantentheorie', *Erkenntnis* **6**, 303–316

Frank, Ph. (1938): *Interpretations and Misinterpretations of Modern Physics*. Paris: Hermann & Cie

Frank, Ph. (1947): *Einstein – His Life and Time*. New York: Knopf

Frank, Ph. (1949): 'Einstein, Mach and Logical Positivism', in P.A. Schilpp *ed.* (1949), Volume I, 269–286

Frank, Ph. (1950): 'Metaphysical Interpretations of Science I, II', *British Journal for the Philosophy of Science* **I**, 60–74, 77–91

Frank, Ph. (1957): *Philosophy of Science*, Englewood Cliffs (N.J.): Prentice-Hall

Franklin, A. (1986): *The Neglect of Experiment*. Cambridge: Cambridge University Press

Franklin, A. (1995): 'Laws and Experiment', in F. Weinert *ed.* (1995), 191–207

Franklin, A. (1999): 'The Roles of Experiment', *Die Naturwissenschaften* **1**, 35–53

Fraser, J.T. *ed.* (1966): *The Voices of Time*. George Braziller/Allen Lane, The Penguin Press 1968

Fraser, J.T./F.C. Haber/G.H. Müller *eds.* (1972): *The Study of Time*. Berlin-Heidelberg-New York: Springer

French, S.. (2003): 'Scribbling on the Blanck Sheet: Eddington's Structuralist Conception of Objects', *Studies in History and Philosophy of Modern Physics* **34** B(2), 227–59

French, S./J. Ladyman (2003): 'Remodelling Structural Realism: Quantum Physics and the Metaphysics of Structure', Synthese **136**, 31–56

Freundlich, E. (1916): 'Die Grundlagen der Einsteinschen Gravitationstheorie', *Die Naturwissenschaften* **4**, 363–392

Freundlich, Y. (1973): '"Becoming" and the Asymmetries of Time', *Philosophy of Science* **40**, 496–517

Friedman, M. (1974): 'Explanation and Scientific Understanding', *Journal of Philosophy* 71, 5–19

Friedman, M. (1981): 'Theoretical Explanation', in R. Healey *ed.* (1981), 1–16

Friedman, M. (1983): *Foundations of Space-Time Theories*. Princeton (N.J.): Princeton University Press

Fürth, R. (1969): 'The Role of Models in Theoretical Physics', in Boston Studies in the Philosophy of Science **V**, ed. by R.S. Cohen/M.W. Wartofsky. Dordrecht: Reidel, 327–40

Gale, R. *ed.* (1968): *The Philosophy of Time*. London: MacMillan

Gale, R. (1968): 'The Static Versus The Dynamic Temporal', in R. Gale *ed.* (1968), 65–85

Galison, P. (1979): 'Minkowski's Space-Time: From Visual Thinking to Absolute World', *Historical Studies in the Physical Sciences* **10**, 85–121

Galison, P. (1987): *How Experiments End*. Chicago/London: The University of Chicago Press

Galison, P. (1995): 'Theory Bound and Unbound: Superstrings and Experiments', in F. Weinert *ed.*, (1995), 369–408

Galison, P. (1997): *Image and Logic*: A Material Culture of Microphysics. Chicago/London: The University of Chicago Press

Galison, P. (2000): 'Einstein's Clocks: The Place of Time', *Critical Inquiry* **26**, 355–89

Galison, P. (2003): *Einstein's Clocks, Poincaré's Maps*: Empires of Time. New York: W.W. Norton

Galison, P./D.J. Stump *eds.* (1996): *The Disunity of Science*. Stanford (California): Stanford University Press

Gamow, G. (1965): *Mr. Tompkins in Paperback*. Cambridge: Cambridge University Press

Gaspar, P./M.E. Briggs/M.K. Francis/J.V. Sengers/R.W. Gammon/J.R. Dorfman/R.V. Calabrese (1998): 'Experimental Evidence for Microscopic Chaos', *Nature* **394**, 865–8

Gehrcke, E. (1913): 'Die gegen die Relativitätstheorie erhobenen Einwände', *Die Naturwissenschaften* **1**, 62–6

Geiger, H. (1909–10): 'The Scattering of the α-Particles by Matter', *Proceedings of the Royal Society of London*. Series A **83**, 493–504

Geiger, H./E. Marsden (1913): 'The Laws of Deflexion of α Particles through Large Angles', *The Philosophical Magazine* **XXV**, 604–623

Gerstein, M./M. Levitt (1998): 'Simulating Water and the Molecules of Life', *Scientific American* (November 1998), 74–9

Gilbert, L. (1914): *Das Relativitätsprinzip*-Die jüngste Modenarrheit der Wissenschaft. Brackwede i. W.: Dr.W. Breitenbach

Godfrey-Smith, W. (1979): 'Special Relativity and the Present', *Philosophical Studies* **36**, 233–44

Gödel, K. (1949): 'An Example of a New Type of Cosmological Solutions of Einstein's Field Equations of Gravitation', *Reviews of Modern Physics* **21**, 447–50

Gödel, K. (1949): 'A Remark about the Relationship between Relativity Theory and Idealistic Philosophy', in P.A. Schilpp *ed.*, Volume II, 557–62

Gold, T. (1962): 'The Arrow of Time', *American Journal of Physics* **30**, 403–10
Gold, T. (1966): 'Cosmic Processes and the Nature of Time', in R.G. Colodny *ed.* (1966), 311–29
Gold, T. *ed.* (1967): *The Nature of Time*. New York: Cornell University Press
Gold, T. (1974): 'The World Map and the Apparent Flow of Time', in B. Gal-Or *ed.*, *Modern Developments in Thermodynamics*. New York: John Wiley & Sons, 63–72
Goswami, A. (2001): *The Physicist's View of Nature*, Part I: From Newton to Einstein. Kluwer
Greene, B. (1999): *The Elegant Universe*. New York: W.W. Norton and Company, Inc.
Griffiths, R.B. (2001): *Consistent Quantum Theory*. Cambridge: Cambridge University Press
Grünbaum, A. (1967): *Modern Science and Zeno's Paradoxes*. London: George Allen and Unwin
Grünbaum, A. (1967): 'The Anisotropy of Time', in T. Gold *ed.* (1967), 149–86
Grünbaum, A. (1968): 'The Status of Temporal Becoming', in R. Gale *ed.* (1968), 322–54
Grünbaum, A. (1969): 'Are Physical Events Themselves Transiently Past, Present and Future? A Reply to H. A. C. Dobbs', *British Journal for the Philosophy of Science* **20**, 145–62
Grünbaum, A. (1970): 'Space, Time and Falsifiability', *Philosophy of Science* **37**, 469–588
Grünbaum, A. (1971): 'The Meaning of Time', in E. Freeman/W. Sellars *eds.*, *Basic Issues in the Philosophy of Time*. Open Court (1971), 196–227; reprinted in M. Čapek *ed.* (1976), 471–500
Grünbaum, A. (1973): 'Is there a "Flow" of Time or Temporal "Becoming"?', in *Philosophical Problems of Space and Time* ([2]1973), 314–29
Habermas, J. (1988): *Nachmetaphysisches Denken*. Frankfurt (a. M.): Suhrkamp
Hacking, I. (1983): *Representing and Intervening*. Cambridge: Cambridge University Press
Haldane, J.B.S. (1936): 'Some Principles of Causal Analysis in Genetics', *Erkenntnis* **6**, 346–356
Hall, M.B. (1966): *Robert Boyle on Natural Science*. Indiana University Press
Harig, G./J. Schleifstein *eds.* (1960): *Naturwissenschaft und Philosophie*. Berlin: Akademie Verlag
Harré, R. (1997): 'Is there a Basic Ontology for the Physical Sciences?', *Dialectica* **51**, 17–34
Harris, E.E. (1968): 'Simultaneity and the Future', *British Journal for the Philosophy of Science* **19**, 254–6
Harvey, W. (1651): 'Anatomical Exercises on the Generation of Animals', *The Works of William Harvey, translated from the Latin with a Life of the Author* by Robert Willis, London 1847
Harwit, M. (1981): *Cosmic Discovery*. Brighton (Sussex): Harvester Press
Hawking, St./R. Penrose (1996): 'The Nature of Space and Time', *Scientific American* (July 1996), 44–9

Hazard, P. (1961): *La Crise de la Conscience Européene 1680–1715*. Paris: Fayard
Healey, R. *ed.* (1981): *Reduction, Time and Reality*. Cambridge: Cambridge University Press
Heath, A.E. (1928): 'Contribution to Symposium on Materialism in Light of Modern Scientific Thought', *Proceedings of the Aristotelian Society*, Volume **8**, 130–42
Heilbron, J.L. (1968): 'The Scattering of α and β Particles and Rutherford's Atom', *Arch. Hist. Exact Sci* **4**, 247–307
Heilbron, J.L. (1981): *Historical Studies in the Theory of Atomic Structure*. New York: Arno Press
Heilbron, J.L. (2000): *The Dilemmas of an Upright Man*. Max Planck and the Fortunes of German Science. Boston: Harvard University Press
Heilbron, J.L./Th.S. Kuhn (1969): 'The Genesis of the Bohr Atom', *Historical Studies in the Physical Sciences* **1**, 211–290
Heisenberg, W. (1926): 'Quantenmechanik', *Die Naturwissenschaften* **14**, 989–94
Heisenberg, W. (1926–27): 'Schwankungserscheinungen und Quantenmechanik', *Zeitschrift für Physik* **40**, 501–6
Heisenberg, W. (1927): Über den anschaulichen Inhalt der quantentheoretischen Kinematik und Mechanik', *Zeitschrift für Physik* **43**, 172–198
Heisenberg, W. (1927): 'Über die Grundprinzipien der "Quantenmechanik", *Forschungen und Fortschritte* **3**, 83; reprinted in W. Heisenberg (1984–5), Band I: Physik und Erkenntnis 1927–1955, 21
Heisenberg, W. (1928): 'Erkenntnistheoretische Probleme der modernen Physik', in W. Heisenberg (1984–5), Band I: Physik und Erkenntnis 1927–1955, 23–8
Heisenberg, W. (1929): 'Die Entwicklung der Quantentheorie 1918–1928', *Die Naturwissenschaften* **26**, 490–6
Heisenberg, W. (1930): *The Physical Principles of the Quantum Theory*. London: Dover
Heisenberg, W. (1931): 'Die Rolle der Unbestimmtheitsrelationen in der modernen Physik', *Monatshefte für Mathematik und Physik* **38**, 365–72; reprinted in W. Heisenberg (1984–5), Band I: Physik und Erkenntnis 1927–1955, 40–7
Heisenberg, W. (1931): 'Kausalgesetz und Quantenmechanik', *Annalen der Philosophie* **2**, 172–82; reprinted in W. Heisenberg (1984–5), Band I: Physik und Erkenntnis 1927–1955, 29–39
Heisenberg, W. (1934): 'Atomtheorie und Naturerkenntnis', *Unversitätsbund Göttingen, Mitteilungen* **16**, 9–20; reprinted in W. Heisenberg (1984–5), Band I: Physik und Erkenntnis 1927–1955, 62–73
Heisenberg, W. (1934): 'Die Entwicklung der Quantenmechanik', *Die Moderne Atomtheorie*. Die bei der Entgegennahme des Nobelpreises 1933 in Stockholm gehaltenen Vorträge. Leipzig: S. Hirzel, 1–18; reprinted in W. Heisenberg (1984–5), Band I: Physik und Erkenntnis 1927–1955, 74–91
Heisenberg, W. (1936): 'Prinzipielle Fragen der modernen Physik', *Neuere Fortschritte in den exakten Wissenschaften*. Leipzig/Wien: Franz Deuticke, 91–102; reprinted in W. Heisenberg (1984–5), Band I: Physik und Erkenntnis 1927–1955, 108–119

Heisenberg, W. (1948): 'Der Begriff "Abgeschlossene Theorie" in der Modernen Naturwissenschaft', *Dialectica* **2**, 331–6

Heisenberg, W. (1951): '50 Jahre Quantentheorie', *Die Naturwissenschaften* **38**, 49–55; reprinted in W. Heisenberg (1984–5), Band I: Physik und Erkenntnis 1927–1955, 354–60

Heisenberg, W. (1952): 'Atomphysik und Kausalgesetz', *Merkur* **6**, 701–11; reprinted in W. Heisenberg (1984–5), Band I: Physik und Erkenntnis 1927–1955, 376–86

Heisenberg, W. (1955): *Das Naturbild der heutigen Physik*. Hamburg: Rowohlt

Heisenberg, W. (1956): 'Die Entwicklung der Deutung der Quantentheorie', *Physikalische Blätter* **12**, 289–304; reprinted in W. Heisenberg (1984–5), Band I: Physik und Erkenntnis 1927–1955, 434–449

Heisenberg, W. (1958): 'Die Plancksche Entdeckung und die philosophischen Grundlagen der Atomlehre', *Die Naturwissenschaften* **45**, 227–34; reprinted in W. Heisenberg (1984–5), Band II: Physik und Erkenntnis 1956–1968, 205–12

Heisenberg, W. (1959): 'Die Plancksche Entdeckung und die philosophischen Probleme der Atomphysik', *Universitas* **14**, 135–48; reprinted in W. Heisenberg (1984–5), Band II: Physik und Erkenntnis 1956–1968, 235–48

Heisenberg, W. (1959): 'Grundlegende Voraussetzungen in der Physik der Elementarteilchen', *Martin Heidegger zum siebzigsten Geburtstag*. Pfullingen: Neske, 291–7; reprinted in W. Heisenberg (1984–5), Band II: Physik und Erkenntnis 1956–1968, 249–5

Heisenberg, W. (1959): *Physik und Philosophie*. Frankfurt (a. M.)/Berlin/Wien: Ullstein

Heisenberg, W. (1967): 'Philosophische Probleme in der Theorie der Elementarteilchen', reprinted in W. Heisenberg (1984–5), Band II: Physik und Erkenntnis 1956–1968, 410–22

Heisenberg, W. (1969): The Concept of "Understanding" in Theoretical Physics', H. Mark/S. Fernbach *eds.*, *Properties of Matter Under Unusual Conditions*. New York/London/Sidney/Toronto: Interscience Publishers, 7–10; reprinted in W. Heisenberg (1984–5), Band III: Physik und Erkenntnis 1969–1976, 335–38

Heisenberg, W. (1970): 'Änderungen der Denkstruktur im Fortschritt der Wissenschaft', *Studium Generale* **23**, 808–16; reprinted in W. Heisenberg (1984–5), Band III: Physik und Erkenntnis 1969–1976, 350–8

Heisenberg, W. (1970): 'Abschluß der Physik', Süddeutsche Zeitung (6. Oktober), 44; reprinted in W. Heisenberg (1984–5), Band III: Physik und Erkenntnis 1969–1976, 385–92

Heisenberg, W. (1973/^{3}1976): *Der Teil und das Ganze*. München: Piper

Heisenberg, W. (1974): 'The Philosophical Background of Modern Physics', *Encyclopedia Moderna* **28**, 133–41; reprinted in W. Heisenberg (1984–5), Band III: Physik und Erkenntnis 1969–1976, 496–506

Heisenberg, W. (1975): 'Bemerkungen über die Unbestimmtheitsrelation', *Physikalische Blätter* **31**, 193–6; reprinted in W. Heisenberg (1984–5), Band III: Physik und Erkenntnis 1969–1976, 514–7

Heisenberg, W. (1977): *Tradition in der Wissenschaft*. München: Piper

Heisenberg, W. (1984–85): *Gesammelte Werke/Collected Works* Abteilung C, Bd. I–III. *eds.* W. Blum, H.P. Dürr/ H. Rechenberg), München/Zürich: Piper
Helmholtz, H. v. (1847): *Über die Erhaltung der Kraft*, in: Ostwalds Klassiker der Exakten Wissenschaften, Band 1, Thun/Frankfurt (a. M.), 1996, 3–60
Helmholtz, H. v. (1854): 'Über die Wechselwirkung der Naturkräfte und die darauf bezüglichen Ermittelungen der Physik', in Helmholtz (1871), 99–136
Helmholtz, H. v. (1854): 'The Interaction of Natural Forces', in Helmholtz (1881), 137–174
Helmholtz, H. v. (1862–3): 'Über die Erhaltung der Kraft', in Helmholtz (1871), 137–80
Helmholtz, H. v. (1862–3): 'On the Conservation of Force', in Helmholtz (1881), 277–317
Helmholtz, H. v. (1869): 'Über das Ziel und den Fortschritt der physikalischen Wissenschaft', in Helmholtz (1871), 181–211
Helmholtz, H. v. (1869): 'Aim and Progress of Physical Science', in Helmholtz (1881), 319–48
Helmholtz, H. v. (1871): *Populäre Wissenschaftliche Vorträge* II. Braunschweig: Vieweg 1871
Helmholtz, H. v. ([2]1881): *Popular Lectures on Scientific Subjects*. London: Longmans
Herfel, W.E./W. Krajewski/I. Niiniluoto/R. Wójcicki *eds.* (1995): *Theories and Models in Scientific Processes*. Amsterdam/Atlanta: Rodopi
Herschel, J. (1830): *A Preliminary Discourse on the Study of Natural Philosophy*. New York/London: Johnson Reprint Corporation (1966)
Hertz, G. (1929): 'Die Bedeutung der Planckschen Quantentheorie für die Experimentalphysik', *Die Naturwissenschaften* **17**, 496–98
Hertz, H. (1894): *Die Prinzipien der Mechanik in neuem Zusammenhange dargestellt*, in: Ostwalds Klassiker der exakten Wissenschaften, Band 263, Thun/Frankfurt (a. M.) 1996
Hertz, P. (1936): 'Regelmäßigkeit, Kausalität und Zeitrichtung', *Erkenntnis* **6**, 412–420
Hesse, M.B. (1961): *Forces and Fields*. London: Nelson
Hesse, M.B. (1966): *Models and Analogies in Science*. Notre Dame (Indiana): University of Notre Dame Press
Hesse, M.B. (1980): *Revolutions and Reconstructions in the Philosophy of Science*. Brighton (Sussex): Harvester Press
Hoffmann, B. (1972): *Albert Einstein*. New York: Viking Press
Holton, G. (1965): 'The Metaphor of Space-Time Events in Science', *Eranos Jahrbuch* **34**, 33–78
Holton, G. (1973): *Thematic Origins of Scientific Thought* – Kepler to Einstein. Cambridge (Mass.)/London: Harvard University Press
Holton, G. (1978): *The Scientific Imagination*: Case Studies. Cambridge: Cambridge University Press
Holton, G. (1986): 'Do Scientists Need a Philosophy?', in G. Holton (1986), 163–78
Holton, G. (1986): 'Einstein's model for constructing a scientific theory', in G. Holton (1986), 28–56

Holton, G. (1986): *The Advancement of Science and its Burdens*. Cambridge: Cambridge University Press

Holton, G. (2000): *Einstein, History and Other Passions*. Cambridge (Mass.)/London: Harvard University Press

Hooke, R. (1705): 'A General Scheme or Idea of the Present State of Natural Philosophy', *The Posthumous Works of Robert Hooke*, published by Richard Waller, London: Sam. Smith and Benj. Walford, 1–65

Hooker, C.A. (1991): 'Projection, Physical Intelligibility, Objectivity and Completeness: The Divergent Ideals of Bohr and Einstein', *British Journal for the Philosophy of Science* **42**, 491–511

Horwich, P. (1987): *Asymmetries in Time*. Boston: MIT

Horwitz, L.P./R.I. Arshansky/A. Elitzur (1988): 'On the Two Aspects of Time: The Distinction and its Implication', *Foundations of Physics* **18**, 1159–93

Hughes, R.I.G. (1989): *The Structure and Interpretation of Quantum Mechanics*. Cambridge (Mass.)/London: Harvard University Press

Hughes, R.I.G. (1989): 'Bell's Theorem, Ideology and Structural Explanation', in J.T. Cushing/E. McMullin *eds.* (1989), 195–207

Hughes, R.I.G. (1999): 'The Ising Model, computer simulations and universal physics', in M. Morgan/M. Morrison *eds.* (1999), 97–145

Humboldt, A. v. (1844): *Kosmos. Entwurf einer physikalischen Weltbeschreibung*. Gesammelte Werke Band I–IV. Stuttgart: Verlag der J.G. Cotto'schen Buchhandlung

Hund, F. (1970): 'Zeit als physikalischer Begriff', *Studium Generale* **23**, 1088–1101; reprinted in J.T. Fraser *et al.* (1972), 39–52

Hund, F. ([2]1978): *Geschichte der physikalischen Begriffe*, 2 Bände. Mannheim: Bibliographisches Institut

Huntington, E.V. (1912): 'A New Approach to the Theory of Relativity', *The Philosophical Magazine* **XXIII**, 494–513

Huxley, T.H. (1874): 'On the Hypothesis that Animals are Automata, and its History', in *Collected Essays* Volume I, London: Macmillan 1893, 199–250

Huxley, T.H. (1876): 'Lectures on Evolution', in *Collected Essays* Volume IV, London: Macmillan 1898, 46–138

Jammer, M. (1960/[2]1969): *Concepts of Space*. New York: Harper

Jammer, M. (1966): *The Conceptual Development of Quantum Mechanics*. Reading (Mass.): Addison Wesley

Jammer, M. (1974): *The Philosophy of Quantum Mechanics:* The Interpretation of Quantum Mechanics in Historical Perspective. New York: John Wiley

Jammer, M. (1979): 'A Consideration of the Philosophical Implications of the New Physics', in: G. Radnitzky/ G. Andersson *eds.*, *The Structure and Development of Science*. Dordrecht: Reidel (1979), 41–62

Jeans, J. (1921): 'The General Physical Theory of Relativity', *Nature* **106**, 791–3

Jeans, J. (1936): *Scientific Progress*. London: Allen and Unwin

Jeans, J. (1943): *Physics and Philosophy*. Cambridge: Cambridge University Press

Jeffreys, H. (1921): 'Relativity and Materialism', *Nature* **108**, 568–9

John, Ch. St. (1921): 'The Displacement of Solar Lines', *Nature* **106**, 789–90

Jönsson, C. (1974): 'Electron Diffraction at Multiple Slits', *American Journal of Physics* **42**, 4–11; first published in *Zeitschrift für Physik* **161**, 1961

Joos, E. (1986): 'Why Do We Observe A Classical Spacetime?' *Physics Letters A* **116**, 6–8

Joos, E. (1996): 'Decoherence through Interaction with the Environment' in D. Giulini *et al.* (1996), 35–136

Joos, E. (2000): 'Elements of Environmental Decoherence' in Ph. Blanchard *et al.* (2000), 1–18

Joos, G. ([2]1951): *Theoretical Physics*. London/Glasgow: Blackie & Son

Jordan, P. (1927): 'Kausalität und Statistik in der modernen Physik', *Die Naturwissenschaften* **15**, 105–10

Jordan, P. (1927): 'Philosophical Foundations of Quantum Theory', *Nature* **119**, 566–69, 779

Jordan, P. (1929): 'Die Erfahrungsgrundlagen der Quantentheorie', *Die Naturwissenschaften* **17**, 498–507

Jordan, P. (1934): 'Über den positivistischen Begriff der Wirklichkeit', *Die Naturwissenschaften* **22**, 485–90

Jordan, P. ([2]1960): *Atom und Weltall*: Einführung in den Gedankeninhalt der modernen Physik. Braunschweig: Vieweg 1960

Kannitscheider, B. (1996): *Im Innern der Natur*. Philosophie und moderne Physik. Darmstadt: Wissenschaftliche Buchgesellschaft

Kant, I. (1755): *Allgemeine Naturgeschichte und Theorie des Himmels*, in Kant (1968), Band I, 226–396

Kant, I. (1781, [2]1787): *Kritik der reinen Vernunft*, in Kant (1968), Band III, IV

Kant, I. (1783): *Prolegomena zu einer jeden Künftigen Metaphysik, die als Wissenschaft wird auftreten können*, in Kant (1968), Band V, 112–264

Kant, I. (1786): *Metaphysische Anfangsgründe der Naturwissenschaft*, in Kant (1968), Band IX, 10–135

Kant, I. (1968): *Werkausgabe*, hrsg. von W. Weischedel. Frankfurt (a. M.): Suhrkamp

Kelvin, W.T./P. Tait (1912): *Treatise on Natural Philosophy*. Cambridge: Cambridge University Press

Kennard, E.H./D.E. Richmond (1922): 'On Reflection from a moving mirror and the Michelson-Morley Experiment', *The Physical Review* **XIX**, 572–7

Kennedy, R.J. (1922): 'Another Ether-Drift Experiment', *The Physical Review* **XX**, 26–33

Kennedy, R.J./E.M. Thorndike (1932): 'Experimental Evidence of the Relativity of Time', *Physical Review* **42**, 400–18

Kiefer, C. (1996): 'Decoherence in Quantum Field Theory', in D. Giulini *et al.* (1996), 137–56

Kiefer, C. (1996): 'Consistent Histories and Decoherence', in D. Giulini *et al.* (1996), 157–86

Kiefer, C. (2000): 'Decoherence in Situations Involving the Gravitational Field', in Ph. Blanchard *et al.* (2000), 101–112

Kirchhoff, G. (1859): 'The Fraunhofer Lines'/'Emission and Absorption', reprinted in W.F. Magie, *A Source Book in Physics*. New York: McGraw-Hill (1935), 354–60

Kirchhoff, G./R. Bunsen (1860): 'Chemische Analyse durch Spectralbeobachtungen', *Annalen der Physik und Chemie* **60**, 161–89

Kitcher, Ph./W.C. Salmon *eds.* (1989): *Scientific Explanation.* Minnesota Studies in the Philosophy of Science **XIII.** Minnesota: University of Minnesota Press

Koenig, E. (1888/1890): *Die Entwicklung des Kausalproblems.* Leipzig

Kopff, A. (1921): 'Das Rotationsproblem in der Relativitätstheorie', *Die Naturwissenschaften* **9**, 9–15

Koyré, A. (1957): *From the Closed World to the Open Universe.* Baltmore: John Hopkins University Press

Koyré, A. (1965): *Newtonian Studies.* Chicago: University of Chicago Press

Koyré, A. (1978): *Galileo Studies.* Hassocks (Sussex): The Harvester Press

Krajewski, W. (1977): *Correspondence Principle and Growth of Science.* Dordrecht: D. Reidel

Krane, K. (1983): *Modern Physics.* New York: John Wiley & Sons

Kries, J. v. (1920): 'Über die zwingende und eindeutige Bestimmtheit des physikalischen Weltbildes', *Die Naturwissenschaften* **8**, 237–47

Kries, J. v. (1924): 'Kants Lehre von Zeit und Raum in ihrer Beziehung zur modernen Physik', *Die Naturwissenschaften* **12**, 318–31

Kroes, P. (1983): 'The Clock Paradox, or how to get rid of Absolute Time', *Philosophy of Science* **50**, 159–63

Kroes, P. (1985): *Time* – Its Structure and Role in Physical Theories. Dordrecht: D. Reidel

Kuhn, T.S. ([2]1970): *The Structure of Scientific Revolutions.* Chicago: The University of Chicago Press

Kuhn, T.S. (1978): *Black Body Theory and the Quantum Discontinuity, 1894–1912.* Chicago/London: The University of Chicago Press

Kwiat, P.G. /A.M. Steinberg/R.Y. Chiao (1992): 'Observation of a "quantum eraser": A revival of coherence in a two-photon interference experiment', *Physical Review A* **45/11**, 7729–7739

Ladyman, J. (1998): 'What is Structural Realism?', *Studies in History and Philosophy of Science* **29**, 409–24

Landsberg, P.T. (1970): 'Time in Statistical Physics and Special Relativity', *Studium Generale* **23**, 1108–1158; reprinted in J.T. Fraser *et al.* (1972), 59–109

Landsberg, P.T. (1996): 'Irreversibility and Time's Arrow', *Dialectica* **50**, 247–58

Lang, A. (1904): *Das Kausalproblem.* Köln

Langevin, P. (1923): *La Physique Depuis Vingt Ans.* Paris: Octave Doin

Langevin, P. (1930): *La Science et le Déterminisme.* Paris

Laplace, P. (1774): 'La Probabilité des Causes Par Les Événements', *Œuvres Complètes de Laplace*, Volume VIII. Paris: Gauthier-Villars et Fils (1841), 27–62

Laplace, P. (1820): *Théorie analytique des probabilités.* Paris: Couvier

Laub, J. (1910): 'Über die experimentellen Grundlagen des Relativitätsprinzips', *Jahrbuch der Radioaktivität und Elektronik* **VII**, 405–63

Laue, M. v. ([2]1913): *Das Relativitätsprinzip.* Braunschweig: Vieweg

Laue, M. v. (1932): 'Zu den Erörterungen über Kausalität', *Die Naturwissenschaften* **20**, 915–16

Laue, M. v. (1934): 'Über Heisenbergs Ungenauigkeitsbeziehungen und ihre erkenntnistheoretische Bedeutung', *Die Naturwissenschaften* **22**, 439–41

Laue, M. v. (1960): 'Erkenntnistheorie und Relativitätstheorie', G. Harig/J. Schleifstein *eds.* (1960), 61–69

Layzer, D. (1967): 'The Strong Cosmological Principle, Indeterminacy and the Direction of Time', in T. Gold *ed.* (1967), 111–120

Layzer, D. (1975): 'The Arrow of Time', *Scientific American* **233**, 56–69

Le Poidevin, R. (1996): 'Time, Tense and Topology', *The Philosophical Quarterly* **46**, 467–81

Leibniz, G.W. (1973): *Philosophical Writings*, ed. by G. H. R. Parkinson. London: J.M. Dent & Sons

Lewis, D. (1973): 'Causation', *Journal of Philosophy* **70**, 556–67; reprinted in *Philosophical Papers* II (1986), 159–72

Lewis, D. (1986): 'Postscripts to Causation', *Philosophical Papers* II (1986), 172–231

Lewis, D. (1986): 'Causal Explanation', *Philosophical Papers* II (1986), 214–240

Lewis, D. (1986): *Philosophical Papers* II. Oxford: Oxford University Press

Lewis, G.N. (1908): 'A Revision of the Fundamental laws of Matter and Energy', *The Philosophical Magazine* **XVI**, 705–717

Lewis, G.N./R.C. Tolman (1909): 'The Principle of Relativity and Non-Newtonian Mechanics', *The Philosophical Magazine* **XVIII**, 510–23

Liebscher, D.-E. (1999): *Einsteins Relativitätstheorie und die Geometrien der Ebene*. Stuttgart/Leipzig: B.G. Teubner

Lloyd Morgan, C. (1924): 'Optical Records and Relativity', *Nature* **114**, 577–8, 681–2

Lodge, O. (1920): 'Popular Relativity and the Velocity of Light', *Nature* **106**, 325–6

Lodge, O. (1921): 'The Geometrisation of Physics, and its Supposed Basis on the Michelson-Morley Experiment', *Nature* **106**, 795–800

Lodge, O. (1921): 'Remarks on Simple Relativity and the Relative Velocity of Light I, II', *Nature* **107**, 716–9, 748–51

Lodge, O. (1921): 'Further Remarks on Relativity', *Nature* **107**, 784–5

Lodge, O. (1921): 'Remarks on Gravitational Relativity', *Nature* **107**, 814–8

Lodge, O. (1922): 'Relativity and the Æther', *Nature* **110**, 446

Lodge, O. (1924): 'A Philosopher on Relativity', *Nature* **114**, 318–21

Lorentz, H.A. (1895): 'Der Interferenzversuch Michelsons', in: *Versuch einer Theorie der elektrischen und optischen Erscheinungen in bewegten Körpern*, Leiden (1895), §§89–92; reprinted in H.A. Lorentz *et al.* (1974), 1–5

Lorentz, H.A. (1904): 'Electromagnetic phenomena in a system moving with any velocity smaller than that of light', *Proceedings Acad. Sc. Amsterdam* **6**; reprinted in H.A. Lorentz *et al.* (1974), 6–25

Lorentz, H.A. (1921): 'The Michelson-Morley Experiment and the Dimensions of Moving Bodies', *Nature* **106**, 793–5

Lorentz, H.A./A. Einstein/H. Minkowski ([7]1974): *Das Relativitätsprinzip*. Darmstadt: Wissenschaftliche Buchgesellschaft [English translation: *The Principle of Relativity*. New York: Dover 1952]

Lucas, J.R. (1973): *A Treatise on Time and Space*. London: Methuen

Macauley, W.H. (1910–11): 'Motion, Laws of', *Encyclopaedia Britannica* Volume **XVIII** (11th edition), 906–9

Mach, E. (1883/⁹1933): *Die Mechanik*. Darmstadt: Wissenschaftliche Buchgesellschaft (1976) [English Translation: *The Science of Mechanics*. Open Court Publishing 1915]

Mackie, J.L. (1980): *The Cement of the Universe*. Oxford: Clarendon Press

Majorana, Q. (1918): 'On the Second Postulate of the Theory of Relativity: An Experimental Demonstation of the Constancy of the Velocity of Light Reflected by a Moving Mirror', *The Physical Review* **XI**, 411–20

Mainzer, K. (1996): *Symmetries of Nature*. Berlin: Walter de Gruyter

Margenau, H. (1949): 'Einstein's Conception of Reality', in P.A. Schilpp *ed.* (1949), Volume I, 243–268

Margenau, H. (1950): *The Nature of Physical Reality*. New York: McGraw-Hill

Margenau, H. (1966): 'The Philosophical Legacy of Contemporary Quantum Theory', in R.G. Colodney *ed.* (1966), 330–56

Margenau, H. (1978): 'Probability and Causality in Quantum Physics', in H. Margenau (1978), 21–38

Margenau, H. (1978): 'Meaning and Scientific Status of Causality', in H. Margenau (1978), 39–51

Margenau, H. (1978): *Physics and Philosophy*: Selected Essays. Dordrecht: D. Reidel

Mathews, G.B. (1921): 'Non-Euclidean Geometries', *Nature* **106**, 790–1

Maudlin, T. (1993): 'Buckets of Water and Waves of Space: Why Spacetime is Probably a Substance', *Philosophy of Science* **60**, 183–203; reprinted in J. Butterfield *et al. eds.* (1996), 263–84

Maudlin, T. (1994/²2002) : *Quantum Non-Locality and Relativity*. Malden (Mass.)/ Oxford : Blackwell

Maupertius, P.L. Moreau de (1768): 'Système de la Nature', *Œuvres* II. Hildesheim: Georg Olms (1965), 138–184

Maxwell, J.C. (1873): 'On Faraday's Lines of Force', *The Scientific Letters and Papers of James Clerk Maxwell,* Volume II (1862–1873), edited by P. M. Harman. Cambridge: Cambridge University Press (1995), 790–811

Maxwell, J.C. (1873): 'Manuscript Fragment on Dynamical Principles', *The Scientific Letters and Papers of James Clerk Maxwell,* Volume II (1862–1873), edited by P.M. Harman. Cambridge: Cambridge University Press (1995), 811–12

Maxwell, J.C. (1877): *Matter and Motion*. New York: Dover

Maxwell, N. (1985): 'Are Probabilism and Special Theory Incompatible?' *Philosophy of Science* **52**, 23–43

Maxwell, N. (1988): 'Are Probabilism and Special Relativity Compatible?' *Philosophy of Science* **55**, 640–5

Maxwell, N. (1993): 'Aim-oriented Empiricism and Scientific Essentialism', *British Journal for the Philosophy of Science* **44**, 81–101

Maxwell, N. (1998): *The Comprehensibility of the World*. Oxford: Oxford University Press

Mayr, E. (2000): 'Darwin's Influence on Modern Thought', *Scientific American* (July 2000), 67–71

McCall, S. (1994): *A Model of the Universe*, Oxford: Clarendon Press
McCall, S. (1995): 'Time flow, non-locality, and measurement in quantum mechanics' in St. Savitt *ed.* (1995), 155–72
McClure, E. (1921): 'Relativity and Materialism', *Nature* **108**, 467
McMullin, E. (1985): 'Galilean Idealization', *Studies in History and Philosophy of Science* **16**, 247–73
McMullin, E. (1988): 'The Shaping of Scientific Rationality', in E. McMullin *ed.* *Construction and Constraint.* Notre Dame: University of Notre Dame Press 1988, 1–47
McMullin, E. (2000): 'The Origins of the Field Concept in Physics', *Die Naturwissenschaften* **4**, 13–39
McTaggart, J. (1908): 'The Unreality of Time', *Mind* **18**, 457–84
McTaggart, J. (1968): 'Time', in R. Gale *ed.* (1968), 86–97
Mehra, J. *ed.* (1973): *The Physicist's Conception of Nature.* Dordrecht/Boston: D. Reidel
Mellor, D.H. (1981): *Real Time.* Cambridge: Cambridge University Press
Mellor, D.H. (1995): *The Facts of Causation.* London: Routledge
Mermin, N.D. (1990): *Boojums all the way through*: communicating science in a prosaic age. Cambridge/New York: Cambridge University Press
Meyer, H.H. (1934): 'Kausalitätsfragen in der Biologie', *Die Naturwissenschaften* **22**, pp. 598–601
Michelson, A.A. (1903): *Light Waves and their Uses.* Chicago: Chicago University Press
Mill, J.S. (1843/1898): *A System of Logic.* London: Longmans, Green & Co.
Miller, R. (1987): *Fact and Method.* Princeton: Princeton University Press
Minkowski, H. (1909): 'Raum und Zeit', *Physikalische Zeitschrift* **10**, 104–11; reprinted in H.A. Lorentz *et al.* (^{7}1974), 54–71
Misner, Ch./K.S. Thorne/J.A. Wheeler (1973): *Gravitation.* New York: W.H. Freeman and Company
Mises, R. von (1930): 'Über kausale und statistische Gesetzmäßigkeit in der Physik', *Die Naturwissenschaften* **18**, 145–53; also in *Erkenntnis* **I** (1930), 189–210
Mises, R. von (1931): 'Über das naturwissenschaftliche Weltbild der Gegenwart', *Die Naturwissenschaften* **18**, 885–93
Mittelstaedt, P. (^{4}1972): *Philosophische Probleme der modernen Physik.* Mannheim: B.I.-Wissenschaftsverlag. [English Translation: *Philosophical Problems of Modern Physics,* Reidel 1976]
Mittelstaedt, P. (^{3}1989): *Der Zeitbegriff in der Physik.* Mannheim: B.I.-Wissenschaftsverlag
Mittelstaedt, P. (1998): *The Interpretation of Quantum Mechanics and the Measurement Process.* Cambridge: Cambridge University Press
Mittelstaedt, P. (2000): 'The Problem of Decoherence and the EPR Paradox', in Ph. Blanchard *et al.* (2000), 149–60
Monod, J. (1997/1974): 'On the molecular theory of evolution', reprinted in M. Ridley *ed.* (1997), 389–95
Morgan, C.L. (1924): 'Optical Records and Relativity', *Nature* **114**, 577–9

Morgan, M./M. Morrison *eds.* (1999): *Models as Mediators*. Cambridge: Cambridge University Press

Morrison, M. (1995): 'Symmetries as Meta-Laws', in F. Weinert *ed.* (1995), 157–88

Myatt, C.J. *et al.* (2000): 'Decoherence of quantum superpositions through coupling to engineered reservoirs', *Nature* **403** (20 January), 269–73

Nagel, E. (1961): *The Structure of Science*. London: Routledge & Kegal Paul

Nahin, P.J. (1993): *Time Machines*. New York: Springer

Nature (1913): Volume **92**: 'Review of Books on Relativity', 485, 577

Nature (1913): Volume **93**: 'Review of Books on Relativity', 28, 56–7, 187, 532–3

Nature (1914): Volume **94**: 'Review of Books on Relativity', 387

Nature (1914): Volume **95**: 'Review of Books on Relativity', 1, 612

Nature (1916): Volume **97**: 'Review of Books on Relativity', 30

Nature (1921): Volume **106**: 'Bibliography of Relativity', 811–3

Nature (1921): Volume **107**: 'Reviews of Books on Relativity', 422, 578–81

Nature (1921): Volume **109**: Reviews of Books on Relativity, 544–5,770–2

Nature (1921): Volume **110**: Reviews of Books on Relativity, 471–2, 568–70

Nature (1923): Volume **111**: Reviews of Books on Relativity, 697–9

Nehrlich, G. (1982): 'Special Relativity is not Based on Causality', *British Journal for the Philosophy of Science* **33**, 361–88

Nernst, W. (1922): 'Zum Gültigkeitsbereich der Naturgesetze', *Die Naturwissenschaften* **10**, 489–93

Newman, M.H.A. (1928): 'Mr Russell's "Causal Theory of Perception"', *Mind* **37**, 137–48

Newton, I. (1687): *Mathematical Principles of Natural Philosophy* and *System of the World*. F. Cajori edition. Berkeley (California): University of California Press 1960

Newton, I. (1687): *The Principia. Mathematical Principles of Natural Philosophy*. A New Translation by I.B. Cohen and Anne Whitman. Berkeley/Los Angeles/London: University of California Press 1999

Newton, I. (1728): *A Treatise of the System of the World*, in I. Newton (1687), Cajori edition

Newton, I. (1706/^{3}1721, ^{4}1730): *Opticks*. London

Newton-Smith, W.H. (1980): *The Structure of Time*. London/Boston: Routledge & Kegan Paul

Newton-Smith, W.H. (1988): 'Space, Time and Space-Time: A Philosopher's View', in R. Flood/M. Lockwood *eds.* (1988), 22–35

Nielsen, M.A. (2002): 'Rules for a Complex Quantum World', *Scientific American* (November 2002), 67–75

Nor, K.B.M. (1992): 'A Topological Explanation for 3 Properties of Time', *Il Nuovo Cimento* **107B**, 65–70

North, J.D. (1970): 'The Time Coordinate in Einstein's Restricted Theory of Relativity', *Studium Generale* **23**, 203–23; reprinted in J.T. Fraser *et al.* (1972), 12–32

Norton, J.D. (1980): 'Science and Analogy', in M.D. Grmek/R.S. Cohen/G. Cimino *eds.*, *On Scientific Discovery*. Dordrecht: D. Reidel 1980, 115–40

Norton, J.D. (1989): 'The Hole Argument', in A. Fine/J. Leplin *eds.* (1989): *Proceedings of the 1988 Biennial Meeting of the Philosophy of Science Association, Volume Two*, East Lansing, Michigan: Philosophy of Science Association, 56–64; reprinted in J. Butterfield *et al.*, *eds.* (1996), 285–294

Norton, J.D. (1992): 'Philosophy of Space and Time', in M. Salmon *ed.* (1992): *Introduction to the Philosophy of Science*. New Jersey: Prentice Hall, 179–232; reprinted in J. Butterfield *et al., eds.* (1996), 3–57

Norton, J.D. (1996): 'Are Thought Experiments Just What You Thought?', *Canadian Journal of Philosophy* **26**, 333–66

Nowak, L. (1980): *The Structure of Idealization*. Dordrecht: D. Reidel

Nye, M.J. (1972): *Molecular Reality*. New York: Elsevier

Omnès, R. (1999): *Understanding Quantum Mechanics*. Princeton (N. J.): Princeton University Press

Omnès, R. (2000): 'Decoherence as an Irreversible Process', in Ph. Blanchard *et al.* (2000), 291–98

Ornstein, R.E. (1969): *On the experience of time*. Harmondsworth (Middlesex): Penguin Books

Pagels, H. (1984): *The Cosmic Code*. Pelican Books

Pais, A. (1982): *Subtle is the Lord*. Oxford: Oxford University Press

Pais, A. (1986): *Inward Bound*. Oxford: Clarendon Press

Pais, A. (1991): *Niels Bohr's Times*. Oxford: Clarendon Press

Papillon, J.H.Fr. (1876): *Histoire de la philosophie moderne dans ses rapports avec le développement des sciences de la nature* 2 vol., Paris: Lévèque

Partington, J.R. (1920): 'Relativity', *Nature* **106**, 113–4

Pauli, W. (1921): 'Relativitätstheorie', in: *Encyklopädie der mathematischen Wissenschaften*, Volume **19**, [English Translation: *Theory of Relativity*. New York: Dover 1981]

Pauli, W. (1936): 'Raum, Zeit und Kausalität in der Modernen Physik', *Scientia* **59**, 65–76, reprinted in W. Pauli (1964), 737–48

Pauli, W. (1952): 'Der Begriff der Wahrscheinlichkeit und seine Rolle in den Naturwissenschaften', *Verhandl. Schweiz. Naturforsch. Ges.* 76–9, reprinted in W. Pauli (1964), 1196–98

Pauli, W. (1954): 'Wahrscheinlichkeit und Physik', *Dialectica* **8**, 112–24; reprinted in W. Pauli (1964), 1199–1211

Pauli, W. (1957): 'Phänomen und Physikalische Realität', *Dialectica* **11**, 36–48; reprinted in W. Pauli (1964), 1350–1361

Pauli, W. (1961): *Aufsätze und Vorträge über Physik und Erkenntnistheorie*. Braunschweig: Vieweg

Pauli, W. (1964): *Collected Scientific Papers*, Volume II. Ed. by K. Kronig/V.F.Weisskopf. New York: John Wiley

Pearson, K. (1892/^{3}1911): *The Grammar of Science*. London: W. Scott

Penrose, R. (1988): 'Big Bangs, Black Holes and Time's Arrow', in R. Flood/M. Lockwood *eds.* (1988), 36–62

Penrose, R. (1990): *The Emperor's New Mind*. Oxford/New York: Vintage

Perrin, J. (1909): 'Mouvement Brownien et Réalité Moléculaire', *Annales de Chimie et de Physique* **XVIII**, 5–114
Perrin, J. (1913/⁴1914): *Les Atomes*. Paris: Felix Alcan
Perrin, J. (1916) : *Atoms*. London: Constable & Company
Physics World (1998): Special Issue – Quantum Information (March 1998)
Piaggio, H.T.H. (1922): 'Summary of the Theory of Relativity', *Nature* **110**, 432–4
Piaggio, H.T.H. (1922): 'Space-Time Geodesics', *Nature* **110**, 699
Pfeiffer, A. (1960): 'Zwei Auffassungen des Kausalitätsbegriffs', in G. Harig/J. Schleifstein *eds.* (1960), 183–93
Planck, M. (1900): 'Über eine Verbesserung der Wien'schen Spectralgleichung', in Planck (1972), 3–5
Planck, M. (1900): 'Zur Theorie des Gesetzes der Energieverteilung im Normalspectrum', in Planck (1972), 6–14
Planck, M. (1908): 'Die Einheit des physikalischen Weltbildes', in Planck (1975), 28–51
Planck, M. (1910): 'Die Stellung der neueren Physik zur mechanischen Naturanschauung', in Planck (1975), 52–68
Planck, M. (1913/⁴1921): *Das Prinzip der Erhaltung der Energie*. Berlin
Planck, M. (1913): 'Neue Bahnen der physikalischen Erkenntnis', in Planck (1975), 69–80 [English translation 'New Paths of Physical Knowledge', *Philosophical Magazine* **XXVIII** (1914), 60–70]
Planck, M. (1914): 'Dynamische und Statistische Gesetzmäßigkeit', in Planck (1975), 81–94
Planck, M. (1920): 'Die Entstehung und bisherige Entwicklung der Quantentheorie', in Planck (1975), 125–138
Planck, M. (1923): 'Kausalgesetz und Willensfreiheit', in Planck (1975), 139–168
Planck, M. (1924): 'Vom Relativen zum Absoluten', in Planck (1975), 169–82
Planck, M. (1926): 'Physikalische Gesetzlichkeit', in Planck (1975), 183–205
Planck, M. (1927): 'Die physikalische Realität der Lichtquanten', *Die Naturwissenschaften* **15**, 529–31
Planck, M. (1929): 'Das Weltbild der neuen Physik', in Planck (1975), 206–227
Planck, M. (1932): 'Die Kausalität in der Natur', in Planck (1975), 250–269
Planck, M. (1933): 'Ursprung und Auswirkung wissenschaftlicher Ideen', in Planck (1975), 270–284
Planck, M. (1933): 'Zur Geschichte der Auffindung des physikalischen Wirkungsquantums', in Planck (1975), 15–27
Planck, M. (1935): 'Die Physik im Kampf um die Weltanschauung', in Planck (1975), 285–300
Planck, M. (1937): 'Determinismus und Indeterminismus', in Planck (1975), 334–349
Planck, M. (1959): *The Philosophy of Physics*. New York: Meridian Books
Planck, M. (1972): *Planck's Original Papers in Quantum Physics*, German and English edition, annotated by H. Kangro. London: Taylor & Francis
Planck, M. (1975): *Vorträge und Erinnerungen*. Darmstadt: Wissenschaftliche Buchgesellschaft

Plessner, H. (1930): 'Das Problem der Natur in der gegenwärtigen Philosophie', *Die Naturwissenschaften* **18**, 869–75

Poincaré, H. (1905): *La Valeur de la Science* [English translation *The Value of Science*, Dover 1958]

Polzik, E. (2002): 'Atomic entanglement on a grand scale', *Physics World* **15** (September 2002), 33–7

Pooley, O./H.R. Brown (2002): 'Relationism Rehabilitated? I: Classical Mechanics', *British Journal for the Philosophy of Science* **53**, 183–204

Popper, K.R. (1950): 'Indeterminism in Quantum Physics and Classical Physics', *British Journal for the Philosophy of Science* **1**, 617–33, 673–95

Popper, K.R. (1956): 'The Arrow of Time', *Nature* **177**, 538

Popper, K.R. (1965): 'Time's Arrow and Entropy', *Nature* **207**, 233–4

Popper, K.R. (1982): *Quantum Theory and the Schism in Physics*. London/New York: Routledge

Popper, K.R. (1982): *The Open Universe:* An Argument for Indeterminism. London/New York: Routledge

Popper, K.R. (1990): 'Two New Views of Causality', in K.R. Popper, *A World of Propensities* (1990). Bristol: Thoemmes, 3–26

Price, H. (1996): *Time's Arrow and Archimedes' Point*. Oxford: Oxford University Press

Prigogine, I. (1973): 'Time, Irreversibility and Structure', in J. Mehra *ed.* (1973), 561–93

Prigogine, I. (1997): *The End of Certainty*. New York/London/Toronto/Sydney: The Free Press

Prigogine, I./I. Stengers (1979): *La Nouvelle Alliance*. Métamorphose de la Science. Paris: Gallimard [English translation: *Order out of Chaos*, 1984]

Putnam, H. (1967): 'Time and Physical Geometry', *Journal of Philosophy* **64**, 240–7; reprinted in *Philosophical Papers* I, Cambridge: Cambridge University Press 1975, 198–205

Rae, A. (^{2}1986): *Quantum Physics*. Bristol/New York: Adam Hilger

Rae, A. (1986): *Quantum Physics:* Illusion or Reality? Cambridge: Cambridge University Press

Rankine, W.J.M. (1880): *Miscellaneous Scientific Papers*. London

Rashevsky, N. (1921): 'Light Emission from a Moving Source in Connection with the Relativity Theory', *The Physical Review* **XVIII**, 369–76

Redhead, M. (1980): 'Models in Physics', *British Journal for the Philosophy of Science* **31**, 145–63

Redhead, M. (1987): *Incompleteness, Nonlocality, and Realism*. Oxford: Clarendon Press

Redhead, M. (1989): 'Nonfactorizability, Stochastic Causality, and Passion-at-a-Distance', in J. Cushing/E. McMullin *eds.* (1989), 145–153

Redhead, M. (1989): 'The Nature of Reality', *British Journal for the Philosophy of Science* **40**, 429–41

Redhead, M. (1995): *From Physics to Metaphysics*. Cambridge/New York: Cambridge University Press

Rees, M. (2000): *Just Six Numbers*. London: Phoenix
Regt, H.W. de (1996): 'Philosophy and the Kinetic Theory of Gases', *British Journal for the Philosophy of Science* **47**, 31–62
Reichenbach, H. (1920): 'Philosophische Kritik der Wahrscheinlichkeitsrechnung', *Die Naturwissenschaften* **8**, 146–53
Reichenbach, H. (1931): 'Das Kausalproblem in der Physik', *Die Naturwissenschaften* **19**, 713–22
Reichenbach, H. (1949): 'The Philosophical Significance of the Theory of Relativity', in P.A. Schilpp *ed*., Volume I. (1949), 287–312
Reichenbach, H. (1958): *The Philosophy of Space and Time*. New York: Dover [Translation of *Philosophie der Raum-Zeit-Lehre*, 1928]
Rey, A. (1907, ^{2}1923): *La théorie de la physique chez les physiciens contemporains*. Paris: Alcan
Ridley, M. *ed*. (1997): *Evolution*. Oxford: Oxford University Press
Riebesell, P. (1916): 'Die Beweise für die Relativitätstheorie', *Die Naturwissenschaften* **4**, 97–101
Rietdijk, C.W. (1966): 'A Rigorous Proof of Determinism Derived from the Special Theory of Relativity', *Philosophy of Science* **33**, 341–4
Rietdijk, C.W. (1976): 'Special Relativity and Determinism', *Philosophy of Science* **43**, 598–609
Riezler, K. (1928): 'Die Krise der "Wirklichkeit"', *Die Naturwissenschaften* **16**, 705–12
Robb, A.A. (1914): *A Theory of Time and Space*. Cambridge: Cambridge University Press
Robb, A.A. (1914): 'The Principle of Relativity', *Nature* **93**, 454
Robb, A.A. (1922): 'Relativity and Physical Reality', *Nature* **110**, 572
Roman, P. (1969): 'Symmetry in Physics', in Boston Studies in the Philosophy of Science **V**, ed. by R.S. Cohen/M. W. Wartofsky. Dordrecht: Reidel, 363–69
Rosen, J. (1995): *Symmetry in Science*. New York, Berlin, Heidelberg: Springer-Verlag
Rosenfeld, L. (1942): 'The Evolution of the Idea of Causality', reprinted in L. Rosenfeld (1979), 446–64
Rosenfeld, L. (1967): 'Observation and the Direction of Time', in T. Gold *ed*. (1967), 187–95
Rosenfeld, L. (1971): 'Unphilosophical Considerations on Causality in Physics', reprinted in L. Rosenfeld (1979), 666–80
Rosenfeld, L. (1972/1974): 'Statistical Causality in Atomic Theory', reprinted in L. Rosenfeld (1979), 666–80
Rosenfeld, L. (1979): *Selected Papers of Leon Rosenfeld*, *ed*. R.S. Cohen/J.J. Stachel, Boston Studies in the Philosophy of Science **21**. Dordrecht: Reidel
Rovelli, C. (1999): 'Quantum spacetime: What do we know?' in C. Callender/N. Hugget *eds*. (1999), 101–122
Russell, B. (1912): 'On the notion of cause', *Proceedings of the Aristotelian Society* **XIII** (1912–13), 1–26; reprinted in Russell (1965), 163–186
Russell, B. (1927): *The Analysis of Matter*. London: George Allen & Unwin
Russell, B. (1925/^{3}1969): *The ABC of Relativity*. London: George Allen & Unwin

Russell, B. (1948): 'Space-Time and Causality', in *Human Knowledge*, 1948, Chapter X, reprinted in Russell (1965), 122–136
Russell, B. (1965): On *the Philosophy of Science*, edited, with an Introduction, by Charles A. Fritz. Indianapolis/New York/Kansas City: The Bobbs-Merrill Company
Rutherford, E. (1911): 'The Scattering of α and β Particles by Matter and the Structure of the Atom', *Philosophical Magazine* **XXI**, 669–88
Rutherford, E./F. Soddy (1902): 'The Cause and Nature of Radioactivity I, II', *Philosophical Magazine* **IV**, 370–96, 569–85
Rynasiewicz, R. (1996): 'Absolute versus Relational Space-Time: An Outmoded Debate?', *The Journal of Philosophy* **93**, 279–306
Rynasiewicz, R. (2000): 'On the Distinction between Absolute and Relative Motion', *Philosophy of Science* **67**, 70–93
Saint Augustin (1961): *Confessions*. Harmondsworth: Penguin
Salecker, H./E.P. Wigner (1958): 'Quantum Limitations of the Measurement of Space-Time Distances', *Physical Review* **109**, 571–7
Salmon, W.C. (1984): *Scientific Explanation and the Causal Structure of the World*. Princeton: Princeton University Press
Salmon, W.C. (1998): *Causality and Explanation*. Oxford: Oxford University Press
Samson, R.A. (1920): 'Relativity and Reality', *Nature* **105**, 708
Savitt, St. F. (1994): 'The Replacement of Time', *Australasian Journal of Philosophy* **72**, 463–74
Savitt, St. F. (1996): 'The Direction of Time', *British Journal for the Philosophy of Science* **47**, 347–70
Savitt, St. F. *ed.* (1995): *Time's Arrow Today*. Cambridge: Cambridge University Press
Schilpp, P.A. *ed.* (1949): *Albert Einstein: Philosopher-Scientist*, 2 Volumes. La Salle (Ill.): Open Court
Schjelderup, K. (1923): 'The Theory of Relativity and its Bearing upon Epistemology', *Scandinavian Scientific Review* **I**, 14–65
Schlegel, R. (1966): 'Time and Thermodynamics', in J.T. Fraser *ed.* (1966), 500–26
Schlesinger, G.N. (1980): *Aspects of Time*. Indianapolis: Hackett
Schlick, M. (1917): 'Raum und Zeit in der gegenwärtigen Physik', *Die Naturwissenschaft* **5**, 162–67, 177–86
Schlick, M. (1920): 'Naturphilosophische Betrachtungen über das Kausalprinzip', *Die Naturwissenschaften* **8**, 461–74
Schlick, M. (1931): 'Die Kausalität in der gegenwärtigen Physik', *Die Naturwissenschaften* **19**, 145–62
Schlick, M. ([3]1920): *Space and Time in Contemporary Physics*, transl. Henry L. Brose. Oxford: Oxford University Press/New York: Dover 1963
Schottky, W. (1921): 'Das Kausalproblem der Quantentheorie als eine Grundlage der modernen Naturforschung', *Die Naturwissenschaften* **9**, 492–496, 506–511
Schottky, W. (1922): 'Zur Krisis des Kausalitätsbegriffes', *Die Naturwissenschaften* **10**, 982
Schrödinger, E. (1928): 'Conceptual Models in Physics and their Philosophical Value', in E. Schrödinger (1957), 148–65

Schrödinger, E. (1929): 'Die Erfassung der Quantengesetze durch kontinuierliche Funktionen', *Die Naturwissenschaften* **17**, 486–9
Schrödinger, E. (1929): 'Was ist ein Naturgesetz?', *Die Naturwissenschaften* **17**, reprinted in E. Schrödinger (1987), 9–17
Schrödinger, E. (1930): 'Die Wandlung des physikalischen Weltbildes', in E. Schrödinger (1987), 18–26
Schrödinger, E. (1931): 'Indeterminism in Physics', in E. Schrödinger (1957), pp. 52–80
Schrödinger, E. (1935): 'Die gegenwärtige Situation in der Quantenmechanik', *Die Naturwissenschaften* **23**, 807–49
Schrödinger, E. (1947): 'Die Besonderheit des Weltbilds der Naturwissenschaft', in E. Schrödinger (1987), 27–85
Schrödinger, E. (1950): 'Was ist ein Elementarteilchen?' in E. Schrödinger (1987), 121–143
Schrödinger, E. (1952): 'Unsere Vorstellung von der Materie', in E. Schrödinger (1987), 102–120
Schrödinger, E. (1957): *Science Theory and Man*. London: George Allen and Unwin
Schrödinger, E. (1967): *What Is Life? & Mind and Matter*. Cambridge: Cambridge University Press
Schrödinger, E. (^{4}1987): *Was ist ein Naturgesetz?* München: Oldenbourg
Sciama, D. (1988): 'Time "Paradoxes" in Relativity', in R. Flood/M. Lockwood *eds.* (1988), 6–21
Science (2002): Manipulating Coherence, Vol. **298** (15 November 2002), 1353–77
Scientific American (2002): A Matter of Time. Special Issue (September 2002)
Scientific American (2002): The Once and Future Cosmos (Special Edition)
Scully,O./B.-G./Englert/H. Walther (1991): 'Quantum optical tests of complementarity', *Nature* **351**, 111–16
Seager, W. (1996): 'A Note on the "Quantum Eraser"', *Philosophy of Science* **63**, 81–9
Seddon, K. (1987): *Time* – A Philosophical Treatment. London/New York/Sidney: Croom Helm
Sellers, D. (2001): *The Transit of Venus*. Leeds: MagaVelda Press
Sellers, W. (1962): 'Time and the World Order', in *Scientific Explanation, Space and Time,* ed. H. Feigl/G. Maxwell. Minneapolis: University of Minneapolis Press 1962, 527–616
Sexl, R./H.K. Schmidt (1978): *Raum-Zeit-Relativität*. Reinbek: Rowohlt Taschenbuch
Segré, E. (1980): *From X-rays to Quarks*. San Francisco: W.H. Freeman
Shallis, M. (1988): 'Time and Cosmology', in R. Flood/M. Lockwood *eds.* (1988), 63–79
Shapin, St./S. Schaffer (1985): *Leviathan and the Air Pump*. Princeton: Princeton University Press
Shimony, A. (1988): 'The Reality of the Quantum World', *Scientific American* **258**, 36–43
Shimony, A. (1998): 'Implications of Transience for Spacetime Structure', in S.A. Hugget *et al.* (1998), *The Geometric Universe*. Oxford/New York: Oxford University Press, 161–172

Silberstein, L. (1918): 'General Relativity without the Equivalence Hypothesis', *The Philosophical Magazine* **XXXVI**, 94–128

Silberstein, L. (1920): 'On the Measurement of Time', *The Philosophical Magazine* **XXXIX**, 366–72

Silberstein, L. (1922): *The Theory of General Relativity and Gravitation.* Toronto: University of Toronto Press

Silberstein, L. (1933): *Causality: a Law of Nature or a Maxim of the Naturalist?* London

Simon, H.A. (1977): 'On the Definition of the Causal Relation', in *Models of Discovery.* Dordrecht: Reidel, pp. 81–92

Simon, H.A. (1977): 'Spurious Correlation: A Causal Interpreation', in *Models of Discovery.* Dordrecht: Reidel, pp. 81–92

Simpson, G.G. (1997/1961): 'One hundred years without Darwin are enough', reprinted in M. Ridley *ed.* (1997), 369–78

Sklar, L. (1974): *Space, Time and Spacetime.* Berkeley: University of California Press

Sklar, L. (1981): 'Time, Reality and Relativity', in R. Healey *ed.* (1981), 129–142

Sklar, L. (1992): *Philosophy of Physics.* Oxford: Oxford University Press

Sklar, L. (1993): *Physics and Chance.* Cambridge: Cambridge University Press

Sklar, L. (1995): 'Time in experience and in theoretical description of the world', in: St.F. Savitt *ed.* (1995), 217–229

Sklar, L. (2000): *Theory and Truth.* Oxford: Oxford University Press

Skyrms, B. (1984): 'EPR: Lessons for Metaphysics', *Midwest Studies in Philosophy* **IX**, 245–5

Smart, J.J. (1949): 'The River of Time', *Mind* **58**, 483–94

Smart, J.J. (1955): 'Spatialising Time', *Mind* **64**, 239–41; reprinted in R. Gale *ed.* (1968), 163–7

Smart, J.J. (1963): 'The Space-Time World', in *Philosophy and Scientific Realism.* London: Routledge & Kegan Paul (1963), 131–48

Smart, J.J. (1967): 'The Unity of Space-Time: Mathematics versus Myth-Making', *Australasian Journal of Philosophy* **25**, 214–17

Smart, J.J. (1980): 'Time and Becoming', in P. van Inwagen *ed.*, *Time and Change.* Dordrecht: D. Reidel 1980, 3–15

Smolin, L. (2001): *Three Roads to Quantum Gravity.* New York: Basic Books

Sommerfeld, A. (1910): 'Zur Relativitätstheorie', *Annalen der Physik* **32**, 749

Sommerfeld, A. (1924): 'Grundlagen der Quantentheorie und des Bohrschen Atommodells', *Die Naturwissenschaften* **12**, 1047–49

Sommerfeld, A. (1927): 'Zum gegenwärtigen Stande der Atomphysik', *Physikalische Zeitschrift* **28**, 231–9

Sommerfeld, A. (1929): 'Einige grundsätzliche Bemerkungen zur Wellenmechanik', *Physikalische Zeitschrift* **30**, 866–71

Spielberg, N./B.D. Anderson ([2]1995): *Seven Ideas That Shook The Universe.* New York: John Wiley & Sons

Stamp, Ph. (1995): 'Time, decoherence, and "reversible" measurements', in St. Savitt *ed.* (1995), 107–54

Stark, J. (1930): 'Die Axialität der Lichtemission und Atomstruktur', *Annalen der Physik* **4**, 710–24

Stebbing, L.S. (1937): *Philosophy and the Physicists*. London: Methuen & Co.

Stein, H. (1968): 'On Einstein-Minkowski Space-Time', *Journal of Philosophy* **65**, 5–23

Stein, H. (1970): 'On the paradoxical time-structures of Gödel', *Philosophy of Science* **37**, 589–601

Stein, H. (1970): 'Newtonian Space-Time', in R. Patter *ed.* (1970): *The Annus Mirabilis of Sir Issac Newton* 1666–1966. Cambridge (Mass.): MIT Press, 238–84; reprinted in J. Butterfield *et al.* (1996), 79–106

Stein, H. (1991): 'On Relativity Theory and the Openness of the Future', *Philosophy of Science* **58**, 147–67; reprinted in J. Butterfield *et al.* (1996), 239–262

Stewart, O.M. (1911): 'The Second Postulate of Relativity and the Electromagnetic Emission Theory of Light', *Physical Review* **XXXII**, 418–28

Strauss, M. (1936): 'Komplementartät und Kausalität im Lichte der logischen Syntax', *Erkenntnis* **6**, 335–338

Strauss, M. (1960): 'Quantentheorie und Philosophie', in G. Harig/J. Schleifstein *eds.* (1960), 167–76

Strong, E.W. (1957): 'Newton's "Mathematical Way"', in P.P. Wiener/A. Noland *eds.*, *Roots in Scientific Thought*. New York: Basic Books, 412–32

Tegmark, M./J.A. Wheeler (2001): '100 Years of Quantum Mysteries', *Scientific American* **284** (February 2001), 54–61

Teichman, R. (1993): 'Time and Change', *The Philosophical Quarterly* **43**, 158–77

Tendeloo, N.Ph. (1913): 'Die Bestimmung von Ursache und Bedingungen: Ihre Bedeutung besonders für die Biologie', *Die Naturwissenschaften* **1**, 153–56

Terhal, B./M.M. Wolff/A.C. Doherty (2003): 'Quantum Entanglement: A Modern Perspective', *Physics Today* (April 2003), 46–52

Terrell, J. (1959): 'Invisibility of the Lorentz Contraction', *Physical Review* **116**, 1041–45

Thirring, H. (1921): 'Über das Uhrenparadoxon in der Relativitätstheorie', *Die Naturwissenschaften* **9**, 209–212

Thirring, H. (1925): 'Relativität und Aberration', *Die Naturwissenschaften* **13**, 445–7

Thirring, H. (1926): 'Neuere experimentelle Ergebnisse zur Relativitätstheorie', *Die Naturwissenschaften* **14**, 111–6

Thorne, K.S. (1995): *Black Holes and Time Warps*. London: Papermac

Tipler, F.J. (1974): 'Rotating Cylinders and the Possibility of Global Causality Violations', *Physical Review* **D9**, 2203–6

Tipler, F.J. (1976): 'Causality Violation in Asymptotically Flat Space-Times', *Physical Review Letters* **37**, 879–82

Tipler, F.J. (1977): 'Singularities and Causality Violation', *Annals of Physics* **108**, 1–36

Tipler, F.J. (1980): 'General Relativity and the Eternal Return', in F.J. Tipler *ed.*, *Essays in General Relativity*. New York: Academic Press, 21–37

Tipler, P.A. (^{2}1982): *Physics*. New York: Worth Publishers, Inc.

Tolman, R.C. (1910): 'The Second Postulate of Relativity', *Physical Review* **XXXI**, 26–40

Tolman, R.C. (1912): 'Non-Newtonian Mechanics - The Mass of a Moving Body', *The Philosophical Magazine* **XXIII**, 375–80

Tonnelat, M.A. (1971): *Histoire du Principe de Relativité*. Paris: Flammarion

Torretti, R. (1999): *The Philosophy of Physics*. Cambridge: Cambridge University Press

Toulmin, St. (1959): 'Criticism in the History of Science: Newton on Absolute Space, Time and Motion I, II', *The Philosophical Review* **68** (1959), 1–29, 203–227

Toulmin, St./J. Goodfield (1965): *The Discovery of Time*. London: Hutchinson

Treisman, M. (1999): 'The Perception of Time', in J. Butterfield *ed.* (1999), 217–46

Tyndall, J. (1868): *Faraday as a Discoverer*. London: Longmans, Green

Unruh, W. (1995): 'Time, gravity and quantum mechanics', in St. Savitt *ed.* (1995), 23–65

Van Fraassen, B. (1970): *Introduction to the Philosophy of Time*. Columbia University Press

Van Fraassen, B. (1989): *Laws and Symmetry*. Oxford: Clarendon Press

Van Fraassen, B. (1991): *Quantum Mechanics*. Oxford: Clarendon Press

Van der Waerden, B.L. *ed.* (1968): *Sources of Quantum Mechanics*. New York: Dover

Vignier, J.P./C. Dewdney/P.R. Holland/A. Kyprianidis (1987): 'Causal particle trajectories and the interpretation of quantum mechanics', in: B.J. Hiley/F.D. Peat *eds.* (1987), *Quantum Implications*. London: Routledge, 169–204

Warren, R.M./R.P. Warren (1968): *Helmholtz on Perception*. New York/London: John Wiley

Watanabe, S. (1966): 'Time and the Probabilistic View of the World', in J.T. Fraser *ed.* (1966), 527–62

Watanabe, S. (1970): 'Creative Time', *Studium Generale* **23**, 1057–87; reprinted in J.T. Fraser *et al.* (1972), 159–89

Webster, Ch. (1975): *The Great Instauration*. London: Duckworth

Weinberg, St. (1993): *The Discovery of Subatomic Particles*. Harmondsworth: Penguin Books

Weinberg, St. (1993): *Dreams of a Final Theory*. London: Vintage

Weinert, F. (1982): 'Tradition and Argument', *The Monist* **65**, 88–105

Weinert, F. (1984): 'Contra Res Sempiternas', *The Monist* **67**, 374–94

Weinert, F. (1992): 'Vicissitudes of Laboratory Life', *British Journal for the Philosophy of Science* **43**, 423–9

Weinert, F. (1993): 'Laws of Nature: A Structural Approach', *Philosophia Naturalis* **30**, 147–71

Weinert, F. (1994): 'The Correspondence Principle and the Closure of Theories: Two Incompatible Aspects of Heisenberg's Philosophy of Science', *Erkenntnis* **40**, 303–23

Weinert, F. (1995): 'Laws of Nature – Laws of Science', in: F. Weinert *ed.* (1995), 3–64

Weinert, F. *ed.* (1995): *Laws of Nature* – Essays on the Philosophical, Scientific and Historical Dimensions. Berlin: de Gruyter

Weinert, F. (1995): 'Wrong Theory – Right Experiment: The Significance of the Stern-Gerlach Experiments,' *Studies in the History and Philosophy of Modern Physics* **26**, 75–86

Weinert, F. (1995): 'The Duhem Quine Problem Revisited,' *International Studies in the Philosophy of Science* **9**, 147–156

Weinert, F. (1998): 'Fundamental Physical Constants, Null Experiments and the Duhem-Quine Thesis,' *Philosophia Naturalis* **35**, 225–52

Weinert, F. (1999): 'Theories, Models and Constraints.' *Studies in History and Philosophy of Science* **30**, 303–333

Weinert, F. (1999): 'Habermas, Science and Modernity', in A. O'Hear *ed. German Philosophy Since Kant.* Royal Institute of Philosophy Supplement **44**. Cambridge: Cambridge University Press (1999), 329–55

Weinert, F. (2000): 'The Construction of Atom Models: Eliminative Inductivism and its Relation to Falsificationism', *Foundations of Science* **5**, 491–531

Weingard, R. (1972): 'Relativity and the Reality of Past and Future Events', *British Journal for the Philosophy of Science* **23**, 119–21

Weingard, R. (1977): 'Space-Time and the Direction of Time', *Nous* **11**, 119–31

Weingard, R. (1979): 'General Relativity and the Length of the Past', *British Journal for the Philosophy of Science* **30**, 170–2

Weingard, R. (1979): 'Some Philosophical Aspects of Black Holes', *Synthese* **42**, 191–219

Weingard, R. (1999): 'A philosopher looks at string theory', in C. Callender/N. Hugget *eds.* (1999), 138–151

Weinstein, M.B. (1913): *Die Physik der bewegten Materie und die Relativitätstheorie.* Leipzig

Weisskopf, V. (1960): 'The Visual Appearance of Rapidly Moving Objects', *Physics Today* **13**, 24–7

Weizsäcker, V.F. von (1917): 'Empirie und Philosophie', *Die Naturwissenschaften* **5**, 669–73

Weizsäcker, C.F. von ([7]1970): *Die Geschichte der Natur.* Göttingen: Vandenhoeck & Ruprecht

Weizsäcker, C.F. von (1973): 'Classical and Quantum Descriptions', in J. Mehra *ed.* (1973), 635–67

Weizsäcker, C.F. von (1973): 'Physics and Philosophy', in J. Mehra *ed.* (1973), 736–46

Weizsäcker, C.F. von (1976): *Zum Weltbild der Physik.* Stuttgart: Hirzel

Weizsäcker, C.F. von ([5]1979): *Die Einheit der Natur.* München: Hanser

Wendorff, R. ([3]1985): *Zeit und Kultur.* Opladen: Westdeutscher Verlag

Wentscher, E. (1921): *Geschichte des Kausalproblems in der neueren Philosophie.* Leipzig: F. Meiner

Weyl, H. (1918): 'Gravitation und Elektrizität', Sitzungsberichte der Preußischen Akademie der Wissenschaften; reprinted in H.A. Lorentz *et al.* (1974), 147–59

Weyl, H. (1921): 'Electricity and Gravitation', *Nature* **106**, 800–2

Weyl, H. ([4]1921): *Raum Zeit Materie* [English translation *Space, Time, Matter.* New York: Dover 1952]

Weyl, H. (1927): 'Quantenmechanik und Gruppentheorie', *Zeitschrift für Physik* **46** (1927), 1–46

Weyl, H. (1931): 'Geometrie und Physik', *Die Naturwissenschaften* **19**, 49–58

Weyl, H. (1976): 'The Open World', in M. Čapek *ed.* (1976), 561–6

Wheeler, J.A. (1967): 'Three-Dimensional Geometry as a Carrier of Information about Time', in T. Gold *ed.* (1967), 90–110

Whitehead, A.N. (1920/1964): *The Concept of Nature*. Cambridge: Cambridge University Press

Whitehead, A.N. (1922): *The Principle of Relativity with Applications to Physical Science*. Cambridge: Cambridge University Press

Whitrow, G.J. (1966): 'Time and the Universe', in J.T. Fraser *ed.* (1966), 563–81

Whitrow, G.J. (1967): 'Reflections on the Natural Philosophy of Time', in *Interdisciplinary Perspectives of Time*, ed. R. Fischer, Volume **138**, Art. 2

Whitrow, G.J. (1970): 'Reflections on the History of the Concept of Time', *Studium Generale* **23**, 498–508; reprinted in J.T. Fraser *et al.* (1972), 1–11

Whitrow, G.J. (1976): 'Becoming and the Nature of Time', in M. Čapek *ed.* (1976), 525–32

Whitrow, G.J. (1980): *The Natural Philosophy of Time*. Oxford: Oxford University Press

Whitrow, G.J. (1989): *Time in History*. Oxford: Oxford University Press

Whittaker, E.T. (1927): 'The Outstanding Problems of Relativity', *Nature* **120**, 368–71

Whittaker, E.T. (1949): *From Euclid to Eddington*: A Study of Conception of the External World

Wiener, Ch. (1863): *Grundzüge der Weltordnung*. Leipzig: C.F. Winter

Wigner, E. (1967): *Symmetries and Reflections*. Bloomington/London: Indiana University Press

Williams, D. (1951): 'The Myth of Passage', *Journal of Philosophy* **48**, 457–72; reprinted in R. Gale *ed.* (1968), 98–116

Williams, L.P. *ed.* (1968): *Relativity Theory:* Its Origins and Impact on Modern Thought. New York: John Wiley & Sons

Wilson, M. (1993): 'There's a Hole and a Bucket, Dear Leibniz', in P.A. French *et al., eds., Midwest Studies in Philosophy* **XVIII**, Notre Dame: University of Notre Dame Press (1993), 202–41; reprinted in J. Butterfield *et al.*, *eds.*(1996), 165–204

Winnie, G. (1970): 'Special Relativity Without One-Way Velocity Assumptions I, II', *Philosophy of Science* **70**, 81–99, 223–38

Winnie, G. (1977): 'The Causal Theory of Space-Time', in J. Earman *et al., Foundations of Space-Time Theories*. Minnesota Studies in the Philosophy of Science, Volume 8, Minneapolis: University of Minnesota Press

Witten, E. (1999): 'Reflections on the fate of spacetime', in C. Callender/N. Hugget *eds.* (1999), 125–137

Worral, J. (1989): 'Structural Realism: The Best of Both Worlds?', *Dialectica* **43**, 99–124

Wright, G.H. von (1971): *Explanation and Understanding*. Ithaca (New York): Cornell University Press

Wright, G.H. von (1974): *Causality and Determinism*. New York/London: Columbia University Press

Wrinch, D./H. Jeffreys (1921): 'The Relation between Geometry and Einstein's Theory of Gravitation', *Nature* **106**, 806–9

Wrinch, D. (1922): 'The Theory of Relativity in Relation to Scientific Method', *Nature* **109**, 381–2

Young, Th. (1845): *A Course of Lectures on Natural Philosophy and the Mechanical Arts*. London: Taylor and Walton

Yourgrau, P. (1961): 'Some problems concerning fundamental constants in physics', in H. Feigl/G. Maxwell *eds.* (1961), *Current Issues in the Philosophy of Science*. New York: Holt, Rinehart and Winston, 319–41

Yourgrau, P. (1991): *The Disappearance of Time*. Cambridge: Cambridge University Press

Zeh, H.D. (1970): 'On The Interpretation of Measurement in Quantum Theory', *Foundations of Physics* **1** (1970), 69–76; reprinted in *Quantum Theory and Measurement*, edited by J.A. Wheeler/W.H. Zurek, Princeton University Press (1983), 342–49

Zeh, H.D. (1986): 'Emergence Of Classical Time From A Universal Wavefunction' *Physics Letters A* **116**, 9–12

Zeh, H.D. ([2]1992/[4]2001): *The Physical Basis of The Direction of Time*. Berlin/New York: Springer

Zeh, H.D. (1996): 'The Program of Decoherence: Ideas and Concepts', in D. Giulini *et al.* (1996), 5–34

Zeh, H.D. (2000): 'The Meaning of Decoherence', in Ph. Blanchard *et al.* (2000), 19–42

Zeh, H.D. (2000): 'Was heißt: es gibt keine Zeit?', web essay: www.zeh-hd.de

Zeilik, M. ([5]1988/[9]2002): *Astronomy*. New York: John Wiley & Sons

Zeilinger, A. (1998): 'Fundamentals of Quantum Information', *Physics World* **11** (March 1998), 35–40

Zeilinger, A. (2002): 'Quantum Teleportation', *Scientific American* (April 2000), 32–41

Zeilinger, A. (2003): *Einsteins Schleier*. Die neue Welt der Quantenphysik. München: C.H. Beck

Zilsel, E. (1927): 'Über die Asymmetrie der Kausalität und die Einsinnigkeit der Zeit', *Die Naturwissenschaften* **15**, 280–6

Zou, X.Y./L.J. Wang/L. Mandel (1991): 'Induced Coherence and Indistinguishability in Optical Interference', *Physical Review Letters* **67/3**, 318–321

Zwart, P. (1976): *About Time*. North-Holland Publishing Company

List of Figure Sources

Robert Boyle (1627–1691) p. 13: Chemical Achievers Copyright 2000
Immanuel Kant (1724–1804) p. 24 and p. 151: Engraving by J.L. Raab, Photo: Deutsches Museum Munich
Alexander von Humboldt (1769–1859) p. 27: Staatliche Museen zu Berlin – Nationalgalerie, Photo: Joerg P. Anders/bpk
Charles Darwin (1809–1882) p. 30: Photo Deutsches Museum Munich
Hermann von Helmholtz (1821–1894) p. 34: Verlag Harri Deutsch, Oswalds Klassiker der exakten Wissenschaften 79/80
James C. Maxwell (1831–1879) p. 38: Engraved by G.J. Stodart from a photograph by Fergus of Greenock, courtesy AIP Emilio Segre Visual Archives
Michael Faraday (1791–1867) p. 44: Photo by John Watkins, courtesy AIP Emilio Segre Visual Archives
Heinrich Hertz (1857–1894) p. 46: Verlag Harri Deutsch, Ostwald Klassiker der exakten Wissenschaften 263
Galileo Galilei (1564–1642) p. 110: Portrait by Leoni; Photo: Deutsches Museum Munich
Isaac Newton (1642–1727) p. 112: Courtesy AIP, Emilio Segre, Visual Archives, Physics Today Collection
Gottfried Wilhelm Leibniz (1646–1716) p. 117: Photo Deutsches Museum Munich
Ernst Mach (1838–1916) p. 118: Photo Deutsches Museum Munich
Albert Einstein (1879–1955) p. 120: Nobel Foundation
Pierre-Simon, Marquis de Laplace (1749–1827) p. 195: AIP Emilio Segre, Visual Archives, Physics Today Collection
Max Planck (1858–1947) p. 212: Nobel Foundation
Ernest Rutherford (1871–1937) p. 216: Nobel Foundation
Niels Bohr (1885–1962) p. 218: Nobel Foundation
Werner Heisenberg (1901–1976) p. 230: Nobel Foundation

Name Index

Subject Index